Writing Built Environment Dissertations and Projects

Writing Built Environment Dissertations and Projects
Practical Guidance and Examples

Dr Peter Farrell MSc FRICS FCIOB
Reader and Programme Leader
MSc Construction Project Management
University of Bolton
UK

with

Dr Fred Sherratt MCIOB C.BuildE MCABE FHEA
Senior Lecturer
Anglia Ruskin University
UK

and

Dr Alan Richardson MSc FCIOB MInstCES PGCEd
Reader and Programme Leader
BEng Civil Engineering
University of Northumbria
UK

SECOND EDITION

WILEY Blackwell

This edition first published 2017
© 2011, 2017 by John Wiley & Sons, Ltd.

Registered Office
John Wiley & Sons, Ltd, The Atrium, Southern Gate, Chichester, West Sussex, PO19 8SQ, United Kingdom.

Editorial Offices
9600 Garsington Road, Oxford, OX4 2DQ, United Kingdom.
The Atrium, Southern Gate, Chichester, West Sussex, PO19 8SQ, United Kingdom.

For details of our global editorial offices, for customer services and for information about how to apply for permission to reuse the copyright material in this book please see our website at www.wiley.com/wiley-blackwell.

Library of Congress Cataloging-in-Publication data applied for

ISBN: 9781118921920

A catalogue record for this book is available from the British Library.

Wiley also publishes its books in a variety of electronic formats. Some content that appears in print may not be available in electronic books.

Cover image: Gettyimages/OJO_Images

Set in 10/12pt Minion by SPi Global, Pondicherry, India

Printed in Singapore by C.O.S. Printers Pte Ltd

1   2017

Contents

Author biographies ix
Preface x
About the companion website xii

1 Introduction 1
  1.1 Introduction 1
  1.2 Terminology; nomenclature 2
  1.3 Document structure 3
  1.4 Possible subject areas for your research 7
  1.5 Professional bodies and the non-technical or technical dissertation or project 8
  1.6 Qualitative or quantitative analysis? 10
  1.7 The student/supervisor relationship and time management 14
  1.8 Ethical compliance and risk assessments 17
  1.9 House style or style guide 22
  1.10 Writing style 23
  1.11 Proofreading 27
  1.12 Extra support? 29
  1.13 A research proposal 29
  1.14 A viva or viva voce 30
  Summary 31
  References 31

2 The introduction chapter to the dissertation or project 33
  2.1 Introduction contents 33
  2.2 Articulation or description of the problem and provisional objectives 35
  Summary of this chapter 37
  References 38

3 Review of theory and the literature 39
  3.1 Introduction 39
  3.2 Style and contents of a literature review 41
  3.3 Judgements or opinions? 43
  3.4 Sources of data 44

3.5   Methods of finding the literature   48
3.6   Embedding theory in dissertations and projects   49
3.7   Referencing as evidence of reading   53
3.8   Citing literature sources in the narrative of your work   54
3.9   References or bibliography or both?   58
3.10   Common mistakes by students   59
3.11   Using software to help with references   60
3.12   Avoiding the charge of plagiarism   62
Summary of this chapter   64
References   64

4   **Research goals and their measurement**   **67**
4.1   Introduction   68
4.2   Aim   70
4.3   Research questions   71
4.4   Objectives   71
4.5   Variables   74
4.6   A hypothesis with one variable   75
4.7   A hypothesis with two variables: independent and dependent   77
4.8   Writing the hypothesis: nulls and tails – a matter of semantics   81
4.9   'Lots' of variables at large, intervening variables   83
4.10   Ancillary or subject variables   83
4.11   No relationship between the IV and the DV   88
4.12   Designing measurement instruments; use authoritative tools and
       adapt the work of others   89
4.13   Levels of measurement   93
4.14   Examples of categorical or nominal data in construction   95
4.15   Examples of ordinal data in construction   96
4.16   Examples of interval and ratio data in construction   97
4.17   Types of data   98
4.18   Money and CO_2 as variables   102
4.19   Three objectives, each with an IV and DV: four variables to measure   103
4.20   Summarising research goals; variables and their definition   104
Summary of this chapter   105
References   105

5   **The Methodology chapter; analysis, results and findings**   **107**
5.1   Introduction   107
5.2   Approaches to collecting data   110
5.3   Data measuring and collection   112
5.4   Issues mostly relevant to just questionnaires   120
5.5   Ranking studies   129
5.6   Other analytical tools   131
5.7   Incorporating reliability and validity   132
5.8   Analysis, results and findings   137
Summary of this chapter   138
References   139

6   Laboratory experiments   **140**
    6.1   Introduction   141
    6.2   Test methodology   142
    6.3   Sourcing test materials   143
    6.4   Reliability and validity of findings   143
    6.5   Sample size   145
    6.6   Laboratory recording procedures   145
    6.7   Dissertation/project writing (introduction, methodology and results)   146
    6.8   Health and safety in the laboratory; COSHH and risk assessments   149
    6.9   Role of the supervisor   151
    6.10   Possible research topics for technical dissertations or projects,
           construction and civil engineering   153
    6.11   Examples of research proposals   153
    6.12   Research objectives and sample findings by the author   154
    Bibliography   163

7   Qualitative data analysis   **165**
    7.1   Introduction   165
    7.2   The process of qualitative data collection   166
    7.3   Steps in the analytical process   168
    Summary of this chapter   175
    References   176

8   Quantitative data analysis; descriptive statistics   **177**
    8.1   Introduction   177
    8.2   Examples of the use of descriptive statistical tools   178
    8.3   Ancillary variables   186
    8.4   Illustration of relevant descriptive statistics in charts   190
    8.5   Normal distributions; Z scores   191
    8.6   A second variable for descriptive analysis; an IV and a DV   197
    Summary of this chapter   201
    References   202

9   Quantitative data analysis; inferential statistics   **203**
    9.1   Introduction   204
    9.2   Probability values and three key tests: chi-square, difference
           in means and correlation   206
    9.3   The chi-square test   210
    9.4   Determining whether the dataset is parametric or non-parametric   220
    9.5   Difference in mean tests; the t-test   223
    9.6   Difference in means; the unrelated Mann–Whitney test   225
    9.7   Difference in means; the related Wilcoxon t-test   230
    9.8   Difference in means; the parametric related t-test   232
    9.9   Correlations   236
    9.10   Using correlation coefficients to measure internal reliability and
           validity in questionnaires   243

9.11 Which test?   243
9.12 Confidence intervals   247
9.13 Summarising results   250
Summary of this chapter   250
References   250

10 Discussion, conclusions, recommendations and appendices   **251**
10.1 Introduction   251
10.2 Discussion   252
10.3 Conclusions and recommendations   253
10.4 Appendices   255
10.5 The examiner's perspective   256
10.6 Summary of the dissertation or project process   258
Summary of this chapter   259
References   259

List of appendices   **260**
Appendix A: Glossary to demystify research terms   261
Appendix B: Research ethics and health and safety examples   268
Appendix C: An abstract, problem description and literature review   272
Appendix D: Eight research proposals   279
Appendix E: Raw data for a qualitative study   309
Appendix F: Statistical tables   340

Index   **350**

Author biographies

Peter Farrell

Peter Farrell is a reader in construction management at the University of Bolton, UK, and programme leader for the university's MSc construction project management. He has delivered undergraduate and postgraduate modules in construction management, commercial management and research methods for 20 years. His industry training was in construction planning and quantity surveying and his post-qualification experience was working as a contractor's site manager.

Fred Sherratt

Fred Sherratt is a senior lecturer in construction management at Anglia Ruskin University, UK. She has over 12 years' experience in the construction industry and worked her way up from site secretary, through construction planning to the position of construction manager for a large UK contractor. Fred has attained numerous awards for her research.

Alan Richardson

Alan Richardson is a reader in civil engineering at Northumbria University, UK, and programme leader for the BEng in civil engineering. He has over 90 publications mainly based upon technical studies of materials. There are two main streams of his current research, one relating to the use of bacteria in cementitious materials to improve long-term durability and reduce life cycle costs. This work is being undertaken in conjunction with RILEM. The other is researching fibre use to improve impact and blast resistance in concrete. His industry experience is 26 years as managing director of an SME construction company.

Preface

There are many changes between the first and second editions. Most important, are welcome contributions from Dr Fred Sherratt and Dr Alan Richardson. Fred has strengthened sections of the text related to qualitative research and methodology, and has also added a glossary of research terms in Appendix A. Alan has added chapter 6, which examines in greater detail technical civil engineering projects. There are eight exemplar research proposals included in Appendix D that cover the fields of building and civil engineering. The authors of these proposals are acknowledged.

The word 'projects' has been added to the title, such that it now reads 'Writing Built Environment Dissertations *and Projects*'. Most universities use the term 'dissertation' for building degrees and 'projects' for civil engineering. The content of the book has been updated to ensure that it does indeed embrace the needs of civil engineers.

There is emphasis on the difference between 'non-technical' work mostly found on building programmes and 'technical' civil engineering projects. Some examples from the first edition are retained, but many are updated and changed. Exemplar datasets in tables are produced in Excel, since spreadsheets are useful for collating and sorting raw data; also for performing analysis. Some examples are screenshots from Word. It is acknowledged that you may use spreadsheets and word processors other than Microsoft. Many new figures and tables are introduced to help support explanations.

The aim of the text is to provide practical guidance on the preparation of undergraduate dissertations and projects in the built environment. Students doing research at masters and PhD level, may also find the text useful. It is hoped that it will give students the platform to attain the maximum possible mark. Some sections of the book may contribute towards enhanced performance in other modules. For example, suggestions about how to develop theory and use literature as part of a critical appraisal are common to many subjects in the built environment and indeed other disciplines. The book is ordered around a structure that may be useful for a research document; that is, it starts with material that should be contained in an introduction chapter and finishes with material that should be in the conclusion. Embedded throughout the book are issues around study skills and ethics. There are many examples included, using a variety of methodological designs in which students are encouraged to consider the concepts of reliability and validity. A key difference between dissertations/projects and other courseworks is that the middle of the document should include a data collection process and some analysis. Suggestions are made about how to collect data and how to do analysis. The analytical chapters cover qualitative and quantitative approaches. The qualitative chapter demonstrates how to include some rigour in the analytical process, rather than is often the case, where students rely on simplistic browsing of material. The quantitative chapter attempts to avoid some of the complexity in statistical work without devaluing its usefulness. The book encourages students to undertake a process of self-reflection at the end of their research, and to include a section on limitations and criticisms of their own studies. It is hoped that the examples used will stimulate ideas about how students can develop their chosen topic area into dissertation format.

Acknowledgement: the authors are grateful to those referees who gave valuable feedback on the first edition; thank you.

About the companion website

A website is provided to support this book (www.wiley.com/go/Farrell/Built_Environment_Dissertations_and_Projects). There are many tables and figures, particularly in the statistical analysis sections, that include large data sets. The web page will allow you to open these tables and perform statistical tests or produce charts using Excel on the same raw data used in the text. Alternatively, you may substitute your raw data in the templates provided, and copy and paste them into your research documents. You may copy the files onto your own computer and adjust font sizes and so on to suit your own requirements. Tables 8.3, 8.5, 8.6, 8.7 and 9.16 in this book have data 'hidden' to allow reproduction on the book page; on the website all the data is 'unhidden'. Many of the references at the end of each chapter are available on the web. While they may be found through search engines, or typing in the full web address, you may find it easier to locate publications through the links on the website. You might find the templates in Appendix B useful. These are available for download. Appendix E includes transcripts of interviews with two site managers. There were eight interviews in total for the given research project; transcripts of the other six are on the website, together with the later stages of the analytical process. Finally, the statistical tables in Appendix F are available for download.

1 Introduction

The titles and objectives of the sections of this chapter are the following:

1.1 Introduction; to set the scene and describe the dissertation process
1.2 Terminology and nomenclature; to emphasise the importance of the objective
1.3 Document structure; to provide a template
1.4 Possible subject areas for your dissertation; suggest topic areas and encourage early reading
1.5 Professional bodies and the non-technical or technical dissertation or project; to distinguish between these two different types
    1.5.1 The difference between non-technical and technical
1.6 Qualitative and quantitative analysis; to distinguish between the two analytical schools
1.7 The student/supervisor relationship and time management; to provide templates
1.8 Ethical compliance and risk assessments; to identify ground rules for compliance with codes of practice
    1.8.1 Physical or emotional harm; laboratory risk assessments
    1.8.2 Confidentiality and anonymity
    1.8.3 Generally
1.9 House style or style guide; to promote consistency and provide a template
1.10 Writing style; to identify potential pitfalls
1.11 Proofreading; to encourage it, as a process, using independent help if necessary
1.12 Extra support?; to describe help available from university disability support units
1.13 A research proposal; what to do if you are required by your university to do a proposal
1.14 Viva or viva voce; to describe what it is and how to prepare

1.1 Introduction

In some universities the dissertation or project may carry as much as one quarter weighting towards the final year degree classification. It is the flagship document of your study. It is the document that external examiners will look at with greatest scrutiny. You may want to take it to your employer and/or prospective employers. You will hopefully be proud to show it to members of your family, and it will sit on your bookshelf so that you can show it to your grandchildren. It is a once-in-a-lifetime journey for most; it is to be enjoyed and remembered. Though it does not happen often, with the help of supervisors, some students may develop

Writing Built Environment Dissertations and Projects: Practical Guidance and Examples, Second Edition.
Peter Farrell, Fred Sherratt and Alan Richardson.
© 2017 John Wiley & Sons, Ltd. Published 2017 by John Wiley & Sons, Ltd.
Companion Website: www.wiley.com/go/Farrell/Built_Environment_Dissertations_and_Projects

their research into a publication. That may involve condensing the work into about ten pages for delivery at a conference or even for inclusion as a journal paper. It is one thing to get a degree qualification on your CV; quite another for you to be a published author.

One of the key criteria for the research is that it must have some originality. That is, not to discover something new but perhaps to look at an area that has already been investigated, and to take a different perspective on it or to use a different methodology. It is more than an assignment – the research process must seek the information, analyse it and offer conclusions. Modest objectives are adequate. Better dissertations and projects have robust methods of analysing qualitative data or some basic statistical analysis.

Dissertations and projects have assessment criteria. To achieve marks in the upper echelons (70%+), criteria often require that work should demonstrate 'substantial evidence of originality and creativity', 'very effective integration of theory and practice', 'excellent grasp of theoretical, conceptual, analytical and practical elements', and 'all information/skills deployed'.

There are two separate strands to your research. The first is that you must develop your knowledge in your chosen topic so that you become 'expert'. One of the reasons you may have chosen your subject is that you may want to learn more about it. Indeed, it is very important that you do this. The second is that you must conduct a piece of research, employing appropriate research methodology. In your document you must explain and substantiate your methodology; it must stand up to scrutiny. The method that you use must include the collection and analysis of data. The two strands go hand in hand. It is not to say that the weighting is 50:50, or any other percentage, but there must be substantial evidence of both in your dissertation. You must demonstrate that you have produced a piece of research in the true meaning of the word 'research'; it is not adequate that your document is a 'mere' report.

1.2 Terminology; nomenclature

Clarity in research is absolutely critical; the plethora of terminology used by academics can be unhelpful, fuzzy and for some misleading. That is just the way it is. It may be useful for you to employ your own rigid definitions of such terminology, or at the very least be consistent in the language you use in your work.

Georg Christoph Lichtenberg (1742–99) a professor of physics at Göttingen University, cited on the Quotations Page (2015), wrote 'One's first step in wisdom is to question everything'. Your research should start with a question, from which you will develop an objective in which you will 'do' something that will enable you to answer the question. What you will 'do' may involve testing a hypothesis. The research question, objective and hypothesis should all match each other, for example:

Research question: How well do UK contractors comply with best practice in health and safety? (note the question mark)

Objective: To determine how well UK contractors comply with best practice in health and safety.

Hypothesis: The compliance of UK contractors with best practice in health and safety is excellent (or in a different context to your research you may write 'not good enough').

You need to make it clear in your introduction that you have a research question, objective and hypothesis that match, but when you communicate with people in industry and also when you find the need to repeat yourself in your document it may be best to do so using the term 'objective'. People in industry are likely to be familiar with the word 'objective', but less familiar with research questions and hypotheses. An objective is a statement of what you will 'do' in your research.

When describing what a research project will 'do', students often express this by using words other than 'objective'. Some examples are: 'the focus of the study', 'the reason for the study', 'the study looks into', 'the study tries to', 'the study examines', 'purpose', 'goal', 'direction', 'intention' or 'seeks to'. Perhaps use of these phrases should be discouraged.

It must be recognised that universities and individual academics will have their own preferences, and students must be able to adapt flexibly to work with supervisors, and also to understand the writing of others who use different language. Most supervisors will be comfortable that you 'hang' the whole of your study around objectives; put more clearly, objectives, objectives and objectives.

1.3 Document structure

A suggested structure/template for a dissertation or project is:

No number	Preliminary pages
Chapter 1	Introduction
Chapter 2	Theory and literature review
Chapter 3	Research design and methodology
Chapter 4	Analysis, results and findings
Chapter 5	Discussion
Chapter 6	Conclusions and recommendations
No number	References and bibliography
No number	Appendices

This is not written in tablets of stone, but is merely a framework around which your structure may be designed. It is for individual researchers to design their structure and to agree it with their supervisor. These may be considered as chapter titles, but they should be 'flavoured' by words relevant to your study area, e.g. 'The development of theory and literature about money as a motivator for construction craftspeople'.

The weight of each chapter, or the number of words, does not necessarily lend itself to one sixth in each. There is an argument for saying that the first two chapters, as the opening to the document, could be about one third weight. The middle two chapters comprising the methodology and analytical framework could be about one third weight. Finally the last two chapters, closing off the document, could be about one third weight. Often it is the last part where students lose marks; they simply run out of time after completing the analysis. The consequence is that documents were heading for really good marks only achieve mid-range marks.

Each chapter should open with an introduction – there should even be an introduction to the introduction chapter – and close with a summary. Students often do not like writing either introductions or summaries, and question their value for the reader. The introduction to each chapter need only be a few paragraphs. It is not for readers to embark on a voyage of discovery as they read each chapter. The 'introduction to the introduction' may start with the aim of the study. It may tell the reader that the introduction chapter will provide a background to the topic area and description of the problem, give a historical perspective, give the research goals (including the objectives), describe briefly the methodology, give an outline of the remaining parts of the document and summarise the chapter. But do not write it as mechanically as the above. Ensure that it is flavoured by your topic area, e.g. a historical perspective of PFI as a procurement method. The writing style of a summary is different from the writing style of an introduction. It does exactly what its name implies: it summarises what has gone before. It should not say 'this chapter has outlined the problem'. It should summarise in the narrative the key points of the problem in a few lines. You need to say what the problem is. A useful tactic when writing a summary is to read each page and condense it into one or two carefully selected sentences. The reason for a summary is that readers who have taken the journey through your chapter, may need some moments of thought and reflection about what they have just read, before going on. They may indeed have forgotten what they read at the beginning of the chapter by the time they get to the end. Also, readers may not read the whole document in one sitting. When they come to recommence reading, the summary can refresh their minds before continuing.

The whole document should be in report numbering format. Start with the introduction chapter as chapter 1. The introduction to the introduction is 1.1., 1.2 definitions of important phrases, 1.3 background to the topic area etc. Try to avoid too many subsections, but if they are needed they become, e.g. 1.3.1, 1.3.2 etc.

Page number the whole document, except the cover page. By convention, preliminary pages are numbered with Roman numerals, that is (i), (ii) etc. The first page is a declaration, numbered Roman numeral (i). People with dyslexia may find it hard to distinguish between Roman numerals; therefore alternatively consider letters, (a), (b), (c) etc. Pages after the preliminary pages, starting with the cover page to chapter 1, use Arabic numerals 1, 2, 3 etc. The cover page to chapter 1, thus starts at page 1. Page numbering with Arabic numerals continues into the reference section and the appendices. Separate parts of the appendices are labelled by letters not numbers; that is appendix A may be a covering letter to a questionnaire, appendix B may be the questionnaire itself and so on. If appendices are related, perhaps use letters and numbers e.g. A1 and A2 have the same theme, B1, B2, B3 ditto etc., as we have done in this book.

The preliminary pages to a research document should include the following separate parts:

(a) unnumbered: a cover page with the document title, name of author, name of university, year and degree title.

(b) declaration using words prescribed by the university such as 'I declare that this research has not been submitted to any other university or institution of learning, and the work included is entirely my own except where explicitly cited in the text'. You will be using and citing the work of others, as described in chapter 3.

(c) an acknowledgements page: it is usual to thank people who have contributed to the research through their time or sponsorship, employers, friends or members of your family and supervisors. Only a short statement is usual.

(d) abstract: the abstract is a very concise summary and should be written very carefully.

Readers may be initially attracted to documents by titles, but these can be misleading, and more information is required. So the purpose of the abstract is to allow readers to make a quick decision about whether they wish to read further sections of the document, or alternatively they may be able to make a sensible judgement that the document is not relevant to their needs. Often readers who are browsing previous research will read abstracts and decide not to read on; that is fine. They have been able to quickly make an informed decision based upon a full and concise summary of the document. Since you have a limited number of words, and you may wish to entice people into the document, each part must be measured carefully. External examiners will read some, but cannot read all documents. Given a choice of which to read, they may be attracted by research with a well-articulated abstract. In academic publications abstracts are often 200–250 words in length, but in dissertations perhaps a larger word count is acceptable. An abstract confined neatly to one A4 page of text, single line space, 12 size font, perhaps three or four paragraphs with a line space between, would be about 500 words. Try to avoid going onto a second page, even for one line. This is your opportunity to sell your work. In research terms, it would be a serious failing if subsequent researchers picked up your document with the idea to further their knowledge in your field, but because of a lack of clarity in the abstract, were led to think that your work was not relevant. If a sentence, or indeed a single word, is not necessary to convey the message required, it should be taken out. The abstract is an art in writing concisely and with precision.

It should: give the topic, state the aim, outline the problem, give the main objectives or hypotheses, summarise the methodology (including population description, sample size if appropriate, method of data collection and analyses) and state the main findings, conclusions and recommendations. It can be written as work proceeds but can only be completed at the end. Students often adopt a writing style for an abstract similar to the following: 'the study will give an objective, and describe the methodology...' etc. This is not an abstract, since it would leave readers without the information required. The abstract must actually state what the objective is, and state the methodology. Some students submit their documents without an abstract; deduct 5 marks!

An example abstract is included in appendix C.

(e)   contents page: this should list the main titles of each chapter. It is not usually necessary to list all subsections of chapters on the main contents page. Subsequently, each chapter should have its own cover page that details the titles of subsections within the chapter.

(f)   list of abbreviations: in your narrative, convention is that at the first point of using each abbreviation in your document it should be spelt out in full, with its abbreviation in brackets, thus: 'The Health and Safety Executive (HSE) is responsible for ...'. At any subsequent need to refer to the HSE you can then just use the abbreviation. If readers later 'forget' what HSE stands for, they can refer to your list of abbreviations at the front of the document. Do not overdo the use of abbreviations; however, the construction industry does use them frequently, and you may reasonably have a list of abbreviations that is about a page long.

(g)   glossary of symbols (if statistical tests are executed): letters of the Greek or Roman alphabet are often used to distinguish between different tests. See sections 8.1 and 9.1 for examples of statistical symbols.

(h)   glossary of terms: this ensures a common understanding even for quite well known terms as well as terms that have a particular meaning in the subject topic of the research. It will include a brief definition of their meaning in the context of the study. Ensure that such definitions are authoritative; that is, from the literature. For example, there may be a need to refer to 'sustainable development in construction' in your document. That may mean different things to different readers, so give an authoritative meaning: 'development that meets the needs of the present without compromising the ability of future generations to meet their own needs' (Brundtland, 1987). You may need to define many phrases in your document.

(i)   lists of appendices, figures and tables: similar to the format at the beginning of this textbook.

Figures may be pie charts, histograms, graphs, or diagrams. Tables may contain results of experiments, or summarise data. Do not overdo pictorial representation of data just to get some colour into your document. A small table, for example, may better show the age profile of people, rather than a brightly coloured pie chart using half a page of space. Figures and tables should be numbered, and prefixed by the number of the chapter in which they appear, e.g. figure 2.3 will be the third figure in chapter 2. The title and content of figures or tables should be such that they can be understood on a stand-alone basis. The reader should not have to browse other sections of text to gain an understanding of a figure or table. Do not refer in your text to 'the figure above' or 'the table below'. Figures and tables should be introduced in your text, and then inserted in your document in the first subsequent convenient position, perhaps at the end of that paragraph or on the next page if that position is close to the bottom of a page. By convention, the titles of figures appears under the figure, and the titles of tables above the table; in both cases the figures/tables themselves and the text for titles can be centred on the page.

If figures or tables are produced in Excel or other software, they can be imported into Word using the 'snipping tool' available in the 'search all programmes and files' box of the start menu, as illustrated in figure 1.1. Alternatively use the print screen, paste, format and crop functions in Word.

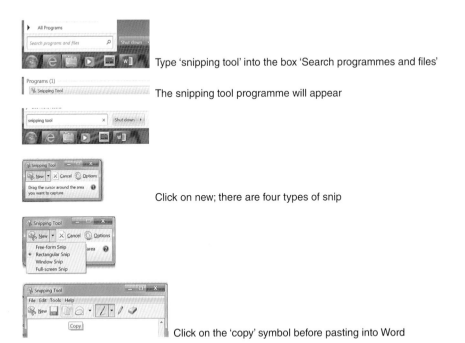

Figure 1.1 Using the 'snipping tool' to cut and paste figures, tables or images in Word.

1.4 Possible subject areas for your research

The topic area that you choose for your work should ideally be related to the specialism that you are studying within construction. You should consider all parts of the construction process from and including inception (clients with ideas that require projects) through to construction, maintenance, refurbishment, demolition and recycle. Most disciplines are interested to use their skills to improve the service provided to clients at all stages of the process. In practice, modern methods of procurement integrate the supply chain, and therefore all professionals are now involved both earlier and later in the process than has traditionally been the case. You may consider issues from the perspective of any party in the supply chain, e.g. clients, end users, consultants, contractors, subcontract specialists, suppliers, manufacturers or indeed other stakeholders such as investors or the public. If you are a civil engineer, you may need to do a 'technical' piece of research, as described in section 1.5.1.

Non-technical topic areas often include soft people issues, such as human resource management, job satisfaction, grievances, employee turnover or quality of life measures. Resources such as subcontractors, plant, material and capital (money) are also popular. You may want to specialise in finance, planning, legal issues or contracts, procurement methods, health and safety, quality, design aesthetics, planning, building information modelling, maintenance, business ethics or use of information technology and software. In the context that you may wish to consider variables in your study, popular dependent variables align with key performance indicators promoted by Constructing Excellence in the Built Environment (Constructing Excellence, 2015), such as client satisfaction, cost predictability, time predictability, quality or safety. Sustainability issues driven by the climate change agenda are often researched. There is great potential for studies in many areas related to sustainability, such as the UK's Building Research Establishment Environmental Assessment Method (BREEAM, 2015) or renewable energy. Defining and measuring best practice in a given field may be the basis of a useful study. The definition of best practice could be an objective of your study met by the literature review. You may find investigating best practice useful to you personally, since it is a valuable way to enhance your own knowledge in the field. The measurement of compliance with best practice by organisations or individuals may then be the basis for another objective, to be met by the main data collection process in the middle part of your study. When Paul Morrell came to the newly created post of UK government chief construction advisor in November 2009, he stated 'we're going to need to start counting carbon as rigorously as we count money, and accepting that a building is not of value if the pound signs look okay, but the carbon count does not' (Richardson, 2009) – lots of opportunities, therefore to measure carbon. The outcome of your research should not be a 'project' of a descriptive kind or a report or the design of a structure. The emphasis is on data collection and analysis, around objectives. It may be management, technology or science based. In July 2013, the UK government launched its publication 'Construction 2025: industrial strategy for construction – government and industry in partnership' (BIS, 2013). It provides many potential subjects for research, for example its vision for 2025 around people, the digital economy (Building Information Modelling et al.), low carbon, industry growth and leadership.

Most often, part-time students select a problem from their workplace; talk to your colleagues at work. Alternatively, you may select something that is current in industry or academia. Full-time students may seek out a mentor from industry – very often practitioners will be delighted to 'put something back' into the education system they have gone through themselves. You should have been reading about current issues throughout your study, so as you are selecting the topic area for your research, you should speed that reading up. The lead sources to look for current issues are websites and conferences of your professional bodies, other academic conferences such as ARCOM, the weekly construction press and construction academic journals. You should be reading each week at least one of *Construction News*, *New Civil Engineer* or *Building*. Download the apps or log on to the websites of *The Construction Index* or *Construction Enquirer*. Find all these sources through a web search engine. To ensure that your study has academic credibility, if you start from a practical perspective, you will need to take it back to its theoretical roots. Alternatively, you may start with a theory and take it forward to its practical application; for example, flagship theories in management, such as leadership and motivation.

1.5 Professional bodies and the non-technical or technical dissertation or project

Undergraduate degree programmes are often accredited by professional bodies. Accreditation is very important to universities, and also very important to you as students. Accreditation means that degree programmes are approved by the relevant professional body, and depending on the level of accreditation, successful students are deemed to have achieved the minimum educational requirements of that professional body. Attaining your degree does not mean you immediately become a full member; there is usually a requirement for a period of practice in industry or research. You will then need to demonstrate your competence against a range of criteria.

It is important for you to become a member of a professional body. Passing your degree demonstrates you are good at academic work, but in your later career you need to demonstrate to employers your competence in practice or, as required by the Construction Design and Management Regulations (HSE, 2015), that you have appropriate skills, knowledge and experience. Your membership of a professional body indicates a commitment to keep yourself up to date with current developments in construction through your continuing professional development (CPD). Also, you will be signed up to a code of professional conduct and comply with the highest ethical standards. The best employers and construction clients need evidence that they are dealing with people who are up to date, professional and ethical. Professional body membership is that evidence, a passport to employment.

The key professional bodies in the built environment, in alphabetical order and spilt between the loose classifications of building and civil engineering, are as follows:

Building
Chartered Association of Building Engineers (CABE)
Chartered Institute of Architectural Technologists (CIAT)
Chartered Institute of Building (CIOB)
Royal Institution of Chartered Surveyors (RICS)
Royal Institute of British Architects (RIBA)

Civil engineering
Institution of Civil Engineers (ICE)
Institute of Highway Incorporated Engineers (IHIE)

The Institution of Highways and Transportation (IHT)
The Institution of Structural Engineers (IStructE)

You should join at least one professional body while you are studying, as a student member; if appropriate, more than one. Many have free membership for students. You should use their libraries and attend CPD events during your study period. After you have completed your degree, you should aspire to become a full member. Depending on the institution, that may be possible within say three years, but some institutions require a masters degree level qualification. Full membership of most of the above professional bodies brings with it 'chartered' status, a prestigious title.

As an entire year of students leave their course of study with their degree, there are 'x' number of students in the job market competing for available jobs. The question to students then will be: 'What will differentiate your application from the next person?' CPD attendance displays to employers, a commitment to being involved in their chosen career pathway. That can make the difference between employment or not! There are other criteria beyond CPD attendance, but if you have already networked you may perform much better in a job interview because of CPD attendance.

When professional bodies accredit programmes, they stipulate generically, the content of degrees. The building professional bodies may be happy with either non-technical or technical research. The civil engineering professional bodies are likely to favour only technical research. Civil engineers often call the final research document a project, not a dissertation. Some universities have a BSc (Hons) degree in civil engineering, others a BEng (Hons) civil engineering. It may be possible for BSc (Hons) programmes to undertake non-technical work, but BEng programmes, which are on the path towards chartered engineer status, will require a technical project. That is a requirement of the professional bodies that accredit civil engineering degrees.

1.5.1 The difference between non-technical and technical

Non-technical work may involve many of the 'soft' or 'subjective' management issues. The data collection and analytical processes may be qualitative, or indeed quantitative and involve statistical calculations or mathematics based on the soft data collected.

Technical subjects or technologies are likely to involve mathematical/quantitative work and science/chemistry; it will involve 'hard' or 'objective' data. It may involve an appraisal of what variables influence behaviour in the civil engineering specialist fields of structures, geotechnics, hydraulics and drainage, materials science, geomatics or land surveying and transportation. It may also be possible to classify your work in one or more of the three key strands of civil engineering work: design, sustainability and health and safety. Your project may not be exclusively technical. Your introduction and the literature review may be partly qualitative, but the literature review may also bring to the table the latest up-to-date technical and scientific position in your subject area. It is also good that you do some qualitative work by getting out and communicating with individual practitioners and companies about your project, and if you do so, mention it in your methodology chapter. However, in the middle part, there must be strong focus on analysis of a technical dataset, and that will very likely involve some mathematical tools. It is unlikely that your method would include a data collection tool such as an electronic questionnaire; yes, do question people but that may be best done as part of some supplementary qualitative work and networking to support the validity of the problem that you will investigate or interpret using some of your results and findings.

What happens if you do non-technical research on a technical programme? Anecdotally, a really good non-technical dissertation may get a 55% mark from your university and a 'glum' look from the accreditation body, while a similarly good technical document might get 65% and a smile.

The next key issue is 'where should the dataset come from'? There is the possibility that you could take the data from the literature, and then, to meet the requirement of your work being original, perform some type of analysis that has not been done before. Perhaps the data collection will involve some experimentation in laboratories or some fieldwork or some computer modelling – though not all students will use these types of methods. Getting your own data supports a view that it is good to demonstrate that you have been proactive in your research, not passive.

While the project may include some sketches, drawings or photographs, the requirement for some analysis excludes the possibility of engineers, architects or architectural technologists submitting a design portfolio or such like.

The building disciplines also welcomes technical research; the content of building degree programmes often involves problems related to such issues. There may be a perception, hopefully unfounded, that building students are less likely to take on the challenge of maths and/or science. If as a building student you can take on these challenges, you may decide to 'go for it'. An issue such as climate change can be addressed from a non-technical or technical perspective. Non-technical research may be around human perceptions or behaviours, while technical work may involve scientific issues.

1.6 Qualitative or quantitative analysis?

The middle of the document should include some analysis – taking one element of a problem, breaking it down and establishing relationships or causes and effects. Robust analysis involves the application of some kind of academic tool, although some academic tools may be considered more robust than others. The way you go about collecting data for analysis, and the way you do the analysis, is one facet of a dissertation or project that can distinguish it from more conventional assessments or courseworks.

Often, analysis is the most challenging part of a dissertation or project. Students beginning to read about research are often faced with a myriad of new terms and complex ideas: ontology, epistemology, positivist, interpretivist and many more. This can make research very daunting and often very confusing for students new to the process – but do not panic! Knowledge and understanding of such 'research paradigms', while beneficial for an undergraduate dissertation, may not be essential. They do, however, become more relevant for MSc students, and certainly for those undertaking a PhD.

These terms all relate to philosophy and the questions around 'what' can be known about our world (ontology) and 'how' can we know it (epistemology). All research is based on such ideas. Much research is not 'bothered' by such thinking. Take for example testing concrete strength; this is an objective fact that can be measured empirically and scientifically – the underlying philosophy of this is well established and does not need restating. Research involving people is more complicated – it can be less objective and more subjective, and this is where research paradigms become more relevant. The terms objective and subjective can be used to illustrate research paradigms at the ends of what is termed the 'research continuum'. Objective or positivist research lies at one end, and seeks to establish *facts* about the world. Subjective or interpretivist research lies at the other, and seeks to explain *why* things are as they are. There are many other approaches to research that lie between these extremes – and indeed some that overlap – which is why students often feel overwhelmed.

Another way of looking at research paradigms is from the perspectives of the two major analytical schools: quantitative and qualitative research. Crudely speaking, qualitative methods involve analysing words (interpreting the *why*) and quantitative methods involve the analysis of numbers (establishing the *what*). Some people may be able to use both methods, but sometimes a person is specialist in one or the other. In their approach to a problem, researchers may lean towards methods that they understand best, but it should be noted that neither approach is 'easier' and both require rigour and significant efforts on the part of the researcher. A simple way of looking at research paradigms is shown in figure 1.2. More explanation of some common research terms can be found in a glossary at appendix A.

Whether to use a qualitative or quantitative approach must be driven by the nature of the problem and the objective to hand. The objective must drive the choice of method, not the other way round. Ideally, you therefore need to have at least an appreciation of each objective to allow you to select the best method to meet that objective. Mindful of your limited time, it may be that your data collection should be only one of qualitative or quantitative. It does not matter which, provided there is some type of analysis of some type of data. In business terms, objectives are often non-negotiable; business executives or politicians will set objectives. They will then select the people with the appropriate methodological expertise who can meet those objectives. In your dissertation or project, it is possible that while objectives should drive

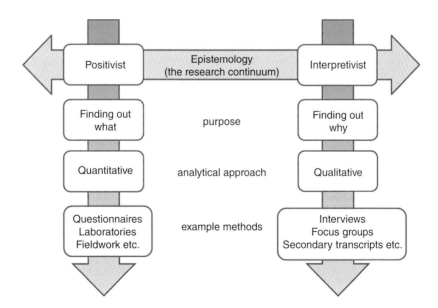

Figure 1.2 Research paradigms.

methods, in reality it could be the other way round, and if you are setting the objective, you can set it to suit your strengths.

For example, in a management research project, you may wish to research motivation. You may set an objective to determine the most effective ways of motivating the workforce. You could adopt a positivist research paradigm, and use a quantitative survey to establish what different measures are used in different organisations, and seek to quantify and measure their effectiveness. This may involve setting a hypothesis – for example that financial motivation is the most effective method – and then testing this using your quantitative survey data, possibly using an electronic tick box format. In this case, you would be seeking to establish facts about motivation. Alternatively you may set an objective to explore why some motivation methods work better than others. You could adopt an interpretivist research paradigm and a qualitative approach, interviewing managers and employees to understand their experiences and attitudes to motivation. In this approach, you have to interpret the answers given in the interviews and develop your findings through analysis, rather than simply presenting them as facts.

The boundary between the qualitative and quantitative can be blurred in questionnaires, by taking qualitative responses to closed questions, and coding them with numbers that have real quantitative value. For example, questions about product or service satisfaction may give participants four qualitative labels as possible answers: (a) very satisfied (b) satisfied (c) not satisfied and (d) not at all satisfied. The analytical process could allocate numbers 3 to 0. The answers given are clearly qualitative; they express the feelings of participants using a carefully selected word, such as 'I am very satisfied with this product'. The passion intended by the qualitative response is diluted by the quantitative number. There are instances where boundaries between qualitative and quantitative are not blurred. Unstructured interviews, whereby interviewees speak freely, are clearly qualitative, and it would inappropriate to allocate numbers to that data.

Qualitative analysis aims to gain insights and understand people's perceptions of the world; people may deliberate with themselves and have to give careful thought to their answers. Beliefs, understandings, opinions and views of people are examined. They may 'pour their heart out to you' about a particular problem. Expression and tone may reveal as much as the actual words. The data you receive is 'rich' and gives valuable insight. The data will be unstructured in its raw form. Such rich data could never be obtained in a tick box electronic survey. If your study was to seek out causes of dissatisfaction among some parts of the workforce, that is probably best done with at least some element of qualitative work. Data needs to be filtered, sorted and manipulated if analytical techniques are to be applied. Qualitative analysis may be merely findings derived from reading other work or from reading transcripts of interviews conducted by others; alternatively, it may be from transcripts of interviews conducted personally. Analysis may involve simple content analysis, whereby there is counting of key words, and deriving data meaning from high frequency hits. A more rigorous approach may involve some kind of coding and comparison of findings through thematic analysis of data. Examples of this can be found in chapter 7. The qualitative analyst will labour meticulously over transcripts. There could be analogies with criminal investigations where prosecutors labour over police interview transcripts; what did the suspect really mean when he/she said '...'? A famous example is the trial of Derek Bentley (BBC, 2005), who told his gun-wielding accomplice to 'let him have it' when a policeman apprehended them. But did he mean let him have the gun or a bullet? A manual approach may be used to analyse data, whereby there is coding, photocopying and then sorting by 'cut and shuffle'. Alternatively, analysts may use specialist qualitative data management software such as NVivo, which enables electronic coding and analysis, or by using standard word processing software.

Quantitative analysis at its simplest level may be a mere comparison of figures. It may involve some descriptive statistics, such as calculation of means. It may include some medians, modes or standard deviations. More rigorous statistical analytical methods involve inferential statistics. Most students should be able to understand and execute the simpler inferential tests; to do the tests is part of hypothesis testing. It requires that you understand the concept of variables; tests seek to determine causes and effects, and whether an independent variable (IV) influences a dependent variable (DV). Some people may have an inherent fear of quantitative analysis or statistics. Be mindful that the use of statistical data is part of everyday life, at its simplest level, dealing with money. If you have a non-negotiable position that you will not do statistics, that is fine, but, be sure that if you do go down the qualitative route, you use robust data collection and analytical techniques as far as is possible.

Constructing Excellence may be used as an example of where qualitative and quantitative data merge or melt into each other. It collects an array of qualitative data to measure client satisfaction. The construction industry needs to know whether its clients are satisfied, and it needs a measurement system to help it monitor its own performance. An output is needed that summarises data, gives quick comparison and allows executives to quickly pick up on areas that need corrective action. The way chosen to do this is to use numbers. Client satisfaction, which has its origins in the collection of qualitative data, is therefore scored on a quantitative scale of 0 to 10. It is important not to allocate a number to client satisfaction 'anecdotally'. A robust set of criteria should be written that would allow a qualitative narrative to be scored within a specified range. The criteria may be in the form of expectations; what would be qualitative expectations of a score in the ranges 0–2, 3–5, 6–7, 8–9 or 10?

It is argued that the best studies comprise the analysis of both qualitative and quantitative data. The qualitative analysis may come first: speaking to people, teasing out issues and problems. The quantitative analysis follows, using numerical data to test hypotheses. The researcher may then revert back to more qualitative data gathering to help in interpreting results and findings from the quantitative tests. The review of the theory and the literature at the early part of the study may be considered to be a qualitative analytical tool, although the review may also include some quantitative analysis. Using both qualitative and quantitative approaches in a study can be called 'mixed methods' or 'methodological triangulation' (Clarke and Creswell, 2008). In the latter case, the three-sided analogy comes from two methods focused towards meeting one objective.

Campbell and Kerlinger quoted in Miles and Huberman (1994) cite two strap lines at opposite ends of the qualitative/quantitative spectrum. Donald Campbell stated, 'all research ultimately has a qualitative grounding', while Fred Kerlinger wrote 'there's no such thing as qualitative data. Everything is either 1 or 0.'

The key thrust of your analytical framework could be quantitative. There are possibilities to flavour your work and to create links into the qualitative school in two ways. The first is the literature review which by default is a qualitative appraisal even though it may include some numerical data. The second is the possibility for you to speak to people, even informally (though better formally in interviews), at various stages of the study. At the early stages it could be around helping you to define the problem and objectives, at the later stages to interpret your results and develop conclusions. You can write up some brief notes that summarise your discussions and place them in an appendix. It would seem very inappropriate for you to jump into a quantitative study, having based it on what you have read; desk bound studies can be insular and misguided.

It would seem reasonable that your work could be substantially, almost exclusively, qualitative throughout. That should not be to the complete exclusion of quantitative data, as though you had a phobia about numbers. You may draw on some quantitative data in the description of the problem or in the literature review, merely by citing some numerical data to perhaps justify the reason for your study. Further qualitative work does not preclude you from needing an appreciation of quantitative methods. Your approach may be to undertake an in-depth qualitative study that investigates a problem, teasing out potential variables. The closing part of the study is a recommendation which sets up a study for another student, perhaps next year. The recommendation should be fully developed, to include the definition of variables and the suggested quantitative analytical tools.

It is not a reasonable expectation that your research should be based on the collection and analysis of substantive qualitative and quantitative data. You can only do what you are able to do, in the given time frame. Research programmes that are well funded may start with an in-depth qualitative investigation, followed by a hypothesis to be tested quantitatively. The whole research programme will close with some more in-depth qualitative work to help in the interpretation of quantitative results, and help in the development of conclusions. It may include a series of pilot studies at each stage, and some intervening qualitative work between the pilot and the main study.

1.7 The student/supervisor relationship and time management

There are lots of factors that influence whether dissertations and projects will be successful or not. A major factor could be the quality of the relationship you have with your supervisor. As far as is possible, it is for students to be in the driving seat. However, supervisors take different views of their role. At one end of the spectrum there are regular meetings, students regularly providing drafts of chapters and supervisors promptly reading and providing feedback, and some supervisors actively chase students. At the other end, are supervisors who do not chase (most will not since they are busy people); they may even refuse to read interim draft submissions, which may be departmental policy. There may be little contact between student and supervisor and a document may be presented by the student that has received no interim feedback.

How much contact there is in a student–supervisor relationship seems to be important. So what level of supervision is appropriate? There is no definitive answer. The volume and type of feedback will vary between supervisors. Most will not identify mistakes in spellings or grammar; feedback may be of a strategic nature, limited to such things as how the overall structure of the documents can be improved, correcting fundamental misunderstandings in research methodology or in the subject area, and identifying gaps. It is absolutely clear that supervisors will not 'write' the dissertation for you. Key learning outcomes of dissertations are that students must demonstrate independent working and use their own initiative in developing their research.

A supervisory grid is illustrated in figure 1.3. You have control over which cell you will be in. Subject to you not having difficulties in your business or personal life, you

Cell A Proactive student Proactive supervisor	Cell B Proactive student Laissez-faire supervisor
Cell C Laissez-faire student Proactive supervisor	Cell D Laissez-faire student Laissez-faire supervisor

Figure 1.3 The supervisory grid and the proactive or laissez-faire relationship. Which cell do you choose?

have absolute control over whether you will be proactive or laissez faire; the latter meaning in the Oxford dictionary 'the policy of leaving things to take their own course, without interference' (Stephenson, 2010). In a research sense, laissez faire may not absolutely mean '... leaving things ...', perhaps giving the impression you are not working; it may be that you can work independently, and if that is the case, excellent. You also have some control over selection of your supervisor; you will know the supervisors who are likely to be proactive and those who are not. You may be quite happy or best suited to a laissez faire supervisor; such supervisors may quite rightly present themselves as such, with a view to promoting your independent working skills. Alternatively, you may need a supervisor who will push, pull and stretch you. One key element to selecting your supervisor, especially if you are in a large university, is to select that person at an early stage in the process. There is a limit to the number of students that supervisors can work with each year. Good supervisors may get fully booked at an early stage. Talk to more than one supervisor to try and tease out who has the expertise and interest in your topic area. Rather than a blank piece of paper and limited ideas, try to take to any meetings a page or so that outlines the problem you may wish to investigate. Demonstrate to supervisors that you have given some initial structured thought to your proposed study.

There are two key elements in selecting your supervisor. First, supervisors should have expertise in the area of your study. Second, you should have some personal chemistry with your supervisor. You may wish to ask your supervisor what the ground rules are in your relationship, such as frequency of contact, potential for feedback, expectation for interim submissions, preferred method of communication, e.g. email.

A possible template that may be agreeable to supervisors is as follows:

- You write a few pages to describe initial ideas and describe what the problem is to be investigated
- You meet with several potential supervisors and select appropriate person.
- You write regularly throughout the process.
- You, submit drafts of your writing.

- Your supervisor provides feedback – you redraft.
- You book face-to-face appointments with your supervisor, possibly every three, four, five, six weeks?
- You then promptly write up notes of these meetings as an aide-memoire for the next meeting.
- You communicate informally with your supervisor, 'regularly'.
- You submit an almost fully complete draft of the document one month before the final submission date, and your supervisor then gives feedback.
- You edit and proofread before final submission.

Note the repeating word 'You'. While some of the above may be 'good practice' and part of the service that universities wish to provide, it may just not be the way you and/or your supervisor want to do it.

Whether or not you make steady progress with your work may be another variable that influences whether your research will be successful or not. Research left to the last minute is less likely to be of good quality. Last minute research is often a tortuous journey for students, that is, it is constantly at the back of the mind as a worry and cause of stress. Research which makes steady progress is most likely better quality and an enjoyable journey. If you are not able to make good progress because of problems elsewhere in your business or personal life, you should keep your supervisor and personal tutor updated. Avoid getting to a late stage in the process before advising of your difficulties and do not just not submit at all on the due date. University systems are generally paternal and supportive in cases of genuine difficulty outside your control.

Figure 1.4 illustrates a possible outcome of a hypothetical research project, the objective of which is 'to determine whether the quality of the dissertation or project process influences

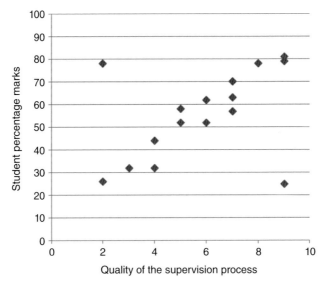

Figure 1.4 The relationship between the quality of the dissertation or project process (IV) and student percentage marks (DV); $n = 16$ (sample size).

student marks'. There could be several concepts wrapped up into 'the quality of the dissertation process'. There is just one measure of 'student marks': the raw number on the scale of zero to 100. If the study were 'real', careful consideration would need to be given to defining each of the variables and to designing a method to measure them. The terminology used later in the text is that the independent variable (IV) is 'the quality of the dissertation process' and the dependent variable (DV) is 'student marks'.

The results indicate that 'the quality of the dissertation or project process does influence student marks'. Marks in the lower left cluster are poor process and poor marks. Marks in the upper right cluster are good process and good marks. There is an outlier at the bottom right; that is a good process but a poor mark – hopefully in reality that would never happen. There is an outlier at the top left; that is a poor process and a good mark. Perhaps this was an extremely talented student, able to work without supervision.

As part of your first proposal to your supervisor you should prepare a bar chart or programme; this can be difficult since to some extent you are going into territory that, for you, is unchartered. Figure 1.5 indicates a typical programme for a dissertation or project that will start towards the end of one academic year and will be completed at the end of the next. It is produced in Excel, since students are often familiar with that software. Specialist planning software could be used too, such as Microsoft Project. Some universities may require that the research be completed in perhaps six months; in such circumstances there is absolutely no time to waste since even a one-year time frame can be tight too.

1.8 Ethical compliance and risk assessments

1.8.1 Physical or emotional harm; laboratory risk assessments

You should ensure that in your study you do not do physical or emotional harm to any person, including yourself. Obviously you will not do that deliberately, but you should not do it accidentally, thoughtlessly or carelessly. Construction is well practised at risk assessments to avoid physical harm. If your research is laboratory based or surveying fieldwork you should undertake a risk assessment and include it in an appendix. That assessment will follow your university procedures, including best practice promoted by the HSE (2014), e.g. avoid, prevent, mitigate: (a) identify the hazards, (b) decide who might be harmed and how, (c) evaluate the risks and decide on precautions, (d) record your significant findings, (e) review your assessment and update if necessary. Section 6.8.4 gives guidance on risk assessments.

The possibility of emotional harm is less well considered in built environment disciplines. Your university will have its own code of practice or similar on its website, detailing ethical standards to be maintained in doing your research. It may be underpinned by three key procedures (adapted from University of Bolton, UoB, 2006):

(1) **Permission-to-do-research form.** For all research, irrespective of the subject area or data collection method, it is likely that you will have to complete a university permission-to-do-research form, and agree and sign it with your supervisor. Appendix A1 includes an exemplar checklist of items that might be included in a permission form. If some items on the checklist identify issues or risks, you will have to describe how you plan to avoid, prevent or mitigate. The form may require that you indicate that you have read the

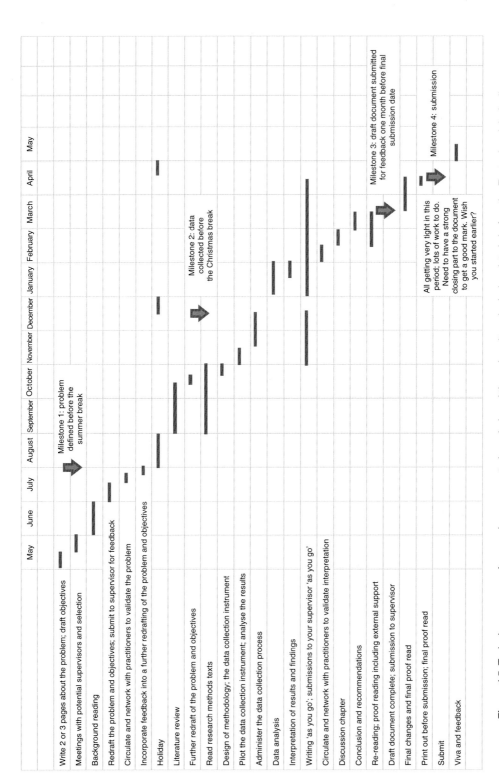

Figure 1.5 Typical programme. Assumed: one-year study period, and the university submission date is before the Easter break in April.

university's ethical code, and that you agree to comply with it. Include your completed form in an appendix to your document, and in your methodology chapter describe the process that you went through to ensure that you complied with ethical rules. If there are some issues which are thought high risk, approval may be needed through an ethics committee. Very unlikely in built environment research, but some disciplines may propose the testing of animals or taking blood from participants. Without very strong arguments to substantiate such methods, permission to do research may be refused.

(2) **The information sheet.** At the point where you come face to face with others in interviews, focus groups (a group discussion) or when observing people, the correct ethical procedure is to give participants an information sheet before data collection begins. In some cases it may be necessary to obtain a signature to evidence informed consent, although it might be sufficient to just make a note in your diary that consent has been given. A typical information sheet, based upon guidelines by the University of Bolton (UoB, 2015), is included in appendix B3. One principle of the information sheet is that it gives potential participants the opportunity to decline involvement in your research; you must not try to force yourself upon unwilling people.

The information sheet should normally include: (i) study title, (ii) identification affiliation and contact details (iii) invitation paragraph (iv) purpose of the study (v) why that person is being invited to take part (vi) what will happen if they agree to take part and how long it will take (vii) possible benefits, disadvantages and risks (viii) assurances of confidentiality and anonymity (ix) what will happen to the results (x) contact details for further information or complaints about the way the research is conducted. Also say thank you.

You will need to give potential participants time to read the information sheet. If you wish to speak to members of the public in a public place, you must introduce yourself politely and step back for a few moments to let them read alone the information sheet. If you are arranging appointments with people, you may be able to send a copy of your information sheet and get consent in advance. If you wish to digitally record any discussions, you must clearly ask for permission, and respect the wishes of participants if they decline; if permission is refused just take some field notes. Always take your student identification card with you and show it to participants.

(3) **The covering letter.** If you propose to collect data from people, though not face to face (e.g. questionnaires that are administered electronically, less often these days by post), you do not need consent forms, since that is implied by returning the questionnaire. However, a covering letter is required in order that potential participants can make a quick informed judgement about whether they wish to reply. You should state how long it will take to participate; if your survey takes five to ten minutes you might get a good response rate, but if it is longer, perhaps not. Section 5.4.3 describes how you should pilot your survey to ascertain a time estimate. Also important is that you should make it clear that participation is voluntary. A sample covering letter is included in appendix B2. It includes information similar to that in an information sheet.

You should not undertake questionnaires or interview or observe people in such a way that it might put participants under pressure, cause anxiety or induce psychological harm. Neither should you do anything that people might consider offensive, even if only mildly offensive, e.g. asking about things that people may consider private, such as family details or personal or company finances. Issues around gender, culture and religion should be treated sensitively.

Obviously you will not convey untruths to people in your research, but making judgements about the issue of deceit can be more difficult. For example, is it appropriate that you covertly observe people in their place of work without advising them you are collecting data for a research project? You will probably be forced to seek permission and notify people if you are a guest in a given situation, but careful judgement needs to be made if your data collection is made alongside your normal work activity, e.g. a study about the effectiveness of site meetings in which you are routinely involved as a participant and in which you now propose to also collect data as a researcher. In this case, should you ask permission of participants to digitally record meetings or take notes, or should you just ask permission of the meeting's chair, or should you just do it without anyone knowing? Since this is not a public meeting, especially if there is to be a digital recording, it is probably best that consent of the meeting is sought through the chair. Also envisage a situation where you may wish to record observations of craftspeople while they work. You may do this as part of your routine employment, or you may spend extra time observing with their knowledge, or perhaps observe from a vantage point without their knowledge. 'Deceiving' people is sometimes justified, providing that the value of the research outweighs the principle of gaining permission. You would have to substantiate this in your permission-to-do-research form. Your argument may be that if participants know that you are collecting data for research, it would unduly impinge upon the study's validity, and prevent tangible benefits being realised. If you are to deceive people it is argued there needs to be a 'greater good for society'. This 'greater good' may be difficult to substantiate at undergraduate level, so openness and asking permission is likely to be the norm.

You must be particularly careful if participants are considered vulnerable – talk to your supervisor. As construction designers, we may legitimately wish to communicate with vulnerable users of buildings and other parts of our infrastructure. You may wish to talk to 14 or 15 year olds about their perceptions of construction as a potential career option. People who are in training or who are accountable to you at work, such as craftspeople, may feel that their job positions are threatened by your wish to collect data. People more senior than you may be fearful about data getting to their superiors or into the public domain. People with a disability or learning difficulties may be reluctant to talk about adjustments they need for their work. People in organisations that are downsizing may be reluctant to talk about anything, and if that is the case, you should respect their wishes.

Part-time students may be able to get data from their own organisations. If it involves other people, permission should be sought from their employer; if it involves taking data from current or archived files, again permission should be sought. If your research involves collecting data in the National Health Service or social services from patients, carers or staff – perhaps about the merits of various building designs in patient recovery – you should gain approval from those agency research ethics committees.

1.8.2 Confidentiality and anonymity

Confidentiality and anonymity are often written of, as though they are the same concept. They are related, but they are different. In a practical sense you will offer participants confidentiality and anonymity, though in your own mind you will recognise there is a difference between the two. Participants do not get complete confidentiality but must always get anonymity.

Confidentiality implies that whatever is written or said will not be shared with other people. That is strictly not the case, since verbatim quotes may be included in documents that will be read by supervisors, external examiners and possibly placed on university library shelves that are accessible to the public. What is important in such circumstances is that there is no attribution to the source; participants cannot be identified and will have their anonymity preserved.

To elaborate the difference between confidentiality and anonymity, consider a situation where you go to your doctor with perhaps a strange phobia. What you tell the doctor is confidential between you and your doctor, though when you leave your appointment the doctor will spend a short time writing down electronically your symptoms and your diagnosis. You may go back to your GP practice some time later with the same problem and see a different doctor; what you said at your first visit is not confidential between you and the first doctor, since the new doctor will most likely read your file before you enter the room. Some medical secretaries in the GP practice may also need to view your file. So what you said initially is not confidential between you and the doctor you initially spoke to, but it is confidential within the bounds of the GP practice. Consider then the situation where several months later one of your doctors is being interviewed in the national media and tells of a young person who came to the practice with a strange phobia. That person is you! The doctor tells the world of your phobia that you thought you were telling the doctor in confidence, and now everyone knows. That is fine. What the doctor does do, is preserve absolutely your anonymity; there will be no small clue that would enable you to be identified.

In the raw data that only you have access to, for example interview transcripts, it is not really necessary to record names of people. In electronic surveys, it would not be usual to ask for names of people. However, consider an example where your supervisor allows you to survey the same group of students on two occasions, with a time gap of perhaps one week. You may be taking the viewpoint of a designer, and wish to measure their learning preference immediately after they have experienced lectures in two different lecture environments. That may involve pairing up scores for each person without using a name; just ask students to label their survey with a code of their choice, known only to them, which would not allow them to be identified personally; but the same code would be used each week.

If names are used anywhere in your research, these documents should be stored securely, with electronic files password protected, just in case pen drives or laptops are lost in public places. Paper files that name people should be shredded on completion of your work. When you refer to this data in your document, anonymity of individuals and companies should be maintained, by typically referring to person A or organisation B. If using a dataset about a company, which is already in the public domain, such as the profit figure for a public liability company, it is acceptable to name names, providing the name cannot be used to find a link to other confidential information. You should offer participants in your study access to a summary of your findings; if they wish to take up that offer, that will require they give you names and contact details which you will tell them you will store securely.

1.8.3 Generally

If you want to observe in a public setting how people use a particular facility, that is ethically acceptable, but if you want to talk to them, you should provide an information sheet. If there are any adverse events during your research, involving either yourself or others, you should report these immediately to your supervisor.

The need to follow ethical rules should not be viewed as a deterrent to you being proactive in your research, and then alternatively encouraging you to undertake passive desk studies. If you consider the checklists in permission-to-do-research forms, that should prevent you making a mistake. Your research permission form is good practice, morally appropriate and is your plan; your information sheet helps your participants and hopefully puts them at ease. If you are able to grasp the principles of research ethics, there is the potential for lots of spin-offs into other spheres of your personal and professional life.

1.9 House style or style guide

Your research document needs to be accessible to all potential readers, and its style needs to be consistent within itself. Some readers have disabilities that make conventionally presented text inaccessible. Some disabilities, such as partial or whole sight loss, affect only a relatively small percentage of people, but dyslexia is relatively common, affecting about 12% of students in higher education. You need to be mindful of how you set out text, font style and font size.

For many people with poor sight, using well-designed printed documents using a minimum of 12 point text is enough, although the Royal National Institute of Blind People recommends 14 point, to reach more people with sight problems. People with dyslexia usually find sans serif fonts such as Arial easiest to read. Left justification is thought to be better than full justification, since the latter necessitates that the spacing of letters within words open to take up the full width of lines. This exacerbates a feeling that some people with dyslexia have, of words merging into one another.

It may be prudent to place a CD ROM version of the document in a plastic wallet stuck to the inside cover. An electronic version allows people to increase font size or use other sophisticated software to read documents to suit their needs. Even if your CD is not used, you would at least be demonstrating to examiners that you have empathy with difficulties experienced by others.

Your university may have a house style that you must follow. Hopefully, it will not be too prescriptive, thus leaving you with a degree of flexibility and personal choice. Some students may find it frustrating to have to follow a style they do not like, and prefer to impose their own tastes on their document. In any case, you should aim for absolute precision in use of the style. Even small differences within the same document can frustrate readers, and distract their attention away from the subject material. Adapting to the discipline of house styles is good practice for industry. Most businesses have styles, so that they have consistency in the way they present information to clients and customers. If you are a part-time student, you may find it frustrating to have to learn two house styles; one in your workplace and one at university. Since the real issues are about accessibility and consistency in documents, and not about the style itself, your university may be happy to let you use your workplace style. If you want to emphasise to examiners that you have not put your document together without regard to accessibility and consistency, write a precept in your preliminary pages stating that you have followed a style, and then put details in an appendix.

If you are not required to follow a style, and do not have one in your workplace, one is suggested below, based on disability literature (ABECAS, 2005). Use the following:

- font type: Arial or Times New Roman
- chapter headings: 14pt bold, sentence case
- subheadings: 14pt bold, sentence case

- main text: 12pt line and a half spacing, sentence case
- table text: 12pt single line spacing, sentence case
- margins 3 cm all round
- one and a half line space between paragraphs; not indented
- one and a half line space between headings and text, text and new heading, and tables/figures and text
- page number at centre bottom of each page, 12pt
- left justification with ragged right edge
- allow generous spacing generally within documents
- black or dark blue print colour

Generally, avoid the following:

- centred text except for headings
- upper case fonts (or capitalisation)
- italic fonts (bold is a better form of highlighting)
- underlining
- roman numerals, e.g. (iii), (iv), (vi), and numbering 3 and 8, 6 and 9, which could be mistaken for each other. Bullet points or (a), (b), (c) are preferred

Unfortunately, the majority of literature in the public domain does not follow accessibility guidelines; it is impossible to make it compulsory.

1.10 Writing style

You are likely to have a word limit or word guide for your document; perhaps 10,000 words. While initially this may seem to be an enormous amount, it is not. A draft may be say 20% higher than your final document, and you will have to work hard to remove irrelevant information. A rule to follow is, 'If what you have written should be in, it should be in; if it should be out, it should be out'. This rule requires a robust editing process. It follows that you must write concisely and with precision. Every sentence must read with absolute clarity – it should mean to readers exactly what it means to you as the writer.

This is not a book about writing style or grammar, but here are some tips arising from common mistakes by construction students:

(1) Do not write in the first or second person. The following first person words should not appear: I, me, mine, my (singular), our, ours, us, we (plural). Nor the following second person: you, yours, your. While they are all acceptable in informal writing, they are not thought to be professional writing styles; a passive approach is required. If you accidentally write in the first person in your first attempt at writing, it may be just that you need to rearrange sentences. For example a first draft may be 'we need to improve the safety performance of the construction industry', but can be rephrased 'there is a need to improve safety performance of the construction industry'. On the occasion that you may need to refer to yourself, you can be author or the writer, not 'I', e.g. 'this study arose from the author's experience working in private practice'.

(2) Avoid writing in the singular; do write consistently in the plural. Therefore, instead of a manager, a contract, a client, a subcontractor, write managers, contracts, clients, subcontractors. This can be really difficult, even with practice, but it does have advantages:

    (a) It helps keep writing concise. A sentence that reads in the singular 'the project manager is responsible for the quality on the site', reads more fluently and with four fewer words in the plural as 'project managers are responsible for quality on sites'.

    (b) It can eliminate inappropriate mixings of singulars and plurals, so that a sentence written 'an architect should be the chair of a meeting and they should ensure …' (a singular architect is not a they, but a 'he' or a 'she') is better as 'architects should chair meetings and they should ensure …'.

    (c) It almost fully eliminates the need to address the gender issue. Some people are quite happy to call a manager 'he', while others may be offended. You do not want to offend anyone, therefore avoid it. In the singular, you may write 'the architect should chair the meeting and he should ensure …'. In the plural this becomes 'architects chair meetings and they should ensure …'. You can write in the singular 'an architect should chair the meeting and he/she should ensure …' but this attracts readers' eyes to the gender issue. It is clearly acceptable to use the male or female gender if you are talking about an individual, e.g. 'Mr Smith chaired the meeting and he …'. In your literature review you may cite Richardson (2009), and 'he stated …'; but be sure it is 'he'. But don't get too hung up about the gender issue. Whether you write for example craftsmen or craftspeople is not likely to make any difference to your mark. UK statute routinely uses the 'he' to also mean 'she', using a precept such as 'any reference to the masculine gender shall be taken to include the feminine'. If you are most comfortable writing of craftsmen, you can do so. Deflect the potential for criticism by including a statement in your preliminary pages or introduction chapter. But otherwise, still try to avoid gender in your writing where possible.

    (d) Writing in the plural promotes consistency in your writing. It some cases it will be inevitable that you write in the plural. It would be a contradiction to write in one paragraph of a singular civil engineer, and then in the next of plural projects.

    (e) Finally, the purpose of research is often to seek out how the population at large behaves. It would seem inappropriate therefore to write of the population as though it comprised a singular company, person or project.

(3) As noted in 2(b), mixing singulars and plurals also occurs when students write of organisations like 'a government', 'a professional body' or 'a company' in terms of the plural 'they'. Each is represented by many singular people coming together as 'they'. While the latter is correct, individually, these organisations have a singular legal identity. A sentence that reads with the plural (they), 'government decided they would change the law', may better read with the singular (it), 'government decided it would change the law'.

(4) Double speech marks for verbatim quotations or citations and single speech marks for highlighting or emphasis or separating from the main text. For example you may cite thus with double speechmarks, the then chief construction advisor to government Peter Hansford, who wrote in the context of skills shortages in construction "Too few teachers know very much about the construction industry. We must create strong, lasting partnerships if we're to develop a talent pipeline of eager, well-informed, inspired new recruits" (Hansford, 2015). Alternatively, use a single speechmark for emphasis or to separate a word from text e.g. the most important word in this text book is 'objectives'. Also use single speechmarks for colloquialisms.

(5)   You should only usually use colloquialisms (not formal or literary), slang and nick-names if you are citing verbatim; otherwise avoid them. You must always write in professional academic language. Some of colloquialisms may arise from the construction industry itself or as local dialect. Not all readers may understand them, e.g. a mobile elevated working platform (MEWP) is informally called a 'cherry picker'. Not everyone may know that, and you should therefore formally call it an MEWP. If the context of your narrative is that you *need to* write of a 'cherry picker' or use slang or nicknames, that is fine, but put them in single speech marks to illustrate that this is not part of your professional writing.

(6)   Avoid using words abbreviated by apostrophes, e.g. can't, won't, don't, shouldn't. Write them out in full as cannot, will not, do not. These sorts of abbreviations are acceptable if you are taking them verbatim from other sources. If the latter is the case, again include them in double speech marks.

(7)   While your role is not that of a censor, it is probably best that you do not use any bad language, mild or strong. In qualitative work it can be important to understand the passion or anger that people may feel about some issues. People may use bad language to express their feelings. Passion or anger must not be lost in analysis. You may type up verbatim transcripts of interviews, and include them in appendices. Some readers may be relaxed about bad language that is left in. Other people may be offended, which may leave you open to criticism. You do not wish to offend anyone. Therefore, whether in the main body of your document or the appendices, it is suggested that you take bad language out. Substitute it with an appropriate number of dashes. Readers will understand.

(8)   Make sure that you can use apostrophes correctly. The two key uses are to indicate possession and to indicate missing letters in the middle of words. It is the possession issue that causes students most problems. For example, 'the house of the architect'; by deleting the word 'of' (the possessor) and switching house and architect around, it becomes 'the architect's house'. The apostrophe is used in place of the word 'of'. If the house is owned by more than two architects, the apostrophe is after the s, thus 'the architects' house'. If you are not sure how to use apostrophes, find out. There are lots of university guides and useful websites that you can locate through search engines. Imagine the discussion that may take place between two examiners who disagree about the mark for a document; one argues for a high mark, while the other suggests a lower mark stating 'this student cannot even do apostrophes'. Imagine employers discarding CVs because candidates cannot even do apostrophes.

(9)   Make consistent use of numbers written as figures or as words, that is the word 'one' or the figure '1'. By convention, anything less than 10 is written as words. Numbers used in calculations should of course be kept as figures; similarly ages, measurements and percentages. Also, 10% is better than 10 per cent.

(10)   Which writing tense to use often causes difficulty. There are four types of tense (a) past tense, which is subdivided into imperfect, perfect and pluperfect (b) present, (c) future and (d) conditional. In simple terms people often refer to only three: past, present and future. Students often write without regard to tense, and switch tenses within adjoining passages of text. You need to be mindful of the tense in which you are writing at each stage of the document. The abstract is written as a concise summary after the whole of the document has been completed. What tense is appropriate?

- present tense: 'the objective is … the method is … it is found that … it is concluded that …'
- past tense: 'the objective was … the method was … it was found that … it was concluded that …'
- a mixture to show that the beginning of the study is past, but the end of the study is present, thus 'the objective was … the method was … it is found that … it is concluded that …'

The introduction chapter will be written as the document proceeds. Some parts will be written at the beginning of the process and some towards the end. The introduction is setting out what will appear in future chapters, but it is also telling the story of what has happened, in a real time frame, in the past. What tense is appropriate?

- 'the objective is … the method was … chapter 2 will outline …'?

The literature review is about work in the past. If writing about work that is well established and dated, the past tense will usually be used, e.g. 'McKay (1943, p. 15) detailed a sketch of a foundation for a one and a half brick wall.' Note 'detailed' in the past tense, not 'details' in the present tense. However, if the work is more recent, it may be written as 'Ashworth and Perera (2015) describe value as 'a comparative term expressing the worth of an item or commodity, usually in the context of other similar or comparable items', or 'Ashworth and Perera (2015) described value as …'; that is 'describe' present tense or 'described' past tense. On the one hand, the methodology chapter tells the story of what you have done in the past, but it also describes what will be presented in the middle of the document. Therefore, it may be written as 'a survey of existing buildings was (past) undertaken, and the data recorded will be (future) analysed in the next chapter.' The analysis chapter may use the present tense; the result is 'this' and 'this is' found. Discussion about the results and findings may be in the past tense – 'in the last chapter the result was … and it was found that …'. Conclusions may start to arise in the discussion chapter, and may be in the future tense 'it will be concluded that …'. As you get to the end of the document you may revert to the present tense 'it is concluded that …'. There is no definitive answer about which tense should be used. The important point is that you should pick your tenses deliberately, recognising in your own mind instances where you have made choices and remaining consistent in those choices.

A useful guide, adapted from Parkin (2005), though not consistent with the above, is thus:

Abstracts: best in the present tense e.g. 'The aim of the study is to …' and 'The results show that …'

Introduction: best in the present tense, but a lapse into the future tense is also possible e.g. 'The literature review demonstrates that …' or '… it will be shown in the literature review that …'

Literature review: could be either in the present or past tense depending on the age of the research and the context and particular sentence structure being employed, e.g. 'Smith (1952) developed the theory of XYZ and Jones (2015) shows that data still supports this theory'.

Methodology, analysis, results, findings, discussion: usually written in the past tense. 'Data from XYZ was analysed and showed that…' Avoid the temptation to write in the future tense, particularly the method, even if you are writing up before carrying it out.

Conclusions: usually in the present tense, e.g. 'It is concluded that …'

1.11 Proofreading

Proofreading is an essential part of all your writing. It must be thorough and meticulous. It must cover content, grammar, spelling, apostrophes, layout, house style and presentation. You should aim for perfection. You should be confident about putting your document on the desk of chief executives or other captains of industry. You should proofread your own work as much as possible. On the one hand you may aim to get chapters 'signed-off' as though floors on a building or phases on a civil engineering project that are being handed over. But, when writing subsequent chapters, you may need to go back to signed-off work to make amendments. Part of the process is that you will be rereading and rereading your work again on a continual basis. You must continue to polish it and polish it again; it will be the result of many iterations. A time gap between reading helps. If you 'finish' a chapter and then go back to it a week or so later, you will no doubt find many things you want to change to make it read with greater clarity, or text to delete or add.

If you are doing this thoroughly, there will come a time when you are just too close to your work. There will be mistakes that you will never find. To get the perfection required will need the independent help of a proofreader. Proofreading is a profession in its own right. It is not expected that you will pay fees to seek professional help. The proofreader may be a member of your family, friend or work colleague. Construction expertise is not necessary, since hopefully you will have been networking with construction professionals and your supervisor throughout the process. There is a line that you must not cross with proofreading by others. Proofreading is proofreading; it is not for someone else to rewrite your work. Students with dyslexia may be able to make a case for greater support in the proofreading process through their university disability advisors. Whatever support you do get with proofreading, declare it in your acknowledgements in the preliminary pages. In the final stages you are building up to the final print-off. In reality there may be several 'final' print-offs. The final print-off may go to the proofreader, who may find some typographical errors. Be mindful that even relatively minor changes can lead to substantive disruption to document presentation. In your final, final document, graphs, pie charts, histograms, photographs or maps should be colour printed.

During the production of your document, you should be repeatedly using the spell and grammar check functions in your software. Perhaps the very last task that you should perform before printing, is to spellcheck your document one final time. At the closing stages when you are doing final editing, there is always the possibility that you will accidentally type something that includes a mistake. Figure 1.6 illustrates the spell check function in Microsoft Word; this function will detect typographical errors other than just spelling mistakes, such as apostrophes, incorrect capitalisation and incorrect spacing between words.

One important element of proofreading is to check references. This chapter has 19 'citations in text' and 19 references; in alphabetical order that is from ABECAS (2005) to UoB (2015). In earlier drafts of the chapter there were fewer citations; some have been added and also some taken out. Care was taken to ensure that as well as changing the main text, citations and references match each other at 19.

If during proofreading you find a mistake that you know will apply to many parts of your document, use the Find and Replace function in Word to locate the other instances, as illustrated in figure 1.7.

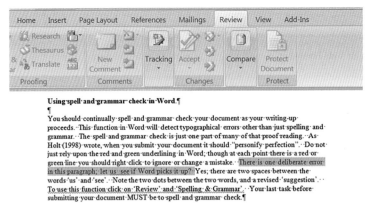

Using spell and grammar check in Word.¶
¶
You should continually spell and grammar check your document as your writing-up proceeds. This function in Word will detect typographical errors other than just spelling and grammar. The spell and grammar check is just one part of many of that proof reading. As Holt (1998) wrote, when you submit your document it should "personify perfection". Do not just rely upon the red and green underlining in Word; though at each point there is a red or green line you should right click to ignore or change a mistake. There is one deliberate error in this paragraph; let us see if Word picks it up? Yes; there are two spaces between the words 'us' and 'see'. Note the two dots between the two words, and a revised 'suggestion'. To use this function click on 'Review' and 'Spelling & Grammar'. Your last task before submitting your document MUST be to spell and grammar check.¶

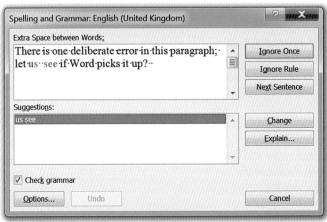

Figure 1.6 The spelling and grammar check facility in Word used to locate typographical errors.

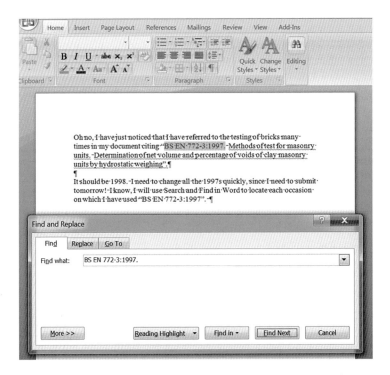

Figure 1.7 The find and replace facility in Word used to locate repeated errors.

1.12 Extra support?

There are examples of students who get to the final stages of their study before they realise that they qualify for extra support from their university. Support may be needed because they have Asperger's syndrome, dyslexia, dyspraxia, mental health difficulties and such like. Some of these students may be quite 'mature', and when they are questioned by academics respond something like, 'Well I knew there was a bit of a problem at school, but I never got around to doing anything about it'. If you think you may have a problem, or are unsure, get yourself assessed by disability specialists in your university. There are many opportunities for you to disclose any problems, for example at enrolment, when speaking to your personal tutor or by approaching central service units in the university. When you do inform them, you are promised confidentiality, and in this context that means only people who need to know are told.

Disability legislation requires that universities make 'reasonable adjustments' in their delivery and assessment to meet your individual needs. In examinations, extra time may be permitted, typically up to 25%. In the context of courseworks and dissertations or projects, there are a host of 'reasonable adjustments' that may be possible: submission dates may be extended or the type of assessment changed. If you have a medical condition that makes it difficult for you to do a presentation to your class peers, perhaps you can give the presentation informally to your tutor? In the UK, you may be eligible for disabled students' allowances (DSA). Such an allowance may enable you to attain laptops, software, specialist tutors, support workers who can be with you in class and take notes, or amanuenses (scribes in examinations).

Students sometimes find themselves in difficulty trying to study and simultaneously act as carers for members of their family, or have one-off problems in a variety of circumstances. You should speak with your personal tutor; as noted in section 1.7, university systems are generally paternal and supportive in cases of genuine difficulty outside your control. You can be given leads to organisations independent of the university to help you.

It may be that you develop personal or family difficulties during your period of study. Being a student can be stressful. Many universities are members of the National Association of Disability Practitioners (NADP), and are able to share good practice among themselves. One recent initiative is the provision of 'life lounges', somewhere to relax and where there is close and free access to specialist services such as counselling, study skills, cognitive behaviour therapy (CBT) and student liaison officers

Students who write and speak English as a second language may need some extra English classes. One goal of study in a UK university may be that on completion of your study you are able to speak and write like a British person. Your aspiration may be that when you do write, readers will not know that English is your second language. Some UK based students may feel they need extra English language support too, or indeed support in other areas such as mathematics. Your university will provide extra support classes free of charge; go and seek them out and use them.

1.13 A research proposal

Some universities require you to submit a research proposal which is marked; perhaps as much as 20% or even 40% weighting of the overall mark. Universities often submit research proposals to seek funding to support postgraduate, doctoral and post-doctoral work over long periods of time. Applications can be for large sums of money. Assessors of these proposals can only come to informed judgements about whether to award money if the proposals are fully detailed.

If the final submission date for your document is towards the end of the academic year, the research proposal may be required before the Christmas break. Your university will guide you about what is needed, but it will usually require that you (i) describe the problem to be investigated, (ii) give your aim and objectives, (iii) execute at least a preliminary literature review, and perhaps most importantly (iv) propose a data collection and analytical method that will enable you to meet your objectives. This last part is particularly tough at relatively early stages of your study; how can you propose some research methods when you are still learning how to do research? The answer to this has to be that you must read research methods texts at an early stage of your study, get involved in any support or seminar classes that the university provides, and meet with your supervisor as often as is reasonable. Appendix D provides eight examples of research proposals.

1.14 A viva or viva voce

Some universities require that you have a viva voce – Latin for an examination by word of mouth – and often abbreviated to viva. It is not a requirement of all universities for undergraduate work, so check your university documentation or ask your programme leader. Where a viva is a requirement, models in universities vary. It may count for perhaps 10% of your mark. Unfortunately some students submit a document that comprises very little of their own original work; for universities, the viva will help to detect whether this has happened.

Hopefully, the viva will be in an informal and relaxed environment, though you might want to dress as though for a job interview. Perhaps there will be two examiners: one the person who acted as your supervisor and a second person with some knowledge of your subject area. To prepare for the viva, it is advisable to reread your document a short time before, since it is likely that there is a time gap of several weeks between submission and the viva. You may be required to give a PowerPoint presentation with a fixed time limit, or you may just be asked speak freely at the beginning of the viva without the support of PowerPoint. Perhaps tell the examiners why you picked your subject area, what the problem is, your aim and objectives, the key literature sources, the methods you used to collect and analyse your data, what you found, what your conclusion is and what you recommend, all of this in perhaps just two minutes. Hopefully you can speak passionately about your subject area.

One view of the viva is that it is truly a verbal examination, a rigorous interrogation of the document. Another is that it brings for you closure to your research, where there is an enjoyable discussion around your work in which supervisors will learn too. From your side, view it as the latter; that will help you relax. In the viva you will be asked questions; the second person may ask most questions since your supervisor will hopefully have a good understanding of what you have done. You may be asked to substantiate your methodology: why you have done it this way and not that, why you favoured this piece of literature, why this method of analysis, why this data collection process, why you wrote this on page x? You should consider yourself to be in a strong position, particularly if you are confident in your work. Nobody knows your document better than you.

A viva can be an advantage, since an examiner may read parts of your work and think 'my best guess is that the student does not really understand what has been written' and will be minded to mark your document down on that basis. No doubt you will be asked about this in the viva, and if you can verbally demonstrate your understanding, your document will be marked up.

At the end of the viva you may be asked to leave the room while the two examiners reflect on some feedback for you and agree a proposed mark. You may be invited back into the room and given that feedback, perhaps also given a mark that will be proposed to the external examiners. If there is disagreement between the two examiners about the mark you should be awarded, a third academic will usually be invited to give a judgement too. The overall duration of the viva will be perhaps 20 minutes with a further 10 minutes for agreeing a mark and feedback.

The above narrative is one way to do it; you need to check with your university which way it does it.

Summary

The dissertation or project is the flagship document in your degree. Language used to describe your research goals can be fuzzy, but clarity in your objectives is essential. The opening part of your document may be an introduction and literature review. The middle part may describe your methodology and present your analysis/results. The closing part may be discussion and conclusions. When selecting a topic area, you should speed up your general reading around current issues. You may choose to do qualitative or quantitative analysis or a mixture of both; whichever is used, the analytical tools must be robust. If you are a civil engineering student, discuss with your programme leader whether there is a requirement for your work to be technical. You should be clear about the way you and your supervisor will work together. The presentation style of your document must be consistent within itself, and you should consider readers who may have partial sight, dyslexia or similar. You must read and understand your university's ethical codes; do not harm or offend any person during your work. You should try to complete your document one month before the final submission date, and use the last weeks to make improvements that may be suggested by your supervisor and for proofreading.

References

ABECAS (2005) *Guidance to higher education institutions: removing barriers and anticipating reasonable adjustments for disabled students in built environment degree programmes.* Farrell, P. and Middlemass, R. The University of Bolton. www.cebe.heacademy.ac.uk/news/abecas/index.php DOI: 10.13140/RG.2.1.5079.9205.

Ashworth, A. and Perera, S. (2015) *Cost Studies of Buildings.* Routledge: Oxon. Chapter 1.5.

BBC (2005) 1953: Derek Bentley hanged for murder. http://news.bbc.co.uk/onthisday/hi/dates/stories/january/28/newsid_3393000/3393807.stm Accessed 19.09.15.

BIS (2013) 'Construction 2025: industrial strategy for construction – government and industry in partnership'. https://www.gov.uk/government/publications/construction-2025-strategy Accessed 26.08.15.

BREEAM (2015) The world's lead design and assessment method for sustainable buildings. BRE. Building Research Establishment. www.breeam.org/ Accessed 27.08.15.

Brundtland, G.H. (1987) *Our Common Future*. Report of the World Commission on Environment and Development. www.worldinbalance.net/intagreements/1987-brundtland.php Accessed 28.08.15.

Clarke, V.L. and Creswell, J.W. (2008) *The Mixed Methods Reader*. Sage, London.

Constructing Excellence (2015) The single organisation driving change in construction. Improving industry performance to produce a better built environment. http://constructingexcellence.org.uk/ Accessed 28.08.15.

Hansford, P. (2015) A letter from the chief construction adviser. Building 16.01.15. www.building.co.uk/ analysis/comment/a-letter-from-the-chief-construction-adviser/5073088.article Accessed 26.09.15.

HSE (2014) *Risk Assessment; a brief guide to controlling risks in the workplace*. Health and Safety Executive. www.hse.gov.uk/pubns/indg163.pdf.

HSE (2015) The Construction (Design and Management) Regulations 2015. www.hse.gov.uk/ Construction/cdm/2015/index.htm Accessed 26.09.15.

McKay, W.B. (1943) *Building Construction*. Vol. 1. Longmans, London. p. 15.

Miles, M.B. and Huberman, A.M. (1994) *Qualitative Data Analysis*. Sage, London.

Parkin, J. (2005) *Notes on the structure of research and the research document*. Unpublished handout. University of Bolton. p. 13.

Quotations Page (2015) Quotation Details. www.quotationspage.com/quote/2933.html Accessed 23.07.15.

Richardson, S. (2009) Morrell takes the stage with a warning to government. *Building* 27.11.09. pp. 10–11 issue 47.

Stephenson, A. (2010) *Oxford Dictionary of English*. 3rd edition. Oxford. Oxford University Press. Available online.

UoB (2006) *Code of Practice for Ethical Standards in research involving Human participants*. University of Bolton. www.bolton.ac.uk/Students/PoliciesProceduresRegulations/AllStudents/ResearchEthics/ Documents/CodeofPractice.pdf Accessed 07.07.15.

UoB (2015) *Example 1. Participant information sheet*. University of Bolton. www.bolton.ac.uk/Students/ PoliciesProceduresRegulations/AllStudents/ResearchEthics/Home.aspx Accessed 07.12.15.

2 The introduction chapter to the dissertation or project

The titles and objectives of the sections of this chapter are the following:

2.1 Introduction contents; identify the parts required in an introduction
2.2 Articulation or description of the problem and provisional objectives; to emphasise that a well developed narrative around a problem provides the best foundation for a study and helps to identify provisional objectives

2.1 Introduction contents

Writers need to focus very carefully on what an introduction should contain. Most academic courseworks contain an introduction. The introduction is not a bit of a chat or the first part of the whole. The Oxford Dictionary defines an introduction as 'a thing preliminary to something else especially an explanation section at the beginning of a book' (Stephenson, 2010).

The introduction chapter should tell readers where the rest of the document or the chapter is going – what items will be covered. Having read the introduction, readers should have a sound platform from which to continue reading. There should be no surprises in remaining chapters, and it is just for readers to continue their read to gain greater understanding and ascertain details. You should start writing your first draft of the introduction chapter at an early stage, but you will not be able to complete it until near the end, since it reports on what you have done in the latter stages. Students often write short introductions; that should not be the case. They should be a substantive part of your document. The introduction chapter should include the following:

(a) introduction to the introduction
(b) definitions of important phrases – this is more than a glossary of terms and is used for one or two key phrases which are fundamental to your work. The definitions should again be drawn from authoritative sources, and may bring together conflicting

Writing Built Environment Dissertations and Projects: Practical Guidance and Examples, Second Edition.
Peter Farrell, Fred Sherratt and Alan Richardson.
© 2017 John Wiley & Sons, Ltd. Published 2017 by John Wiley & Sons, Ltd.
Companion Website: www.wiley.com/go/Farrell/Built_Environment_Dissertations_and_Projects

definitions. For example, research that examines building information modelling (BIM) may draw upon three authoritative definitions:

BIM is essentially value creating collaboration through the entire life cycle of an asset, underpinned by the creation, collation and exchange of shared 3D models and intelligent, structured data attached to them (BIM Task Group, 2015)

… the means by which everyone can understand a building through the use of a digital model. Modelling an asset in digital form enables those who interact with the building to optimise their actions, resulting in a greater whole life value for the asset. (National Building Specification, NBS, 2015)

… a collaborative way of working, underpinned by the digital technologies which unlock more efficient methods of designing, creating and maintaining our assets. BIM embeds key product and asset data and a 3 dimensional computer model that can be used for effective management of information throughout a project life cycle – from earliest concept through to operation (HM Government, 2013).

Give all three or more definitions in your document, and then use them to write your own definition in the context of your research, acknowledging that your definition is adapted from the three authoritative ones.

Alternatively, there may be one well established quotation from the literature that epitomises your work. For example, in the context of a problem around tendering methods, a piece of research might start with the writing of Latham (in Wolstenholme, 2009):

A number of clients are being led by their construction costs consultants to abandon frameworks and go back to lowest price tendering. That is a mistake. Partnering and close collaboration between the client and the whole construction team will mean that the project will come in to quality, time and cost … But if lowest price is demanded by the client, the tender price will not be the actual financial outturn at the end of the project, because the supply side will be looking for claims and variations to make up for what was not in the tender.

(c) background to the topic area and definition/articulation or description of the problem

(d) possibly a historical perspective – this need not be an in-depth narrative, but it can be useful to readers to establish a context for the study. Subject areas are rarely 'new', and they may have origins going back many decades in UK construction or internationally or in other industries. For example, many initiatives in construction have their roots in manufacturing.

(e) research goals – the aim of the study, research questions, objectives and hypotheses. 'Research goals' is a useful umbrella label for all these terms. Identify the variables you will measure and define them in the context of your study. If for example one of your variables is 'quality', that can mean different things to different people. What does it mean in the context of your study?

(f) an outline methodology – this will give a brief summary of the methodology and may include a brief diagram or model of the variables being measured. The model should link the aims, objectives and hypotheses. The methodology, as described in the abstract, might be two or three sentences. The outline methodology as described in the introduction might be two or three paragraphs. But there should be a full chapter devoted to a detailed description of the methodology later in the document.

(g) an outline of the remaining parts of the document – this will tell the reader where the rest of the document will go. It is necessary to explain the intended structure and the route that the work will take. It will briefly tell readers about the contents of each chapter.

(h) a summary of the introduction chapter – one or two sentences to concisely summarise each page or theme

2.2 Articulation or description of the problem and provisional objectives

Albert Einstein (1879–1955) the German-born physicist is quoted as saying 'If I had an hour to solve a problem I'd spend 55 minutes thinking about the problem and 5 minutes thinking about solutions' (Quote Investigator, 2015).

There is agreement that research in construction is often problem-based; this is supported by Creswell (1997, p. 74) and Silverman (2001, p. 5) in the fields of human and social sciences. However, some terminology in texts is variable, for example according to Hart (2001, p. 9) the first part of a research document should be a rationale; Holt (1998, p. 9) asks for a broad discussion, Naoum (2013, p. 12) suggests a purpose, and Fellows and Liu (2008, p. 42) a proposal. This does not matter, since a definition of all these terms usually includes 'articulation of the problem'.

One of the key skills of higher level study is problem-solving; indeed it may be argued that it is 'the' primary skill. The examination of an industrially orientated problem helps to integrate academic and industrial communities; in many cases sponsorship of academic work is only offered if output can be fed back into the market environment. Articulation of the problem is the starting point for a research project. It provides the foundation for the study. A really well thought out narrative at this stage gives an excellent platform for good work. It should be the subject of many iterations, so that improvements are made as the work progresses. It may be based on a modest amount of reading, and will be subsequently redefined as the researcher becomes more knowledgeable. Part-time students may wish to draw on problems from their own work environment. Collecting data for analysis at a later stage of the study can be difficult in research; it follows that if the problem is based in the workplace, you may have relatively easy access to data. Any evidence found to support the articulation of the problem should be cited. Redefinition may result from some exploratory interviews or from the early stages of the literature review. A key concept in research is to 'circulate', that is, get out and talk to people about your study. Do this informally. If they are students or academics, talk to them about their study so that you can learn about broader concepts and methods of research. Talk to people at all stages of the study. Examiners may wish to see that students have been proactive in their research, rather than passive. Do not be shy. Desk bound studies can be labelled dangerous, insular, blinkered and not valid. There are many forums in which to talk informally with knowledgeable people in your profession. For example, professional body seminars often welcome students, even if they are not in membership. Get up, be proactive and travel.

Take the first drafts of your narrative to your supervisor and other people for feedback. Undertake some formal unstructured interviews, perhaps two. The basis of these interviews may be 'Look, here is the problem as I see it. What do you think?' Ask about possible omissions in your work; have you fundamentally misunderstood something? You may get some leads to important publications or other authoritative sources that you would otherwise have missed. Redraft your articulation of the problem after your interviews. When you write up your research methodology, describe how you did some initial reading and writing, and then you undertook some interviews before redrafting your work. This will all lend to supporting the validity of your work.

It is unlikely that students will have the resources to take on both robust quantitative and qualitative analytic methods within their work, that is, triangulation or mixed-methods approach, as suggested by Somekh and Lewin (2005, p. 274). Whether the main analytical method is quantitative or qualitative, the validity of the work can be improved and flavoured by getting out and speaking to people.

It may be useful to put the problem in the middle of a continuum. Before the problem (to the left) are the perceived causes, the things that may have an influence on the problem. In the middle is the problem itself, and the output (to the right) of the continuum is the consequences of the problem – what is it that results, or how does the problem manifest itself, and what are the effects? There is some reflection and early speculation about cause and effect.

Figure 2.1 illustrates the use of safety as an analogy to develop the description of a problem. Possible causes of safety problems originate in a lack of investment. This manifests itself in incidents and accidents that can be supported by some statistics, such as considering the national or international picture, the number of accidents, the number of lost working hours. The consequences are losses of time/money in the production process, negative image labelling for industry and not least, of course, the human consequences for victims and their families. The research process often involves finding evidence to prove that perceived causes of problems are actual causes; it is to seek out the variables. In business terms, if there is evidence that a known cause has a known effect, it may be possible to manipulate that known cause such that there is a change also in the effect. For example, training reduces accidents, so improved training can be 'managed' or 'manipulated'. The proof or evidence found should then be fed

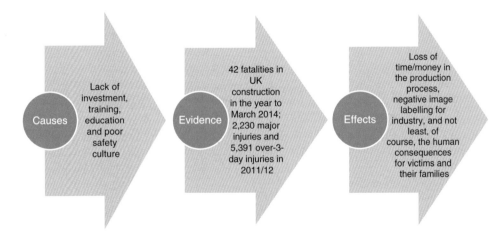

Figure 2.1 Causes, evidence and effects of a problem.

into the research conclusions and recommendations. The thought process at this time may be around the concepts of cause and effect, but as the study progresses, the language of cause and effect may change to variables: independent variables (IVs) and dependent variables (DVs). Appendix D contains eight research proposals as exemplars in the following areas: transportation, historic buildings, timber, structures, rainwater harvesting, health and safety, concrete and underground cable strikes. Each of these research proposals contains a description of the problem to be investigated.

Initial articulation of the problem need not be in too much depth – two or three pages will perhaps suffice, but it can be added to as you enhance your knowledge at the literature review stage. It may be beneficial to conclude the articulation of the problem with one, two or three provisional objectives. These objectives are written mindful that they are likely to change as work progresses, since your knowledge in the topic area is still being developed. They will however give you some early focus, and they are useful for recognising change in the study as it takes place. The objectives should be constantly under review, until that is, the end of the literature review, when they should be 'written in stone'. The articulation of the problem, the final output being the result of many iterations, should also be 'written in stone' at the end of the literature review. The task for researchers beyond the literature review is to 'merely' meet the objectives that are set, employing an appropriate methodology.

The articulation of the problem should come before writing provisional objectives. Students often comment at an advanced stage of their studies, 'I am not sure what my objectives are'. If the articulation of the problem is well thought out and developed, the objectives should 'hit the reader in the face'. They will be obvious. An alternative approach to writing the objective at this stage is to decide what you want to measure. Measurement is an important concept that underpins what you will do in your research. As the study progresses, you may develop the idea of measuring two things – an independent variable and dependent variable; see chapter 4. Perhaps it is sufficient just to think about measuring one thing at the moment, such as compliance with best practice initiatives in health and safety in construction. That would seem a useful thing to measure. How does the UK construction industry score on a scale of 0 to 10? We would hope 9 or 10, but perhaps the answer is only 6 or 7. Do the research and find the answer to this question.

If you rush into research work without properly defining the problem, your work is potentially flawed. Since the description of your problem is at the start of your research, think of it as the foundation to a house. The design specification is 600×150 mm, 28 N/mm^2 in-situ concrete laid on firm ground. You must not lay an undersized foundation with a weaker mix concrete on soft ground. The specification for the description of your problem is a narrative that is well thought out, about three pages long, the result of many iterations and dialogue with your supervisor and industry practitioners. As figure 1.4 suggests, it should be well developed before your summer break. Figure 2.2 illustrates stretcher bond brickwork constructed on weak foundations; analogous to research completed on a poorly defined problem. Is this your research?

Summary of this chapter

The introduction chapter sets the scene and tells the reader where the remaining parts of the document will go. It should be a substantive part of your document. A well-defined problem that is based upon several iterations will provide a strong foundation for the study and help to

Figure 2.2 A crack in a research document founded on a poorly defined problem.

identify objectives. You should take first drafts of the description of your problem and provisional objectives to others, and ask what they think.

References

BIM Task Group (2015) BIM frequently asked questions. www.bimtaskgroup.org/bim-faqs/ Accessed 29.08.15.

Creswell, J.W. (1997) *Qualitative inquiry and research design: choosing among five traditions*. Sage, London.

Fellows, R. and Liu, A. (2008) *Research Methods for Construction*. 3rd edition. Blackwell Science: Oxford, UK.

Hart, C. (2001) *Doing a literature search*. Sage, London.

HM Government (2013) Building Information Modelling. Industrial strategy: government and industry in partnership. https://www.gov.uk/government/uploads/system/uploads/attachment_data/file/34710/12-1327-building-information-modelling.pdf Accessed 29.08.15.

Holt, G. (1998) *A guide to successful dissertation study for students of the Built Environment*. 2nd edition. The University of Wolverhampton, UK.

Naoum, S.G. (2013) *Dissertation research and Writing for Construction students*. 3rd edition. Butterworth-Heinemann, Oxford.

NBS (2015) What is BIM? National Building Specification. www.thenbs.com/bim/what-is-bim.asp Accessed 29.08.15.

Quote Investigator (2015) Exploring the origin of quotations. Albert Einstein. http://quoteinvestigator.com/2014/05/22/solve/ Accessed 24.07.15.

Silverman, D. (2001) *Interpreting Qualitative Data*. 2nd edition. Sage, London.

Somekh, B. and Lewin, C. (2005) *Research methods on social sciences*. Sage, London.

Stephenson, A. (2010) *Oxford Dictionary of English*. 3rd edition. Oxford University Press. Available online.

Wolstenholme, A. (2009) *Never waste a good crisis. A review of progress since Rethinking Construction and thoughts for our future*. Constructing Excellence. https://dspace.lboro.ac.uk/dspace-jspui/bitstream/2134/6040/1/Wolstenholme%20Report%20Oct%202010.pdf Accessed 24.07.15.

3 Review of theory and the literature

The titles and objectives of the sections of this chapter are the following:

3.1 Introduction; to provide a context for the literature review
3.2 Style and contents of a literature review; to describe how to write a review
3.3 Judgements or opinions; to distinguish between the authority of each
3.4 Sources of data; to distinguish between and to identify the most rigorous material
3.5 Methods of finding the literature; to identify the tools to be used when searching
3.6 Embedding theory in research; to encourage students to address theory in their research, and to distinguish between theory and literature
3.7 Referencing as evidence of reading; to emphasise the necessity of reading and citing
3.8 Citing literature sources in the narrative of your work; to illustrate how to 'cite in text', and how to cite verbatim or by paraphrasing
   3.8.1 Verbatim citations
   3.8.2 Paraphrasing
   3.8.3 Secondary citing
   3.8.4 Who to cite in your narrative
   3.8.5 Page numbers and emphasising the authority of the source
3.9 References or bibliography or both; to distinguish between and promote both
3.10 Common mistakes by students; to provide examples
3.11 Using software to help with references; to provide examples of software
3.12 Avoiding the charge of plagiarism; to explain what it is and identify software to help

3.1 Introduction

The literature review is an important part of your research. You need to survey previous work to determine whether similar work has already been executed, otherwise you cannot be assured that your research is original. The literature review focuses almost entirely on the work of others. You need to be looking for (a) similarities to your work – what can you draw on to help you, and what is it that is slightly different that you might do? (b) gaps in the literature, and (c) contentious issues – do authoritative sources disagree? Gaps might be only just

Writing Built Environment Dissertations and Projects: Practical Guidance and Examples, Second Edition.
Peter Farrell, Fred Sherratt and Alan Richardson.
© 2017 John Wiley & Sons, Ltd. Published 2017 by John Wiley & Sons, Ltd.
Companion Website: www.wiley.com/go/Farrell/Built_Environment_Dissertations_and_Projects

one small facet of your topic area that has not been examined before at all. In many spheres of life disagreement occurs frequently in both the hard and soft sciences. Sometimes it will be voiced loudly in the literature, where one source produces evidence with the intention of contradicting evidence produced by another, for example in the climate change debate. Sometimes the disagreements are not so apparent, and you may have to work hard to tease them out.

There is rarely an argument to say that literature reviews are not necessary. That could only be justified in circumstances where your work is closely associated with a previous research project. The literature review may have been completed in that previous project, and the thrust of your work may be to substantially develop one strand of what has already been done. In such a case it would be at least prudent to update the literature review of the earlier project.

A question often asked is 'What percentage of the document should be devoted to the literature review?' The percentage may be about time devoted or words written; students probably ask the question in the context of the latter. There is no definitive answer. It would normally be a substantial piece of work. The suggested structure of a dissertation or project is six chapters. The literature review may be chapter 2 of 6; it may be part of the opening third of the document. Perhaps it should be more than one sixth of the document? Upto a third of the document may not be unusual, with one overriding caveat: quality not quantity.

Often students select an area of study because they want to enhance their knowledge in a field. The literature review may be the part of the research document where you can do this. You may specify a study objective to be met by the literature review such as to review current theories, or to baseline current knowledge, or to establish current best practice. You hopefully become immersed in your subject area as you develop the thirst for knowledge. The literature review involves the compilation of a large number of articles and extensive reading. You may save some articles electronically; for others you may want to develop a well-organised paper file. Students often have lever-arch files of previous papers where they have highlighted important text.

Figure 3.1 illustrates a literature review constructed as though through a funnel. It is adapted from the work of Holt (1998, p. 67), who uses the extremely useful analogy of literature summarised by a wide top part, moving down the aperture of the funnel towards a narrow outlet. To develop his analogy, the wide top part is a receptor for material from a wide variety of sources. It should also consider the topic area in an international setting and across all professions or industries. As you move down through the funnel, material is considered in the context of the country of study, be that the UK or elsewhere, and in the context of the construction industry. Finally, at the narrow outlet there is material that will be related to your study objective. You may find yourself starting at the wide top part, and then you need to work hard to find a narrow field of literature directed towards a modest objective. Or alternatively, you may start at the narrow outlet, and then you will need to work upwards to put the study in the context of broader fields of knowledge. It is important that you address both ends of the funnel. A study that remains at the top is likely to be too superficial to be of value; a study that is only at the bottom will fail to put itself into strategic context.

Consider starting at the top of the funnel with a study to investigate climate change; as a topic area in itself it is too broad for one research project, but the international context must be addressed by reference to international authoritative publications and agreements. The

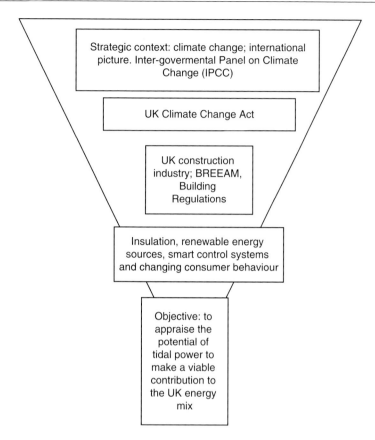

Figure 3.1 The literature review constructed as though through a funnel (adapted from Holt, 1998, p. 67).

position of the UK government can be identified and appraised against the position of a variety of pressure groups. There can be links into the Building Research Establishment Environmental Assessment Method (BREEAM) ratings; also consideration of how insulation, renewable energy sources, smart control systems and changing consumer behaviour can be used to reduce CO_2 emissions. The role of energy generation and transmission industries may be briefly examined. The narrow part of the review may be around the potential use of intelligent energy management systems by speculative developers or a comparison of energy generated by solar and wind or … or …

3.2 Style and contents of a literature review

Readers will be able to deduce from the review what the up-to-date position is in this field, what is the extent of current knowledge and what is happening at the leading edge. It should be a summary of the state of the art. There can be some limited explanatory or descriptive material to provide context for readers, but not too much. Too often literature reviews are a mere description of what has gone before. There is an assumption that readers are construction professionals, but not necessarily experts in the topic area you have selected. The review should bring together common themes and issues in the literature, make intelligent links and demonstrate that the literature has been examined with insight. Most of all, the literature review should be critical and offering judgements; it is these latter concepts that are often most difficult for students to master. It is not adequate to re-present the work of authoritative people, and then merely rubbish it. It is not sufficient that a literature review merely comprises statements extracted from previous work, which are bolted together in a clumsy fashion. The thrust should be to collect evidence from previous work, and pitch the work of one or several sources against others in a critical sort of way. Thus the criticism is informed.

> The critical nature of a literature review may be enhanced by the use of terms that emphasise that comparison is taking place: 'whereas', 'on the other hand', 'alternatively', 'but', 'another view', 'the opposite stance', or 'this is contradicted by'.

As you write, these sorts of terms may help to focus your attention on the need to be critical. But try not to use these phrases mechanically. The University of Manchester (UoM, 2015) provides a useful 'academic phrasebank', which includes tips on being critical, being cautious, comparing and contrasting and so on.

The literature review needs to flow as a narrative – an enjoyable read – and you need to demonstrate your writing skills as you weave it together. Take care not to throw in anecdotal statements, even if, from your practical experience, they seem to be facts, e.g. 'the construction industry is inherently unsafe'. As Fenn (1997) correctly states 'prove it'; where is the evidence for this bold assertion? You should not be discouraged and made afraid to write, since anecdotes can help you to weave the narrative together. You must either find the proof or, if appropriate, use a caveat such as 'it may be argued that the construction industry is inherently unsafe'. Using another example, it may be tempting to write as fact that CO_2 causes climate change, since 'everyone knows that'. However, this type of writing is not acceptable without citing the evidence, so that the reader can go back to the source if necessary and challenge it. What is acceptable is 'carbon dioxide (CO_2) is one of the main greenhouse gases that cause climate change (Energy Saving Trust, 2010)'. As part of your review you should also identify authoritative sources that dissent from this view.

Towards the end of the review you should assess in your words, the implication of the literature on your study – relate it to the aims and objectives. It may be titled 'discussion', 'summary', 'critical appraisal' or 'appraisal'. While the literature will be from a wide variety of sources, it must be written as though it were a funnel, with the output being consolidated and narrow. Legitimate contentions, assertions and arguments for advancing the area of knowledge should be given. The identification of gaps is justification for further research, and should therefore lead to the objectives or hypotheses of the research project. You should identify any fundamental issues that arose. Revisit the research questions, objectives and hypotheses at the end of your literature review. Having done your reading and appraisal, you have transformed yourself from having too little knowledge to being an expert. Provisional objectives set at the start of your study should now be reappraised and adapted as necessary; at the completion of the literature review the objectives should become set in stone. Having laid a strong foundation to your study with a robust description of the problem and your literature review, it is now merely for you to move forward and collect some data to allow you to meet your objectives.

Appendix C includes a sample literature review, prepared by Challender at al. (2014), which includes a description of the problem upon which it is founded.

3.3 Judgements or opinions?

The Oxford dictionary (Stephenson, 2010) defines an opinion as a 'belief or assessment based on grounds short of proof'. Judgement is defined as 'critical faculty, discernment, insight'. It may be helpful to consider the exercise as one whereby all the evidence pertaining to a particular issue is collated and placed on a set of balancing scales. Some of the evidence may be substantive, robust and stem from a research project that has been well resourced and funded. Other evidence may be less substantive, lightweight and merely opinions of important (or unimportant) people taken from the media. We all have opinions on a variety of subject matter and we may enjoy expressing those anecdotally in social circles. But in academic and business terms, decisions must be made on the evidence. That evidence must be weighed carefully. Decision-makers must get themselves into positions of knowledge on a given subject area; they must not be ignorant. You must consider the weight of each piece of evidence, and place it on one side of the balancing scale; add your own evidence. The issue then becomes one of making judgements, very importantly not giving opinions or expressing personal views. The judgement should be made impartially, recognising the weight of the evidence on each side of the scale. Opinions, taken individually, are lightweight, anecdotal and prone to change. You may put yourself in the position of judge in a court: you are going to say that, on the balance of the evidence before you which at this stage of the research project is mostly the literature and the writer's experience, it is likely that 'xyz' is the case. The analogy of the balancing scales is illustrated in figure 3.2. When clients procure projects from the construction

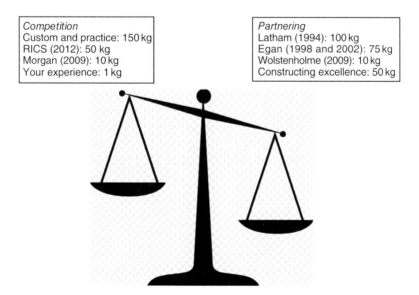

Figure 3.2 Weighing the evidence before coming to judgements; which is best, competition or partnering?

industry, should they seek many bids from many parties in competition with each other or should they seek to negotiate/partner with one or two companies?

The concept of weight is important; the analogy is that a government-sponsored report executed by a team of leading people in the field may weigh 100 kg, while the writer's own experience may weigh less than 1 kg. It is for you as the writer to judge the validity of each source and weigh it accordingly. Do not use the weighing analogy too rigidly, since life is complex. At one point in time, given particular circumstances, the scales may drop most heavily on side 'A' and then at another time with slightly different circumstances, the scale may drop on side 'B'.

You must not be naïve as a writer. You must consider the validity of sources and take account of that when you weigh the literature. For example, a manufacturer will argue that its product is best. There is lots of bias around and lots of motives around for bias: money, job preservation, politics. The political world employs specialist spin-doctors. It is your job as the writer to skilfully weigh the validity of each source; be suspicious. Scientists have said that they never believe anything that is only written down; it has to be validated in the laboratory.

It is important to distinguish between isolated opinions and collective opinions sourced as part of the main research data collection exercise. Research places little emphasis on isolated opinions, especially if they are expressed spontaneously with little thought or reflection. Similarly, research is not interested in *your* opinion as a researcher and analyst. Research is very interested to learn from the opinions of a small number of people during in-depth reflective interviews; these may be analysed qualitatively. Also, research wants to learn from the opinions of many people collected perhaps from a survey by quantitative analysis. These opinions may form part of the main data analysis used to test, if appropriate, hypotheses. In a business sense, the collective opinions of people about products and services are drivers for buyer choice; clearly these opinions are very important. Since research does not want your opinion, do not let this leave you feeling 'worthless' in the process. The research exercise that you are undertaking is you; it savours of you. It is your articulation of the problem, your use of the literature, your objective, your choice of method to meet the objective, your analysis and your interpretation of results to develop findings and conclusions. The research requires you to use your skills and qualities to produce valid conclusions to an important problem. You will not express your opinion, but you will be reflective, use insight and make interpretations. Based upon the evidence, you will make your judgements, as the expert, from a position of knowledge and based upon the data and results that you produce. Use the data, including the literature, to inform your interpretation of findings, or 'let the data do the talking'. Do not give conclusions based upon what you thought before you started your research.

3.4 Sources of data

Some literature is more rigorous than others. The literature review needs to draw on the best evidence. Where a source is published can be an indicator of its worth. One method of making judgements about the rating of journals is to take note of their 'impact factor'. An impact factor is based upon the number of citations that papers in journals receive; well cited journals have higher impact factors. In the construction discipline, impact factors in the range of 1–3 are highly respected. Even some of our best journals may have impact factors of less than 1.0, but in other more mature research disciplines such as medicine and nature, impact factors may be over 30.

Papers in academic journals are rigorously refereed, and are based on robust research methodologies. The refereeing process ensures that only the best research is published; attrition rates vary, but perhaps only one in five or one in ten papers submitted are published. Authors are specialists in their field, and papers are often the output of funded research. They may be considered to be in the top division of written work. Within this top division, there are some journals that may be informally ranked in the upper quartile, some in the middle and some in the lower quartile. It is an accolade for academics to have their work published in leading journals in their field. The analytical framework in the middle of some of these papers can be extremely complex; in those cases it is perhaps just sufficient that you digest the early parts and then the findings/conclusions at the end.

Conference proceedings are often useful sources of data. Some conferences are more prestigious than others. Conferences are not thought to be as prestigious as refereed academic journals, since it is more difficult to achieve publication in journals. Other dissertations, projects and theses can provide useful ideas about document structure and research methodology. They may provide a useful list of references. Reports commissioned by governments or government quangos are often very important. They may represent a turning point in the life of a discipline. Such reports are often well resourced, compiled by leading people in the field and therefore very authoritative – similarly for reports commissioned by learned professional bodies or trade bodies. Textbooks often give leads to, or appraise, other authoritative literature. Some textbooks are written for practitioners, while others are for students to merely explain how systems work. Student textbooks are important to read to bring your own base knowledge to the required level. The most critical material in textbooks may be where one or two people have acted as editors to bring together (though usually in separate chapters) the most authoritative writers in a field.

Weekly magazines, monthly magazines and websites in the discipline must be located and regularly reviewed; they are absolutely essential reading. These are published commercially and by professional bodies. Section 1.4 identifies some magazines, websites and apps. The articles in such papers are sometimes merely gossip, items of news and the like; that is fine. They may therefore be considered, in some parts, to be lightweight and anecdotal. However, they often contain leads to important and current work elsewhere, such as government reports that are out for consultation. At the early stages of research projects they clearly give clues to what the issues of the day are and where problems exist. You must also be mindful of the availability of audio and audio-visual material, which is often supported by written notes.

A literature review that is based on academic journals, conference papers and authoritative reports (flavoured by a sprinkling of textbook and magazine citations) is likely to score a much higher mark than one that is only based on textbooks and magazines. You need to demonstrate that the quality of your reading is at the higher levels.

The Institution of Civil Engineers' virtual library gives access to 47 journals, as listed in table 3.1 (ICE, 2015). The Association of Researchers in Construction Management (ARCOM, 2015) lists and gives links to 50 journals relevant to the field of construction management, civil engineering and architecture, as illustrated in table 3.2. In your research you may look at just several academic journals.

Table 3.1 Journals in the field of civil engineering.

Advances in Cement Research
Bioinspired, Biomimetic and Nanobiomaterials
Civil Engineering Innovation
Dams and Reservoirs
Emerging Materials Research
Environmental Geotechnics
Geosynthetics International
Geotechnical Research
Green Materials
Géotechnique
Géotechnique Letters
ICE Construction Law Quarterly
ICE Engineering Division Papers
ICE Proceedings
ICE Proceedings: Engineering Divisions
ICE Selected Engineering Papers
ICE Transactions
Infrastructure Asset Management
International Journal of Physical Modelling in Geotechnics
Journal of Environmental Engineering and Science
Journal of the ICE
Life of Telford
Magazine of Concrete Research
Minutes of the Proceedings
Nanomaterials and Energy
Proceedings of the ICE – Bridge Engineering
Proceedings of the ICE – Civil Engineering
Proceedings of the ICE – Construction Materials
Proceedings of the ICE – Energy
Proceedings of the ICE – Engineering History and Heritage
Proceedings of the ICE – Engineering Sustainability
Proceedings of the ICE – Engineering and Computational Mechanics
Proceedings of the ICE – Forensic Engineering
Proceedings of the ICE – Geotechnical Engineering
Proceedings of the ICE – Ground Improvement
Proceedings of the ICE – Management, Procurement and Law
Proceedings of the ICE – Maritime Engineering
Proceedings of the ICE – Municipal Engineer
Proceedings of the ICE – Structures and Buildings
Proceedings of the ICE – Transport
Proceedings of the ICE – Urban Design and Planning
Proceedings of the ICE – Waste and Resource Management
Proceedings of the ICE – Water Management
Proceedings of the ICE – Water Maritime and Energy
Proceedings of the ICE – Water and Maritime Engineering
Structural Concrete
Surface Innovations

Source: ICE (2015).

Table 3.2 Journals in the field of Construction Management.

Architectural Engineering and Design Management
Architectural Record
Automation in Construction
Building and Environment
Building Research and Information
Civil Engineering and Environmental Systems
Computer-Aided Civil and Infrastructure Engineering
Construction Innovation: Information, Process, Management
Construction Law Journal
Construction Management and Economics
Cost Engineering Journal (Formerly AACE Bulletin)
Design Studies
Engineering Project Organization Journal
Engineering, Construction and Architectural Management
Facilities
Intelligent Buildings International
International Journal of Construction Education and Research
International Journal of Construction Management
International Journal of Disaster Resilience in the Built Environment
International Journal of Facilities Management
International Journal of Housing Markets and Analysis
International Journal of Law in the Built Environment
International Journal of Project Management
International Journal of Project Planning and Finance
International Journal of Strategic Property Management
International Journal of Sustainable Engineering
Journal of Architecture
Journal of Construction in Developing Countries
Journal of Construction Research
Journal of Corporate Real Estate
Journal of Cost Analysis and Management
Journal of Digital Information
Journal of Engineering, Design and Technology
Journal of Financial Management of Property and Construction
Journal of Housing and the Built Environment
Journal of Information Technology in Construction (ITCon)
Journal of Management Procurement and Law
Journal of Property Investment and Finance
Journal of Property Research
Journal of Property Valuation and Investment
Journal of Real Estate Literature
Journal of Urban Planning and Development
Lean Construction Journal
NICMAR Journal of Construction Management
Organization, Technology and Management in Construction
Proceedings of Institution of Civil Engineers-Civil Engineering
Property Management
Smart and Sustainable Built Environment
Structure and Infrastructure Engineering
Urban Studies

Source: ARCOM (2015).

3.5 Methods of finding the literature

The expectations of a literature review may have increased in recent years, because the process is assisted by electronic searching techniques. You must clearly make best use of the internet, commercial CDs and library search catalogues. You should find out if there is a university that specialises in the topic area you have chosen. It follows that there may be several dissertations or projects at that university that you may be able to access electronically. Direct access is available, whether or not you are a student at that university, to library search catalogues.

Equally as important as the electronic search is to manually browse library shelves. You must not think that all important information in a field is available electronically. If the discipline relevant to the research project is engineering, it is useful to look for titles beginning with 'e' and 'j', such as The Engineering Project Organisation Journal and the Journal of Engineering, Design and Technology. In some subjects also look for 'i', since several journals titles start with 'International'. Several years' editions of a leading journal in a field may be quickly browsed by looking at the index of articles on the front or back cover of the text.

The depth in a literature review is probably best obtained by gaining 'references from references'. Early leads can be obtained by electronic searching and browsing. Once an article is found, at the end of that article is often a list of references to other work in the field. When you obtain these references, you may find more references at the end of the new reference, and so the process goes on. Often such articles may not be available in the local library, so you will have to use inter-library loan processes. The important point here is that it can be a time-consuming process, and so it is important to start the literature review early. Start the browsing process well in advance of the allocated period of study.

At the end of the literature review process you should clearly know, for your field of study, which are the leading: journals, industry magazines, conferences, centres of knowledge/excellence, relevant professional bodies, government departments, government quangos, countries in the world, national and international individuals and websites.

As well as your own university library, you may also use your local public library or another university library. Your own university may not be the closest to you at all times during the year, and another university may be more convenient to travel to, or may contain specific publications that you need to browse. Often universities have reciprocal arrangements whereby students from one institution may visit the library in another. Government agencies often have libraries related to business, the environment and so on. Joining a professional body is important. The UK professional bodies may argue that they are the leading experts in their field in the world. At the very least, they have access through their libraries to the most authoritative work in the world. A lot of the information is available electronically or can be posted to you. It may be necessary to be a student member to obtain material. If you are not a member, it is advisable that you join. Membership may be free to students; the time in your career that you may make most use of services provided by professional bodies is as a student. Most professional people, at least once in their lives, visit the headquarters of their professional body. If you can, go and browse the library shelves of your professional body while you are a student.

3.6 Embedding theory in dissertations and projects

Clough and Nutbrown (2007, pp. 104–105) quote Lewin 'there is nothing so practical as a theory' and Silverman 'without a theory, there is nothing to research'. Kerlinger in Creswell (2003, p. 120) defined theory as 'a set of interrelated constructs (variables), definitions, and propositions that presents a systematic view of phenomena by specifying relations among variables, with the purpose of explaining natural phenomena.' Examples of flagship theories are theories of gravity first developed by Isaac Newton in 1687, Darwin's theory of evolution by natural selection in 1865, and Einstein's general theory of relativity in 1916.

Theory answers the 'why' and 'how' questions. Answers to 'why' questions often commence with 'because …', but the answers to 'why' and 'how' can often not be seen with the naked eye. Explanations underpinning 'why' and 'how' may be extremely complex, and it is not reasonable to expect that they can be understood by all. Laws and principles arise from theories; they can only be written when theories are proven over time. If the theory is proven absolutely, a law can be written, such that 'if this happens, that will definitely happen', e.g. as cited in O'Brien (2010) Boyle's law, 1662, is that 'the volume of a fixed mass of gas is inversely proportional the pressure acting on it'. Principles are written as guides, e.g. again in O'Brien (2010) Archimedes' principle (287–212BC), 'any object, wholly or partly immersed in a fluid, is buoyed up by a force equal to the weight of the fluid displaced by the object'. People who do not understand the 'why' and 'how' explanations, may be happy to accept laws or principles without further question. Laws and principles make it possible to apply theories to practice without having to understand the 'why'.

Figure 3.3 uses a spider's web as an analogy to a theory, with early ideas of a theory expressed as the spider's bridge thread. It spans from branch to branch, but structurally will not stand alone without many further threads below, all meshing together to form a theory which is proven to hold over time and 'watertight'. Theories are meshed into the complex web of knowledge. Nodes on the web are variables, and threads between nodes indicate relationships

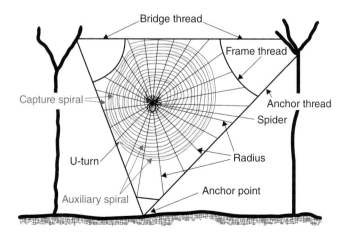

Figure 3.3 The analogy of the spider's web constructing new theories.

between variables: cause and effect. As new research is executed, it may give strength by adding one new thread to the web or by increasing the gauge of an existing thread. Alternatively, new research may break a thread. The broken thread may have been placed in position based on evidence available at that time, but the new research provides insight to discredit earlier beliefs. Threads that are placed in or removed from the web are not usually accompanied by grand announcements, since each individual thread is developed over a long period of time, and tested using multiple methodologies. Theories evolve and there are few 'Eureka' moments. Theories indicate the strength of any relationships and direction. The strength of relationships may be strong, such that movement of IVs will instigate similar movement in DVs, or not so strong relationships (but still significant), whereby large movement in IVs is followed by only small movement in DVs. Direction may be positive or negative: manipulating IVs up may cause DVs to go up, or manipulating IVs up may cause DVs to go down. Rather than the analogy of a web, in construction terms, you may wish to think of your work as merely putting a brick in the huge wall of knowledge, or more modestly, pointing or repointing mortar joints in that wall. Your research may comprise testing of one very small part of a theory, using only one methodology.

Theories in hard sciences are proven to hold, both in laboratories and in practice, using many methodologies. Soft or social sciences involve human behaviour. In these fields there are lots of theories that have been found to hold, but they often have caveats such as a particular theory may only apply given a certain set of circumstances, but if circumstances change, that theory may not apply. Most theories about people behaviour originate from disciplines outside construction. A construction context is sometimes used by researchers from other disciplines to conduct their work. There is a need to know whether theories developed elsewhere apply to construction. Researchers in construction therefore take theories from other disciplines and test them in a construction setting. In soft sciences, theories may compete, conflict and overlap with each other. For example, there are three competing leadership theories; trait, style and contingency.

Some students may start their study with examination of a theory. Qualitative studies may develop hypotheses or theories at the end, and include recommendations for testing by quantitative methods. The theory is the bedrock or foundation for the study, and it should be expressed at the start of a quantitative study. Construction students often start to describe problems from a very pragmatic or practical position. This may be particularly so if part-time students investigate industry-based problems. Research takes place to develop knowledge of society at large, and embed that knowledge such that it is a platform for application and for more research. It is therefore necessary to take descriptions of industry-based problems back to the theoretical foundations from which they originate.

To illustrate the gap between theory and literature, let them be placed on a continuum, with a numerical scale of 0–100. Let the lowest score, zero, for ease of illustration, be mere literature and the highest 100, theory. Let applied and pure research also be placed on this same scale, with applied research (or industry practice) scored at 0, and pure research scored at 100. Literature is not theory; it does not have that status. Items of industry news that appear in the press can be classified as literature, and their value is 0 on this scale, with no real value in research terms. As the academic quality of literature improves, it may have a higher rating, such that, for example, a scholarly reflective article may be scored at 10. Clearly, these numbers are not real numbers; however, continuing the analogy, papers in refereed journals perhaps score 20. The journey required to classify something at the level of theory is a long one, continuing over decades, and on this scale finishes – if it ever finishes – at 100. If you start your description of the problem with a theory in position 100, you can stay in that position if you intend your research to be 'pure' –but there is potential for more marks if you can demonstrate application. If you start your description of the problem from a practically based problem in industry in position zero, you should move along the continuum, and towards 100, to find a theory onto which you can 'hang' your study; what is your 'theoretical coat peg'?

It is not to find a theory just for the sake of finding a theory; your work should, in a very modest way, contribute to knowledge formation. Be assertive in your search for theories, using abstracts in electronic databases such as SCOPUS. This may involve exploration of literature in other disciplines, and will be part of your literature review. A thorough search should leave you overwhelmed with theories related to the problem that you have defined, and while you may review many of those as part of the literature review, select just one key theory for your investigation. The research question will narrow the number of possible theories.

Bothamley (1993) edited the multi-disciplinary text, 'Dictionary of Theories; More than 5000 Theories, Laws, and Hypotheses Described'. Labels ascribed to theories are not always expressed with clarity. Some theories are mentioned more often in the literature than others, for example motivation theories or organisational theories. Some theories are mentioned less often, sitting quietly in the background known only to experts in that field. Some theories have their labels ascribed to the originator's name, e.g. McGregor's Theory X, while others are labelled according to their meaning, e.g. bidding theory, competition theory, decision-making theory, theory of design, economic theory (e.g. supply and demand), theory of planned behaviour.

Some literature or authors use the word 'theory' loosely, claiming mere hypotheses to be theories, such as 'my theory is that quantity surveying is excellent training for a career in project management'. This is a hypothesis worth testing, but it does not have the status of a theory established by testing and re-testing using different methodologies over time. Anecdotally, the media often reports conspiracy theories. These too are not theories, but more like unproven hypotheses. The relevance of some theories may decrease in modern society; some theories are cast in stone for ever, such as classical theories of organisations, now over a century old, emanating from work by Max Weber in 1904. Newer theories, such as those in collaborative working, arguably have their origins over a mere few decades, and they are still under development.

It is possible to take a problem in practice or industry, and to base research on theoretical principles underpinning methodologies, rather than theory underpinning the problem itself. There may be debates about the validity of certain methodologies to test hypotheses, and their appropriateness to given problems. These debates may be around populations, sampling techniques, data collection, analytical methods.

Consider three examples of taking industry-based problems back to theory:

(1) Site waste management plans (SWMPs) were made compulsory in 2008 for all construction projects over £0.3 M. Contractors' problems include administrative and training costs, including uncertainty about whether plans are effective and add value. Site managers complain about the administrative burden passed to them with no additional resource to help. The research questions are (1) do SWMPs reduce waste? (2) do SWMPs reduce cost? As an IV the quality of SWMPs can be expressed on a scale of 0–10 with 0

being no plan, and 10 being a plan and its resulting actions arising from being 'best in class'. There are two DVs: waste and cost. Waste may be measured by the number of skips to landfill or such like. Cost may be a calculation of staff training, time, extra cost of labour sorting and skips. If relationships are proven, quantitatively, and in a positive sense, it would be found that SWMPs do reduce waste and cost.

There are theoretical issues to explore around the decision by government to legislate for SWMPs, when in many spheres self-regulation or voluntary codes of practice are preferred. There are issues in social sciences about the propensity of companies and individuals to do what is right when faced with voluntary codes or legislation. Also, the role of the State in protecting the environment. Perhaps the most appropriate theory is systems theory. SWMPs are systematic of government, local authorities (as enforcers), contractors, subcontractors, suppliers and employees coming together to reduce waste. Systems theory 'predicts that the complexity of organisations, and therefore the role of management, will probably continue to increase – at least for so long as the efficiency-enhancing potential of complexity can continue to outweigh its inevitably increased transaction costs' (Charlton and Andras, 2003). Another research question could be that with the introduction of SWMPs 'has the efficiency-enhancing potential of complexity continued to outweigh increased transaction costs of SWMPs?' Also applicable may be the invisible hand theory developed by Adam Smith (1723 -1790) quoted by Joyce (2001), '... being the managers of other people's money than of their own, it cannot well be expected that they should watch over it with the same anxious vigilance with which partners in a private co-partnery frequently watch over their own. Like the stewards of a rich man, they ... consider attention to small matters as not for their master's honour and very easily give themselves a dispensation from having it.' Is it the case that construction companies delegate control of materials to others, who are motivated not to use those materials prudently but to maximise their own earnings? Not on a theoretical basis, but as part of the government's red-tape challenge, the requirement to produce site waste management plans was repealed on 1 December 2010; however, in a public survey the UK's Department for Environment, Food and Rural Affairs (DEFRA, 2010) reported that '83% of respondents said they would still use some form of tool to record and manage waste on site.'

(2) In research involving materials testing, relevant theories include those around the chemical behaviour of constituent parts. These theories should be examined in the literature review. In tests of concrete, constituent parts can be varied by volume or by type. If new concrete mixes are tested, the results should be evaluated against theory. If a new mix has greater compressive strength, that is fine, but why? The answer lies in the theory. Discussion around possible whys should take place in the closing stage of documents.

(3) You work for a local authority that is reflecting whether to remove speed cameras. Cameras have been in place for a number of years. Accident data is available for periods before and after installation. Articulation of the problem revolves around public and motorists opinions and attitudes towards cameras, but the lead issue is whether they reduce the number of accidents. An initial approach may be to analyse the data using mathematical tools, and look for statistically significant differences in the number of accidents before and after installation of cameras. There is the possibility that findings may indicate either way: cameras do or do not reduce the number of accidents. You may challenge theories with mathematical or statistical tools. There are theories about the law being used as a deterrent to human behaviour.

To find theories, there is no substitute for general reading and browsing literature. If you are able to astutely link your work into the theoretical web, you give yourself the possibility of gaining many more marks. As an alternative to identifying a theoretical foundation for your study, you may place your work in the 'body of knowledge establishing best practice' (Seymour, et al., 1998). There has been significant resource devoted to establishing best practice in many spheres of construction activity since Latham (1994), e.g. health and safety, procurement, culture. That work is coordinated by Construction Excellence in the Built Environment. While this may not gain marks as high as theory-founded research, it does place your work within a structured framework of applied research. It is implicit that best practice draws on theory. Without a base in either theory or best practice, your work is likely to make little contribution and score fewer marks.

3.7 Referencing as evidence of reading

Sir Issac Newton wrote in 1676 (cited by the Royal Mint, 2015) 'if I have seen a little further it is by standing on the shoulders of Giants', which has been interpreted to mean 'today's achievements would not be possible without the discoveries of the past'. In your research, in a very modest way, you are looking to do something new by building on 'discoveries of the past', but you must use referencing as a tool to demonstrate that you have done that. Do not try to reinvent the wheel.

Readers should be able to clearly distinguish between what is your original work and what is the work of others. You must use the work of others, but ensure that it is cited. A valid research project may be to take the published data of one researcher and analyse it in a different sort of way – provided that the original source of the data is acknowledged.

Referencing is not just about getting the technical details of referencing correct, although that is important. It is the tool that you use to demonstrate to examiners and other readers that:

- you have been reading
- you are not passing the work of others off as your own
- you can digest and appraise the work of others in a critical way.

If you do not show evidence of reading in your work, you will not pass. If you pass the work of others off as your own you will be failed (plagiarism). If you do not critically appraise the work of others, you are likely to get lower marks. Being able to reference in a technically correct way is therefore essential. Some students may conscientiously do extensive reading, but neglect to cite their sources. This is not good enough. Using the analogy of courts, judges cannot reach decisions based upon material presented to them that may (or may not) be spurious.

Some students have really good practical experience in some topic areas. Clearly they are in a good position, and can bring the weight of their experience into arguments being made. But to write an academic piece of work purely based on practical experience is not good enough either. After all, who is one student to write authoritatively about a topic area, without having digested the evidence collated from the experience and research of others? Students with practical experience have a strong platform from which to build, but they must integrate that experience with academic tools; that is, they must read, and in their writing blend their practical experience with other evidence from the literature.

3.8 Citing literature sources in the narrative of your work

There are two separate components to referencing; that is, citing literature sources in the narrative of your work, known as 'cite in text', and the full reference details in a separate section towards the end of your document.

The two components briefly illustrated are:

(1)   cite in text: the name of the author or organisation and year in the narrative of your work: e.g. Mitchell and Mitchell (1947) stated that 'Lime, used in building operations as a cementing material, is obtained when naturally occurring forms of calcium carbonate, $CaCO_3$, are suitably calcined.'

(2)   and then in the 'references' section, provide precise details of where this work can be found, thus:

Mitchell, G.A. and Mitchell, A.M. (1947) *Building Construction*. A textbook on the principles and details of modern construction for the use of students and practical men [sic]. 16th Edition Vol. II. Batsford Ltd: London. p. 1

There are hundreds of different referencing styles e.g. APA, Chicago, GOST, IEE. Some UK styles are based on British Standard BS 5605: 1990 'Recommendations for citing and referencing published material' (BSI, 1990). Others are based on the International Standards Organisation (ISO). Many are based on styles adopted by different academic journals in different professional disciplines. The two common styles are normally referred to as Vancouver (numeric) and Harvard (name date) systems. Humanities coursework (e.g. literature, philosophy, history, art and design) normally uses Vancouver. Social sciences or technology (e.g. education, health, sociology, psychology, business, engineering) normally use the Harvard system. You must follow the guidance of your university department; in the built environment it is likely to be Harvard. There may be slightly different interpretations of the Harvard system. Follow to the letter your own university guide. This book is using an adaptation of the British Standard Harvard version.

You need to aim for perfection in writing references/bibliographies, that is, full stops, semicolons, colons, commas, all in the correct place. All information should be in the correct sequence, no information missing and with consistent presentation. Precision is required so that libraries can locate cited articles with ease, without having to ask for more details or clarification. If you were to ask for the 'Oxford' book, is that the name of the book, the author, the publisher or the place of publication? More importantly for you, if you need to go back to the library to relocate a text you have had previously, you need to be able locate it quickly, including relevant page numbers.

As an example, generically, the British Harvard System requires the following information in the reference section:

- author's surname
- publication year
- title
- edition
- publisher
- place of publication

Title of the whole work would be in italic; the title of a textbook would therefore be in italic, but for a paper in a journal, the title of the paper would be in roman (not italic) font and the title of the journal in italic. International Book Standard Numbers (ISBN) or International Standard Serial Numbers (ISSN) are not needed. The full reference details are in a 'section' and not a 'chapter', and should be placed in your document immediately after the conclusion chapter; and before the bibliography and appendices. This text provides examples of how to cite and give references at the end of your writing. For the convenience of readers of this text, references are given at the end of each chapter, but in your document there should only be one section that provides references.

If readers, after reading the narrative, are motivated to seek out a source you have cited in the text, they can turn to the References section. The references will be in alphabetical order, thus if the text by Mitchell and Mitchell (1947) is of interest, the reader can turn to the references section to learn the full reference details.

3.8.1 Verbatim citations

Literature may be cited verbatim; the alternative is to paraphrase as described in 3.8.2. Verbatim is word for word. If you do cite verbatim, you must put that section of text in speech marks. It is the speech marks that distinguishes verbatim text from paraphrasing; paraphrasing is not enclosed in speech marks, but the source of your paraphrasing should still be cited in your text. Be careful that if you cite verbatim, not to do it out of context. Verbatim quotes may most often be just sentences or parts of sentences. The source of the quote may be written as part of the flow of a sentence with only the year given inside of the bracket, thus:

> Wolstenholme (2009) in a review of construction industry progress wrote that 'Since Sir John Egan's Task Force published its report Rethinking Construction in 1998, there has been some progress, but nowhere near enough'.

Alternatively, not written as though part of the flow of a sentence, with the author and year given inside the bracket, thus:

> In a review of construction industry progress it was argued that 'Since Sir John Egan's Task Force published its report Rethinking Construction in 1998, there has been some progress, but nowhere near enough' (Wolstenholme, 2009).

Both methods are acceptable. To make your writing less repetitive, perhaps use both methods interchangeably as you cite different authors.

It is not appropriate for a literature review to have lots of large paragraphs of text taken from other sources verbatim. If there is a whole paragraph that is central to an important argument being made around an objective, then by all means cite it all. When citing paragraphs of text verbatim, the convention is not to use speech marks, but to clearly indent the paragraph, and cite the source at the end, thus:

> Since Sir John Egan's Task Force published its report Rethinking Construction in 1998, there has been some progress, but nowhere near enough. Few of the Egan targets have been met in full, while

most have fallen considerably short. Where improvement has been achieved, too often the commitment to Egan's principles has been skin-deep. In some sectors, such as housing, construction simply does not matter, because there is such limited understanding of how value can be created through the construction process (Wolstenholme, 2009).

3.8.2 Paraphrasing

To paraphrase is to read a section of text, whether that be a paragraph, page, chapter or whole book, then to digest and understand what the writer is saying, and to summarise it more succinctly in your own words. The words that you choose must not misrepresent what the original writer has said, either accidentally or otherwise. Misrepresenting the work of others can lead to the 'Chinese whispers' effect. As work passes from writer 'A' to writer 'B' to writer 'C' etc., what comes out from the last writer is radically different from what was written by the first. You must execute any paraphrasing accurately and fairly. The paragraph from Wolstenholme (2009) as noted in section 3.8.1 may be paraphrased as:

> It is argued that the targets set by Egan in 1998 have not been achieved and that the housing sector does not understand value creation (Wolstenholme, 2009).

Note that there are no speech marks since these precise words did not come from the pen of Wolstenholme. It is likely that a substantial part of your literature review will comprise paraphrased statements, alongside verbatim quotations of short length. It is not to measure percentages for each, but for example it may be 50% short verbatim quotes and 50% paraphrasing.

3.8.3 Secondary citing

Secondary citing is used when you read what one author 'says' another author has 'said', and then wish to cite the first author based on what the second author has written. If we are being suspicious, we could say what the second author alleges the first author has said. The need to secondary cite may occur often. Secondary citing should be a signal for readers to be careful. It is life that all forms of communication are at risk of being misinterpreted as they pass from one source to another. It is part of your task to make sure that you do not contribute to any misinterpretation. It is probably not a good idea to paraphrase something that has itself been paraphrased.

John Ruskin was a 19th century English art critic wrote in 1860:

> It is unwise to pay too much, but it's worse to pay too little. When you pay too much, you lose a little money – that is all. When you pay too little, you sometimes lose everything, because the thing you bought was incapable of doing the thing it was bought to do. The common law of business balance prohibits paying a little and getting a lot – it can't be done. If you deal with the lowest bidder, it is well to add something for the risk you run. And if you do that, you will have enough to pay for something better.

It is highly unlikely that you will have read the 1860 text. Unless your whole study was around his work, there is not a reasonable expectation that you will read it; indeed, it may not be accessible to you. Learning of Ruskin's work through secondary sources is fine. If you have not read the text, it is important not to imply that you have. To secondary cite therefore just write Ruskin (quoted in Ward, 2008) states that 'It is unwise to pay too much, but it's worse to pay too little'. You should not give the full reference details of Ruskin's text in your work; that may mislead examiners into thinking that you have read it first hand. Readers who may wish to follow up the 1860 reference to Ruskin must do so through Ward (2008).

As another example, you may need to refer to authoritative construction industry reports that have gone before. They might be very difficult to locate. If they are in professional body library archives, since they are valuable and out of publication, they may be only available to read in person; not for loan. For instance, the landmark report by Banwell (1964), thus:

> Banwell, H. (1964) The Placing and Management of Contracts for Building and Civil Engineering Work. London, HMSO

This text has not been seen in the writing of this chapter and is not in the reference section. You may alternatively quote, by writing that, in the context of construction industry problems at that time, in 1964 Banwell, cited in Murray and Langford (2003), wrote 'the average year at the time might see 250–300 stoppages per annum, involving up to a total of 50,000 men, losing literally hundreds of thousand working days'.

3.8.4 Who to cite in your narrative

Citations should be to the author. If there are two authors cite them both, e.g. Mitchell and Mitchell (1947) state that … or Ashworth and Perera (2015) argue that … If there are three or more authors, e.g. Griffith, Stephenson and Watson, (2000), cite it in your text as Griffith et al. In your reference section towards the end of your document it would give all three authors' names, e.g. Griffith, A., Stephenson, P. and Watson, P.A. (2000). 'Et al.' is Latin for 'and others'.

If the name of the author is not given, cite the organisation, e.g. BBC (2015) or the source, e.g. The Times (2015). If it is not possible to locate any of these, as a last resort use anon (anonymous), e.g. anon (2015).

If authors have more than one publication in a year that you wish to cite, the first one in your document should be given the letter 'a', then 'b' and so on: HSE (2015a), HSE (2015b). These letters should appear in the narrative and in the references section.

3.8.5 Page numbers and emphasising the authority of the source

It is important to point readers directly to a source with its page number. Readers may often wish to follow up a citation and read more about the surrounding context in the original work. It can be very frustrating to be directed to a textbook of several hundred pages, and then not to be able to locate the relevant part of the text. There is an issue about whether to include page numbers in the narrative part of the text or just in the reference section at the end. As a general rule it is best to keep as much information out of the narrative that would distract readers from the flow of the text. On that basis, put the page numbers in the reference section. However, if a source is cited in the text more than once, it is necessary to give the page numbers at all points cited in the narrative, e.g. Latham (1994, p. 1) and Latham (1994, p. 32). The details of the Latham's publication are then written out in full, only once in the reference section, without pages numbers. Whether you are citing page numbers in text or in references, cite the parts that you have read and digested. If you are citing one page, cite it singularly as say 'p. 10'. If it is a range of pages, and you will paraphrase what had been written, cite it in the plural as say 'pp. 10–15'.

You may need to cite the generic work of previous writers. If the writer has been prolific, that might be just by giving the author's surname, and no references at all to specific texts in the reference section.

If you were citing not just 'any' Powell, but someone who is extremely authoritative, you may wish to emphasise this by saying something like 'Martin Powell, the chief executive of the Institution of Structural Engineers, reported that 'We have created a modern workplace, delivering pragmatic and responsible sustainability solutions wherever possible – while also celebrating structural design' (Powell, 2015). Or to emphasise that this is not just 'any Taylor', and not just any of Taylor's books you may write something like F.W. Taylor, the father of scientific management, in his seminal text 'Principles of Scientific Management' in 1911, stated that '…'. Also, rather than just cite HSE (2005), you may wish to state 'the HSE publication 'Health and Safety Induction for Smaller Construction Companies' (HSE, 2005) states that '…'.

3.9 References or bibliography or both?

Definitions of bibliography and references vary between academic disciplines, universities, university departments and even individual academics in the same department. The two words are used interchangeably, although they are different – or are they? The two key authoritative sources for you must be your university department and your supervisor. Tell supervisors if their definitions conflict with departmental definitions, and seek clarification. The definitions used here are as follows:

> References – everything cited in text in your document
> Bibliography – everything that you have read or browsed that is relevant to your subject area, but has not necessarily been cited in the text

Some authors use a broader definition for a bibliography, to include all material relevant to a subject area, even though the writer has not read or browsed it – but this may not be appropriate for your document.

References may be considered the most important; they are what you have read, digested, understood and used in the appraisal of your subject area. The fact that literature has been cited in the text is the 'evidence' that you have done the reading. If you cite a book, it is not expected that you have read the whole of the book. You may have browsed the whole book, and read relevant parts. Marks are awarded for evidence of reading and for the way you have used the literature in your work. Evidence for this comes partly from verbatim citations in text or paraphrasing and subsequently references. Examiners will usually browse the list of references at the end of your work to make a judgement on the extent and the academic weight of your reading; remember weight comes from the type of material you have cited. Reading and citing weekly construction magazines is to be encouraged, but that will not carry so much weight if it is done to the exclusion of academic papers and other authoritative sources.

A good bibliography can serve two purposes. Firstly, students may wish to demonstrate to examiners that they have read more widely than just the references, and therefore get credit or marks for that reading. It may be that having done the reading, the material was not contentious, not directly relevant to the objectives or not relevant to a wider argument. Secondly, it might be that subsequent researchers may be interested in bibliographies, as they look for leads to other literature.

Examiners may be suspicious of bibliographies. It is possible that having completed research, students merely locate textbook title details electronically and pass them off as though 'read' by putting them in a bibliography. This cannot be done so easily if using references, since within the flow of the narrative work the text will be cited, e.g. you cannot write in your narrative 'McKay (1943, p. 1) states 'bricks are made from clay and shale', if you have not had that text by McKay open at page 1. Perhaps a sensible position is that if you have done little more than casually browse some literature, you should not include it in your bibliography. If you have read a chapter or so, do put it in, and include in the bibliography the pages you have read. You may have a viva at the end of your process; you do not want to be embarrassed if you cannot speak with sincerity about your bibliography.

So to answer the question, 'references or bibliography or both?' Ideally both. Some examiners may not be worried at all if there is no bibliography. However, if there are no citations in text (verbatim or paraphrasing) supported by references, the dissertation or project will fail.

3.10 Common mistakes by students

In the narrative, *do not*:

- give abbreviated or full web addresses; the golden rule is – no 'www' or 'co.uk' or 'org.uk' or 'com' anywhere in the narrative, but do include these web addresses in the 'References' at the end of the document
- give the title of the text, unless you want to emphasise the stature of a particular piece of work
- detach the information from its source by having a full stop in the wrong position. This is the correct way: 'bricks are made from clay and shale' (McKay, 1943). The following is incorrect, since the full stop should be after the close bracket: 'bricks are made from clay and shale'. (McKay, 1943)

In the reference section, *do not*:

- use the abbreviated '&'
- number references; you must put them in alphabetical order of author
- bullet point references
- arrange references under separate headings of textbooks, journals, websites etc
- use the combined title of references and bibliography; keep them separate
- put references at the end of each chapter in the document; towards the end of the document only in one section
- repeat sources in both the reference section and the bibliography; in one of them only.

Citing web pages often causes students difficulties. In bibliographies, students sometimes cite a generic website for an organisation they have been browsing, e.g.

www.constructingexcellence.org.uk/

but this is not appropriate. Whether it be in a reference or bibliography, you must cite a specific web page. References are often given starting with the web address. That is not correct. Always start with the surname of the author or the organisation first:

Constructing Excellence (2015) KPIs and benchmarking. http://constructingexcellence.org.uk/kpis-and-benchmarking/ Accessed 25.07.15.

Abbreviations should be given in full in your references, though not at its first point of mention, that is before the year. In your narrative, you may write thus: 'the HSE (2005) states …'. In your references it will start with 'HSE' and then after the title HSE should be written in full thus:…

HSE (2014) Extendable Scaffolding Loading Bay Gate – use of cable ties to secure loose mesh and unsafe means of operation. Health and Safety Executive. www.hse.gov.uk/safetybulletins/loading-bay-gate.htm. Accessed 25.07.15.

In your references section, set each citation out so that it is easily distinguished from the next with either a line space between each reference or a tab set in of second and subsequent lines.

3.11 Using software to help with references

You have a choice of three methods to compile your references at the end of your document:

(a) do it 'manually'; that is use the word processor to type in all the information required
(b) use tools in word processing software, e.g. the references drop-down menu in Microsoft Word or similar
(c) use specialist web-based software such as Mendeley, RefWorks or other bibliographical software

In many aspects of learning it is necessary to grasp basic underpinning principles. Understanding may be best digested manually before using electronic or software aids. For example, it is better to learn mental arithmetic before using a calculator. Therefore at least some time, using method (a) would seem appropriate.

If you are to record and manage your references manually, it can be difficult to closely follow the requirements of either Harvard or numeric systems with precision. If you decide to do it manually, keep practising and keep your university guide in your briefcase as though a dictionary. It may help if you set up an electronic file to record reference details of the work you have read. Start writing up the references/bibliography at the start of the research. It may be unclear at the start as to whether articles will be a reference or in a bibliography, but at the writing-up stage they can be easily allocated to the appropriate section using cut and paste.

One issue with software is the time required to learn to use it, and whether that time invested will save time later. Referencing is not just for dissertations and projects, but for academic work in all subjects. If you invest your time in learning to use the software in modules completed before your research starts, there are greater potential savings.

Figures 3.4 and 3.5 illustrate the dialogue boxes that will open if you use the Microsoft Word reference facility. Figure 3.4 is located by clicking on 'References' and within the 'Citations & Bibliography' section click on 'Style'. There are 14 different styles are available to you e.g. APA Fifth Edition, APA Sixth Edition etc. Figure 3.4 is located by clicking 'References' then 'Insert Citation' and then 'Add New Source'. You need to complete all the relevant sections. While figure 3.5 illustrates only six 'type of source' options such as 'book' and 'book section', by scrolling down the software, you will note there are a total of 19. To insert references or bibliography in your document, click on 'Bibliography' and 'Insert Bibliography'.

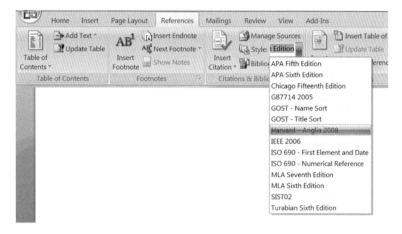

Figure 3.4 Fourteen 'styles' on Word to set out references.

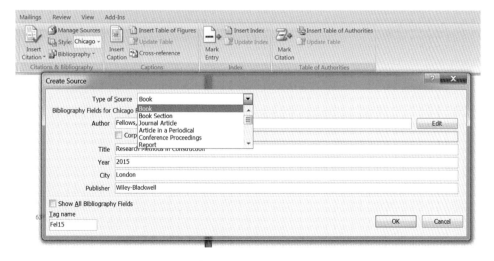

Figure 3.5 Six of the source options in Word.

The specialist software is more sophisticated than the word processing software, and will do things such as:

- does everything that the word processing software
- offers you a wide range (hundreds) of different methods of referencing or 'output styles'
- offers you the format required for more different sources (maybe 30), e.g. journal paper copy, journal electronic
- allows you to hold, copy and paste between folders; perhaps you may have folders for different assessments that you are doing
- allows references from other electronic sources to be copied and pasted without you having to manually type the data into the software
- imports references from the software into your word processed document
- allows you to access your references remotely, over the internet.

You have to make your decision: (a), (b) or (c). If it is only (a) then that leaves you vulnerable to making mistakes, and having to spend more time, probably at a later stage of your study, getting your references into a position where they are as good as you can get them. The most sensible option is clearly (c). To learn to use the software initially is just a matter of an hour or two. Expertise will develop as you use it more, and it will take far less time in use than method (a). The key saving features of specialist web-based software are:

- the dialogue boxes act as 'prompt' facilities – otherwise it is more difficult to distinguish which type of material you need to record for different sources
- importing references from electronic sources – otherwise you have to type in the data manually
- sorting references into order and inserting punctuation at the end of your work – otherwise you have to do this manually.

Library or departmental tutors may be willing to give demonstrations to groups of students. Alternatively, just browse through the help facilities for yourself.

3.12 Avoiding the charge of plagiarism

To plagiarise is defined as 'to take and use (the thoughts, writings, inventions of another person) as one's own' (Stephenson, 2010). There are sophisticated and powerful software tools that universities are routinely using that 'ensure originality as well as use of proper citation' (Turnitin, 2009). In other words, to spot instances of plagiarism or cheating. You may be asked as a matter of routine to submit your document electronically, or if you have submitted only a paper copy of your document, the university may later ask you for an electronic version, if it has some suspicions. The software is a vast repository of data, and holds an electronic record of 'everything that has ever been written' – not literally of course, but with the passage of time it comes ever closer to that. It may take the software a day or so to search your document for matches to information that it holds in its repository. The result of the software search is that it gives a result in percentage terms, e.g. 15% match. Anything above this figure may initiate some concerns. However, a 30% match may be perfectly acceptable, *providing* you have cited

Sample text for Turnitin

This is a sample piece of text uploaded to Turnitin. This first paragraph is original text and does not get a 'similarity hit'. The second paragraph is a well cited piece of work by Ruskin, and it is therefore part of the similarity count; previously submitted elsewhere. Since the source is cited as Ruskin in Ward (2008) it is not plagiarism. The third paragraph is taken verbatim from the executive summary of the report by Wolstenholme in 2009. It would be difficult for a tutor to recognise whether this text is the student's own or not. Since the source is not cited, this is plagiarism; cheating. The fourth 'paragraph' is a reference; including a reference in your work is clearly not plagiarism.

1

"It is unwise to pay too much, but it's worse to pay too little. When you pay too much, you lose a little money – that is all. When you pay too little, you sometimes lose everything, because the thing you bought was incapable of doing the thing it was bought to do. The common law of business balance prohibits paying a little and getting a lot – it can't be done. If you deal with the lowest bidder, it is well to add something for the risk you run. And if you do that, you will have enough to pay for something better" (Ruskin in Ward, 2008).

2

Since Sir John Egan's Task Force published its report Rethinking Construction in 1998, there has been some progress, but nowhere near enough. Few of the Egan targets has been met in full, while most have fallen considerably short. Where improvement has been achieved, too often the commitment to Egan's principles has been skin-deep. In some sectors, such as housing, construction simply does not matter, because there is such limited understanding of how value can be created through the construction process.

Reference

4   **3**

Latham, M. (1994) Constructing the Team. Final Report on the Joint Review of Procurement and Contractual Arrangements in the UK Construction Industry. July. London. HMSO. Available at: http://constructingexcellence.org.uk/wp-content/uploads/2014/10/Constructing-the-team-The-Latham-Report.pdf Accessed 09.09.15.

turnitin   **61%**   --
SIMILAR   OUT OF 100

Match Overview

1 dilbert.com
   Internet source   29%

2 www.dinz.org.nz
   Internet source   23%

3 Submitted to University...
   Student paper   7%

4 www.rics.org
   Internet source   3%

Figure 3.6 Example Turnitin submission; 61% similarity, 23% plagiarism.

the sources of the material in your work. An important point to be clear about here is that if you have cited the source, you have correctly acknowledged that this work, or these words, are not your own. You are legitimately using the work of others to support a point you need to make. Indeed you have to do this; you have to read, cite in text, develop your knowledge and enhance society knowledge from a platform created by others. However, if you do not cite the source, you are passing someone else's work off as your own – cheating. The penalties for plagiarism, particularly in later years of study, can be severe, including failure or even expulsion from the university. Universities may retain a plagiarism register, so that repeat offenders can be identified.

As universities are increasingly using such software, the expanse of repositories increases significantly. Work submitted by students becomes part of the repository. Therefore, student 'A' may cite an original piece of work that was never produced electronically. That may have been as far back as the 1800s, or more recently in the 1990s. When student 'A' submits an assessment, the 1800s or 1990s original work is in the repository. A short time later student 'B' cites the same 1800s or 1990s work; the software will give a match for student 'B' to the work of student 'A'.

The premise of the software is to 'ensure … originality as well as use of proper citation' (Turnitin, 2009). It is for you to check your own work before submission and not for others to check your work. You should be able to submit a draft of your work before submission to your tutor, so that you can identify any potential problems yourself, such as missing citations. Do not view the software with fear. Your view should be that it is a fantastic tool that (a) prompts you to insert citations in drafts of documents that have innocently been missed, and more importantly (b) reinforces underpinning concepts of academia, such as the necessity to read and bring the weight of the evidence of your reading into your writing.

Figure 3.6 illustrates a submission to Turnitin with a 61% match. The third paragraph only is cheating; the words about Sir John Egan's taskforce are being passed off as though the student's own; they are not.

Summary of this chapter

The literature review is essential to establish a baseline for the remaining parts of your study. It is likely to be a substantial part of the document. At the end of the literature review you should be able to make judgements rather than give opinions and convert provisional objectives to firm objectives. You should manually browse paper-based material extensively, as well as using electronic search tools. Academic journals are more rigorous than industry magazines, and you should try to incorporate both in your review. Similarly, theory is more rigorous than mere literature, and you should try to embed the theory in all parts of the document. Citing sources in text, providing references and bibliographies should be done using the correct technical methods. Do not plagiarise your work; if you do, you may fail.

References

ARCOM (2015) Journals in the field of Construction Management. Association of Construction Managers. www.arcom.ac.uk/res-journals.php Accessed 24.07.15.
Ashworth, A. and Perera, S. (2015) *Cost Studies of Buildings*. Routledge: Oxon.

BSI (1990) British Standard BS 5605: 1990 'Recommendations for citing and referencing published material', 2nd Edition.

Bothamley (1993) 'Dictionary of Theories; More than 5000 Theories, Laws, and Hypotheses Described'. Gale Research International, London.

Challender, J., Farrell, P. and Sherratt, F. (2014) Partnering in practice: an analysis of collaboration and trust. *Proceedings of ICE – Management, Procurement and Law.* Vol. 167. Issue 6. pp. 255–264. www.icevirtuallibrary.com/content/article/10.1680/mpal.14.00002 Accessed 26.09.15.

Charlton, B.G. and Andras, P. (2003) The Modernization Imperative; a systems theory account of liberal democratic society, p. 85, Imprint Academic, Exeter.

Clough, P. and Nutbrown, C. (2007) A student's guide to methodology. Sage, London.

Constructing Excellence (2015) KPIs and benchmarking. http://constructingexcellence.org.uk/kpis-and-benchmarking/ Accessed 25.07.15.

Creswell, J.W. (2003) Research design: qualitative, quantitative, and mixed methods approaches. Sage, London.

DEFRA (2010) Site Waste Management Plans scrapped 1st December. Legislation update service. www.legislationupdateservice.co.uk/site-waste-management-plans-scrapped-1st-december/ Accessed 24.07.15.

Energy Saving Trust (2010) Climate change. www.energysavingtrust.org.uk/Climate-Change?gclid=CK G4qe3mpqACFQE8lAoddSeEZQ.

Fenn, P. (1997) Rigour in research and peer review. Construction Management and Economics. July. Vol. 15, Iss. 4. pp. 383–385.

Griffith, A., Stephenson, P. and Watson, P.A. (2000) Management Systems for Construction Longman.

Holt, G. (1998) A guide to successful dissertation study for students of the Built Environment. 2nd edition. University of Wolverhampton, UK

HSE (2005) Health and Safety Induction for Smaller Construction Companies. The Health and Safety Executive. www.hse.gov.uk/construction/induction.pdf Accessed 09.09.09]

HSE (2014) Extendable Scaffolding Loading Bay Gate – use of cable ties to secure loose mesh and unsafe means of operation. Health and Safety Executive. www.hse.gov.uk/safetybulletins/loading-bay-gate. htm Accessed 25.07.15.

ICE (2015) ICE virtual library; Journals. www.icevirtuallibrary.com/content/journals?page=3&perPage=50 Accessed 19.19.15.

Joyce, H. (2001) Adam Smith and the invisible hand. http://pass.maths.org/issue14/features/smith/index.html Accessed 02.09.15.

Latham, M. (1994) Constructing the Team. Final Report on the Joint Review of Procurement and Contractual Arrangements in the UK Construction Industry. July. London. HMSO. http://constructingexcellence.org.uk/wp-content/uploads/2014/10/Constructing-the-team-The-Latham-Report.pdf Accessed 09.09.15.

McKay, W.B. (1943) Building Construction. Vol. 1. Longmans, London. p. 15.

Mitchell, G.A. and Mitchell, A.M. (1947) Building Construction. A textbook on the principles and details of modern construction for the use of students and practical men [sic]. 16th Edition Vol. II. Batsford Ltd: London. p. 1.

Morgan, S. (2009) The right kind of bribe: BAA's Steven Morgan on project roles. Building Magazine. 09 October

Murray, M. and Langford, D. Ed. (2003) Construction Reports 1944–98. Blackwell Science, Oxford.

O'Brien, T. (2010) Brain-powered science. https://books.google.co.uk/books?id=oWhy72Kghy8C&pg= PA322&lpg=PA322&dq=he+volume+of+a+fixed+mass+of+gas+is+inversely+proportional+the+pr essure+acting+on+it+O'Brien&source=bl&ots=PgK4X4aa&sig=Lhl4fSd540ls3lgFHUmABtKryao& hl=en&sa=X&ved=0CCAQ6AEwAGoVChMIiJiUwsiUyAIVC7wUCh1lGgng#v=onepage&q= he%20volume%20of%20a%20fixed%20mass%20of%20gas%20is%20inversely%20proportional%20 the%20pressure%20acting%20on%20it%20O'Brien&f=false Accessed 26.09.15.

Powell, M. (2015) The Institution of Structural Engineers officially opens new international headquarters. https://www.istructe.org/news-articles/2015/announcements/the-institution-of-structural-engineers-officially Accessed 26.09.15.

Royal Mint (2015) Where did the edge inscription 'standing on the shoulders of giants' come from? www.royalmint.com/help/help/standing-on-the-shoulders-of-giants Accessed 29.07.15.

RICS (2012). Contracts in Use. A Survey of Building Contracts in Use During 2010. London: Royal Institution of Chartered Surveyors Publications.

Seymour, D., Crook, D. and Rooke, J. (1998) The role of theory in construction management: reply to Runeson. Construction Management and Economics. Vol. 16, Iss. 1. pp. 109–112.

Stephenson, A. (2010) Oxford Dictionary of English. 3rd edition. Oxford University Press. Available online.

Turnitin (2009) Turnitin original checking. https://www.submit.ac.uk/static_jisc/ac_uk_index.html Accessed 01.02.10.

UoM (2015) Academic Phrasebank. University of Manchester. www.phrasebank.manchester.ac.uk/summary-and-transition/ Accessed 02.10.15.

Ward, G. (2008) The project manager's guide to purchasing: contracting for goods and services. Hampshire: Gower. p. 147. https://books.google.co.uk/books?id=zT1tolYWiWEC&pg=PA147&lpg=PA147&dq=it+unwise+to+pay+too+much+quotations&source=bl&ots=3A3KXHixIg&sig=4WsB5HxFhOUbo5jSGhw6UdO8Q8I&hl=en&sa=X&ved=0CFYQ6AEwCWoVChMI4vCq8Nv0xgIVhrgUCh1ItQx9#v=onepage&q=it%20unwise%20to%20pay%20too%20much%20quotations&f=false. Accessed 24.07.15.

Wolstenholme, A. (2009) Never waste a good crisis. A review of progress since Rethinking Construction and thoughts for our future. Constructing Excellence. https://dspace.lboro.ac.uk/dspace-jspui/bitstream/2134/6040/1/Wolstenholme%20Report%20Oct%202010.pdf Accessed 24.07.15.

4 Research goals and their measurement

The titles and objectives of the sections of this chapter are the following:

4.1 Introduction; to provide a context for defining research goals
4.2 Aim; to define and provide an example
4.3 Research questions; to define and provide an example
4.4 Objectives; to define and provide an example. To emphasise the importance of objectives
    4.4.1 Objectives that 'wobble'
    4.4.2 The literature review as an objective?
    4.4.3 Objectives that do not match was has been done
4.5 Variables; to provide examples
4.6 A hypothesis with one variable; to provide an example
4.7 A hypothesis with two variables; independent and dependent variables
    4.7.1 Which is the IV and which the DV? Variables 'melting' into each other
    4.7.2 Manipulation or observation variables in research?
    4.7.3 A relationship, not a cause: strength of relationships
4.8 Writing the hypothesis: nulls and tails – a matter of semantic
4.9 Lots of variables at large, intervening variables; to define and provide examples
4.10 Ancillary or subject variables; to define and provide examples
4.11 No relationship between the IV and the DV; to explain the consequences
4.12 Designing measurement instruments; use authoritative tools and adapt the work of others; to illustrate examples that do and do not need your own design
    4.12.1 Variable values with high or low numbers as best?
    4.12.2 Measurement scales of 0–10 and 0–100; multiple-item scales
4.13 Levels of measurement; to define categorical, nominal, interval and ratio
4.14 Examples of categorical data in construction; to provide examples
4.15 Examples of ordinal data in construction; to provide examples
4.16 Examples of interval and ratio data in construction; to provide examples
4.17 Types of data; to distinguish between primary and secondary and objective and subjective data
    4.17.1 Primary or secondary data
    4.17.2 Objective or subjective data: hard or soft
    4.17.3 Using subjective data since objective data is too difficult to secure
4.18 Money and CO_2 as variables; to identify how they may be measured
4.19 Three objectives, each with an IV and DV: four variables to measure; to describe our to link independent variables (IVs) to dependant variables (DVs)
4.20 Summarising research goals; variables and their definition

Writing Built Environment Dissertations and Projects: Practical Guidance and Examples, Second Edition.
Peter Farrell, Fred Sherratt and Alan Richardson.
© 2017 John Wiley & Sons, Ltd. Published 2017 by John Wiley & Sons, Ltd.
Companion Website: www.wiley.com/go/Farrell/Built_Environment_Dissertations_and_Projects

4.1 Introduction

Research goals may be used as the umbrella term to encompass all things that you will do in your research: aims, research questions, objectives and hypotheses. Definitions vary in the literature, particularly the definitions of aims and objectives.

Only one aim is appropriate; there may be more than one objective – three objectives that you will meet by your main data collection instrument seems 'doable' in a dissertation or project. More than three may leave your research too superficial in too many areas, lacking depth. However, if you wish to express things that you will do in your literature review as objectives, you may actually list more than three.

Sometimes students do not write research questions, but it is better that they do, since it aids the logical progression of studies and illustrates understanding of how objectives and hypotheses are derived. The sequence should be that the articulation of the problem gives rise to research questions, then the objective follows, as a statement of what you will do to answer the question. A hypothesis is finally written from the objective, in such a form that it is suitable for testing. Sometimes students write objectives that are similar to questions, but similar is not good enough. If they are only similar, it follows that there are differences. Differences can be so distinct that if the research meets objectives, it will not answer the questions that originated from the problem. There should not be any differences; research questions, objectives and hypotheses should imitate each other, with precision. Frequently the justice system is required to interpret the difference in meaning between clauses in contracts; you do not want readers or the examiners of your research to have to spend time interpreting differences in words in your research questions, objectives and hypotheses.

It is suggested that dissertations and projects are bound around objectives; that is because industry and academia may come together best talking about objectives rather than questions or hypotheses. Therefore, when there is a need to refer to a research goal (research question, objective or hypothesis), refer to the objective – but not if you are to refer to a research goal before a statistical test, then perhaps refer to the hypothesis. It is permissible that research is alternatively bound around research questions or hypotheses; you decide. It just becomes a matter of semantics. Better marks can be obtained if you identify and describe the variables in your research; ideally you will have an independent variable and a dependent variable.

Some students may see the possibility of getting some really good data and therefore do the study 'back to front'; that is, they set objectives to suit the data that they can get. While on the one hand, that is arguably not the way to do industry-based research, getting data and having a really good 'poke-around' it, may give some important insights. Researchers may be asked at an early stage of the study what the proposed methodology will be. A legitimate answer for you could be 'I don't know, I will pick the most appropriate methodology to meet the objective and that objective is currently only provisional'. Altering only a few words in an objective could radically alter any proposed methodology. However, it is useful to have a methodology in mind at an early stage of the study. If you are asked to present a research proposal, you may label your methodology provisional.

The description of the problem, aim, research questions, objectives, hypotheses and variables all need to come together under a document title. The title needs to be attractive, to entice potential readers, but also concise and accurate so that it does not mislead. If your document is made available electronically, it is the title that is primary in attracting search engines. The title should be different from the aim, but may typically include a mix of terms

Figure 4.1 The Castle Museum, York. Dense sand and stone plinths represent the description of the problem and literature review. Four columns represent research questions (RQ), objectives (OB) and hypotheses (H), all supporting the aim represented by the roof parapet. In between the objectives are variables (VARS).

from the aim and objectives. It may also give an indication of what was found or concluded in a study. A title set at the beginning of the process can be provisional. Arguably, the aims and objectives are more important than the title at the early stages; the title can be established towards the end of the research. Figure 4.1 is a photograph of the Castle Museum in the historic English city, York; it is used to present an analogy of the relationship between the description of a problem and the development of aims and objectives.

There are four circular stone columns circa 15 m high and 600 mm diameter. The ground was excavated to expose dense sand with a bearing capacity of $600 \, kN/m^2$. Consider the ground to be your description of the research problem, which is the result of many iterations and discussions with practitioners. Stemming from the problem description are some provisional research questions. Placed on the sand are four huge stone plinths acting as pads to support the columns; they are analogous to the literature review supporting what follows. On completion of the literature review, you will then reaffirm or develop your research questions or goals. This is done by erecting the four tall columns, each of which comprises three pieces of stone 5 m high placed carefully upon each other. The first stone for each column represents a research question, the second piece an objective as a statement of what you will 'do' to answer the research question and finally a hypothesis that you will test, again to allow you to answer the research question. There are thus, four research questions/objectives/hypotheses shown. The roof parapet is pitched, representing your aim; the four columns support the roof in the same way that your research goals support your aim. The ridge on the roof parapet is the point of your aim, driving forward your research and giving you direction. Inside or behind your columns, the door and windows are further openings into your research; the variables and their definitions, give you the possibility of attaining better marks. Weaknesses in this structure will cause your research to fail; poorly defined problem or failure to network and circulate (loose, not dense sand), missing important literature (no pad foundations), wobbly or weak objectives (columns defective or not plumb), forgetting your variables (no doors or windows), misdirected aim. In the case of research, you will write a provisional aim, or design the roof, before you form the foundations or start to write about the problem. There is lots of work to be done later to clad the structure and ensure that the research is watertight: sound methodology, reliability and validity, appropriate data collection and robust analysis, strong discussion and conclusions and so on.

Health and safety is used in section 4.20 to illustrate exemplar aims, research questions, objectives and hypotheses. These are based on recommendations arising from the work of Arewa (2013), which is more fully described in appendix D6.

4.2 Aim

The aim is the ultimate goal of the study; it is visionary and a statement at a strategic level. The introduction to the document will often start by saying 'The aim of this study is to investigate...' The aim will identify the context of what is to be attempted; what field are you in? If the aim is carefully written at the outset, it should not change from beginning to end. If you reflect at some later stage of the process that the words used in the aim are not sufficiently succinct, by all means change them but try and do it without altering in your own mind the strategic direction of your research. The aim will give you direction for your objectives, and if objectives change at the early stages of a study, they should remain within the remit of the aim. The aim can be written without consideration of resource limits. Resources will always limit researchers, particularly personal time, and therefore research should not be judged on the yardstick of the aim. Aims will not be achieved. Small studies have one aim. There are analogies with business: mission statements of companies express aims, designed in boardrooms by directors or partners. Objectives are statements at an operational level. They support the mission statement in a business sense, or support the aim in the sense of a dissertation or project. The operational level is a level below the boardroom; that is, managers who are closer to the 'shop floor'.

The retail store Marks and Spencer (2010) had as one of its aims thus: 'Climate change; we'll aim to make our UK and Irish operations carbon neutral within 5 years.' It was acknowledged to be a bold aim, and was supported by twenty objectives, which were statements of what would be done to help achieve the aim. Has the aim been achieved?

Military history across the world can be used to illustrate the relationship between aims and objectives. The aim of a war may be to defeat an enemy, and this aim is supported by a series of military objectives that may involve individual battles on the one hand and intelligence on the other. If the objectives are not met, the aim cannot be achieved. In a construction context a study may start with an aim around customer satisfaction in the context that an industry is losing clients, and then as the research develops, one of the objectives may be around the quality of the product.

In a health and safety context, a study aim may be:

To investigate processes for improvement in UK construction industry health and safety performance.

It could be argued that this is a huge task, with potentially no end, since the construction industry needs to continually improve in this area; it is too tough for an objective.

4.3 Research questions

An early part of the document would normally give an articulation or description of the problem being investigated. Arising from the problem should be research questions. At this point the substance of the research can stand or fall. A problem really well articulated will almost automatically 'spit out' questions. Independent readers of your narrative should be able to tell you what your questions should be. Your early exploratory work, perhaps speaking informally with professionals, could be around matching up the description of the problem with the questions and making sure that your perception of that match mirrors the perception of others. The research questions need to be strong, robust, relevant to the topic area and, in the context of industrial problems, addressing real issues of importance. Weak research questions will mean that the remainder of the research is merely an academic exercise of data collection and analysis. Research questions should state the position for the argument or investigation – written in the form of 'what' or 'how'. Do not make readers work hard to find your research questions buried in the middle of large paragraphs of text. You may find it useful to set them apart from your main text with an indent of about 1 cm. When you write your research question, it can perhaps be labelled RQ (or RQ1, RQ2 etc. if you have more than one) and should clearly end with a question mark, thus:

RQ1: What is the level of compliance of UK contractors with best practice in health and safety?

4.4 Objectives

Objectives must be capable of having an outcome, and the success of the research will be measured against them; they are statements of research hoped for outcomes. Objectives are often preceded by words such as 'to determine', 'to assess', 'to compare', 'to design', 'to determine', 'to develop', 'to establish', 'to evaluate', 'to examine', 'to find out', 'to measure', 'to review', 'to show', 'to survey', 'to test'. Research will be measured against them. They are statements at an operational or tactical level. They take the aim and recognise constraints to translate the aim into 'doable' statements – what the study hopes to achieve or discover. Perhaps a useful strap line is that:

an objective is a statement of what you will do, and you will 'damn well' make sure you do it … lest you fail

Your objective should stem from your research question; it is clearly not appropriate that you 'do' something that will not directly answer your question. Therefore the words should match each other very closely. Your research question and objective (perhaps abbreviated to 'OB') can now be written together, thus:

RQ1: What is the level of compliance of UK contractors with best practice in health and safety?

OB1: To determine the level of compliance of UK contractors with best practice in health and safety

The RQ and objective are written as strap lines and are something you can carry round with you at the forefront of your mind. When somebody asks you what you are doing in your research, your verbatim response is 'I want to determine the level of compliance of UK contractors with best practice in health and safety'. It will 'roll off your tongue'. No stuttering, pauses, 'erm', 'uh', 'I am not sure'; you will be absolutely clear. In conversation with others you should then be able to explain the nature of the problem from which the RQ was derived, and you may have in mind a provisional methodology to meet the objective. The words in the strap line are too few; at a later stage there is a need to define them in the context of your research.

You should work long and hard at your objectives; labour over each word. They may be the result of many iterations. At the early stages of your study, objectives may be considered provisional. Your objectives may change as you develop the early part of the study. Those changes may result from some early exploratory interviews or from new insights gained from the literature. At the end of the literature review, your objectives should become set in stone; it is merely then for you to go out and meet your objectives.

The whole study must revolve around the objectives. At the risk of being monotonous in your document, mention them in each chapter if necessary. The objectives must be in the introduction, and the literature can be rounded off by reaffirming the objectives. The description of the method could have a precursor something like 'This is the method used to meet the objective, thus: …'. Again without being monotonous, the analysis and discussion chapters could have similar statements. In the conclusion the objective strap lines may be given as subheadings.

4.4.1 Objectives that 'wobble'

In many completed studies objectives 'wobble like jelly'. Objectives are not stated with clarity; if they are stated, they change as the document develops. The methods used, perhaps questionnaires, measure some concepts, often poorly, and not related to the objectives. Sometimes objectives disappear part way through the document, and then re-emerge towards the end under a different label (e.g. reason for the study) with some key words changed or shuffled around. The following is an example of a 'wobbly' objective:

Chapter 1: The objective for this study is to determine the level of compliance of UK contractors with best practice in health and safety.

Chapter 2: Having appraised the literature the purpose of the research is to determine how well best practice in health and safety is done.

Chapter 3: This chapter will describe the method to achieve the intention of the study: to find out what clients think of health and safety.

Chapter 4: The analysis in this chapter is undertaken to prove that health and safety in the UK is better than elsewhere in Europe.

Chapter 5: (No objective mentioned; it just disappears!)

Chapter 6: This chapter concludes the examination of the degree of compliance of UK contractors with best practice in health and safety and compares it with Europe.

Be clear; this is not the way to do it. Whatever words are written in chapter 1 to describe your objectives must be repeated verbatim (word for word) as you pass through the document.

> To avoid the wobble or slip of even one word, do not retype objectives as you need to repeat them; copy and paste them from the introduction.

4.4.2 The literature review as an objective?

One reason for your literature review may be to bring your own and your readers' knowledge and understanding up to date. Should that be expressed as an objective? It is arguably implicit that that is what your document will do, and so there is no need to express it as an objective. However, you may want to make sure that you express with clarity specific areas of the literature that you will explore, and that is best done by defining those areas in objectives. You cannot be criticised for the latter. If you do adopt this approach, be sure that you make it clear which objectives are being met by the literature review, and which by the main data collection and analysis. In terms of numbering, it may be logical that the literature objectives are numbered first (say OB1 and OB2), and the objectives to be met by your main data collection exercise in the middle of the document are numbered subsequently (say OB3, OB4 and OB5). Section 4.1 notes that perhaps three objectives are 'doable' in a dissertation or project; that can be perhaps four or five if the method for some objectives is based on the literature review. To reiterate, not too many objectives, otherwise the depth of the study may suffer.

4.4.3 Objectives that do not match was has been done

There are examples of really good studies which hang together really well, but near to the end of the process some hard reflection shows that what has been done in the middle of the document, although excellent, does not reflect with precision the words in the objectives. If you find yourself in that situation, on the one hand that is not good, but it is not unusual. If you had been commissioned by a government minister to research and write a report around some given terms of reference, you cannot change the terms of reference if you get to the end of the process and reflect that you have done something slightly different; ministers will not tolerate that. But your dissertation or project is 'only' training for research; while ideally you should not be changing your objective towards the latter stages of the research, in discussions with your supervisor you may jointly agree that that is the best thing to do. Having realised that you have made a mistake, the strap line could be altered from 'an objective is a statement of what you will do, and you will 'damn well' make sure you do it … lest you fail' to 'an objective is a statement of what you have done … lest you fail'.

4.5 Variables

A variable is any facet that can have more than one value, or that can exist in more than one form; it is capable of moving. In mathematics, a concept that is not a variable is a constant. Variables may exist in quantitative or qualitative studies.

In research we are interested in measuring variables; a useful strap line may be, 'if it moves, measure it'. Businesses are dealing with variables every day without labelling them as such; they do not usually use this type of language. In your research you will be dealing with variables. You should clearly use the language of variables in your document, so label, define and measure them.

Each variable should have a name assigned to it; some software such as SPSS (the Statistical Package for the Social Sciences) calls the name of a variable a 'label'. Since variables are capable of moving, they will be able to have more than one 'value'. It could be just two values, or three or indeed an unlimited number. Examples 1 and 2 illustrate potential 'values' of two variables:

Example 1

Consider potential values of a variable with the name or label 'commitment to best practice in health and safety':

Two values: good and poor
More than two values – three, four, five or perhaps more, but say five: excellent, very good, good, satisfactory and poor
An unlimited number of values: perhaps a percentage scale, with possibilities, though unlikely, for scores to many decimal places

Section 4.17 will describe this variable as being 'subjective'. Section 4.13 will describe its measurement level as being (a) categorical data, (b) ordinal or (c) interval.

Example 2

Consider potential values of a variable with the name or label 'compressive strength of concrete':

Two values: good and poor
More than two values – three, four, five or perhaps more, but say five: excellent, very good, good, satisfactory and weak (there is a classification of concrete called 'weak mix')
Lots of values; though with perhaps a pragmatic range of $10\,N/mm^2$ to $50\,N/mm^2$ (maybe higher), with possibilities for scores to many decimal places

Section 4.17 of this chapter will describe this variable as being 'objective'. Section 4.13 will describe its measurement level as being (a) categorical data, (b) ordinal or (c) interval. Most likely if you are measuring compressive strength of concrete, you will measure it at the interval level.

The National Physics Laboratory (NPL, 2015) identifies the seven SI (Système International d'Unités) base units that can be used as tools in measurement:

ampere (A) – electric current
kilogram (kg) – mass
metre (m) – length
second (s) – time
kelvin (K) – thermodynamic temperature
mole (mol) – amount of substance
candela (cd) – luminous intensity

Some variables may be measured:

scientifically: using these SI units or using one unit as a function of another: e.g. velocity of fluids in pipes in metres/second
numerically: e.g. number of people employed, number of accidents, cost indices, age, percentages
qualitatively: judgements about quality of products or services: excellent, very good, good, satisfactory and poor

In some cases, qualitative judgements may be left as 'words', but quantitative analysts may want to convert them to numbers such as excellent = 4, very good = 3, good = 2, satisfactory = 1 and poor = 0.

Depending on the context of your study, there are many possibilities of things that you might want to measure around construction problems. Tables 4.1 and 4.2 illustrate examples of things that the construction industry is very keen to measure; that is key performance indicators (KPIs), as published by Glenigan (2015) working with Construction Excellence, the CITB and the Department of Business Information and Skills.

4.6 A hypothesis with one variable

There are several definitions of a hypothesis (plural, 'hypotheses'). Just to take one at the moment Holt (1998, p. 16) defines a hypothesis as a 'suggested explanation for a group of facts or phenomena either accepted as a basis for further verification (known as a working hypothesis) or accepted as likely to be true'. Having established an aim and defined the problem, research questions should arise, which are then matched by an objective. The next stage is to add the hypothesis so that it matches the research question and objective, thus:

RQ1: What is the level of compliance of UK contractors with best practice in health and safety?

Table 4.1 Performance variables measured in construction.

			Performance		
KPI name	Measure		2012	2013/14	2015
Client satisfaction	Product	% scoring 8/10 or better	83%	82%	81%
	Service		75%	75%	73%
	Value for money		78%	75%	74%
Contractor satisfaction	Overall performance		75%	74%	69%
	Provision of information		73%	69%	69%
	Payment		80%	79%	81%
Defects	Impact at handover		74%	71%	73%
Cost predictability	Project	% on cost or better	61%	69%	69%
	Design		79%	79%	75%
	Construction		58%	57%	56%
Time predictability	Project	% on time or better	34%	45%	40%
	Design		48%	52%	53%
	Construction		42%	67%	48%
Profitability	%	Median % profit before interest and tax	2.70	2.10	2.80
Productivity	VAPH current values	Median value added / FTE employee (£000)	60.00	63.90	66.00
	VAPH constant 2005 values		52.70	60.90	61.30
Repeat business		Median % turnover from companies worked with previously	79%		

Source: Glenigan (2015)

Table 4.2 Environment variables measured in construction.

			Performance		
KPI	Measure		2012	2013/14	2015
Product performance	Energy use (designed)	Median energy use kg CO_2 / 100 m² gross floor area	2000	2254	1970
	Energy use (designed) – Housing SAP Rating	Median SAP2005 rating	86.5	84.8	86
	Mains water use (designed)	Median water use m³ / 100 m² gross floor area	46.8	54.8	56
	Area of habitat – created / retained – product	% reporting no change or an increase in area of habitat	77	—	—
Construction process performance	Energy use	Median energy use kg CO_2 / £100k project value	196	214	199
	Mains water use	Median water use m³ / £100k project value	6.9	4	4.1
	Waste	Median waste removed from site m² / £100k project value	19.4	22.6	21.6
	Commercial vehicle movements	Median movements onto site / £100k project value	16.1	11.2	20

Source: Glenigan (2015)

OB1: To determine the level of compliance of UK contractors with best practice in health and safety

H1: The level of compliance of UK contractors with best practice in health and safety fails to meet expectations of society.

Note the negative style of this hypothesis: it is written in the context that there is a perceived problem around compliance not being as high as desired. With an RQ, OB and H in place, a methodology will be designed to test the hypothesis, by 'doing' the objective, which will simultaneously answer the research question that has been established from the description of the problem.

4.7 A hypothesis with two variables: independent and dependent

It can be argued that better studies contain at least two variables: an independent variable (IV) and a dependent variable (DV). They can be defined as follows:

IV: the variable whose value is manipulated or controlled, or observed in different settings
DV: the variable whose value changes as a result of movement in the IV

To have just one variable in your study is, however, perfectly acceptable. It would not carry the label of IV or DV, but merely 'variable'. The study objective may be merely to 'find something out'. For example, it is really important that the construction industry knows how satisfied its clients are; otherwise it operates it the dark. However, the analysis of objectives with only one variable may not yield the best marks. Anecdotally, a study with one variable in objectives, that otherwise hangs together well and demonstrates good research, may secure a mark in the 50s. A one-variable study may be considered lacking in terms of demonstrating some academic/intellectual rigour. The data from of a quantitative study with one objective can (only) be used for descriptive analysis and presented as histograms, bar charts, pie charts, means, medians, modes etc. While the qualitative study will not seek to definitely test for relationships, or causes and effects, they may best be directed to teasing out potential relationships.

RQs, OBs and Hs with two variables (IV and DV) opens the possibility of marks in the 60s, 70s and more. Having two variables is also important because it gives you the opportunity to do more robust analysis (see chapters 8 and 9), particularly inferential analysis if you are doing a quantitative study. Inferential analysis is a little more complex than descriptive analysis. If you can do it well, and demonstrate that you understand the principles underpinning some of these statistical tests, you have the opportunity to score more marks.

An IV and a DV are also better in studies because life, politics and business are often about answering the 'what if' question. What will happen to 'this' if we change 'that'. If 'A' is manipulated or moved, what will happen to 'B'? The change is for a purpose, not just for change's sake. At a personal level we are looking to change things, for example to change the way we work to save time expended on doing certain tasks.

An academic exercise which examines industry problems may not solve them. It can, however, identify causes or identify what the variables are, particularly what the IVs are.

Developing theories about 'how' and 'why' things happen is assisted by establishing causes. Conclusions may indicate a need to change or manipulate IVs, and recommendations may suggest how this manipulation could take place.

Section 4.6 defined a hypothesis as a 'suggested explanation for a group of facts or phenomena either accepted as a basis for further verification (known as a working hypothesis) or accepted as likely to be true' (Holt, 1998, p. 16). Now add three further definitions. Kinnear and Gray (1994) describe a hypothesis as a 'provisional supposition that an independent variable has a causal effect upon a dependent variable'; and then the same authors in 2008 'often a hypothesis states that there is causal relationship between two variables'. Fellows and Liu (2008, p. 127) say that a hypothesis is 'a statement, a conjecture, a hunch, a speculation, an educated guess … which is a reasonable suggestion of a (causal) relationship between the IV and the DV'. The best studies will collect data about an IV and a DV, and seek to determine whether the IV (the cause) influences (effects) the IV. In the context of studies where the IV and DV are identified, the definition by Kinnear and Gray may be most appropriate.

With an IV and a DV, in the context of health and safety and a new variable 'profitability', the research goals may then become:

RQ1: Does the level of compliance of UK contractors with best practice in health and safety influence profitability?

OB1: To determine if the level of compliance of UK contractors with best practice in health and safety influences profitability

H1: The level of compliance of UK contractors with best practice in health and safety influences profitability.

4.7.1 Which is the IV and which the DV? Variables 'melting' into each other

Sometimes students have two variables in their study, but have difficulty in identifying which is the IV and which the DV. The golden rule is to establish a time gap between the variables; in a time frame the IV must come first. In experimental work, the IV is manipulated, since we have control over it and at some later point in time, be it short- or long-term, this movement will impact on the DV. It may be helpful to present the relationship as a diagram in your introduction chapter, as illustrated in figure 4.2.

In observation studies, the issue as to which is the IV and which the DV is complicated a little by the notion that IVs and DVs can 'melt' or 'merge' into one another, that is, they are acting on each other simultaneously. It could be argued in one study that 'commitment to health and safety improves profitability'. However, construction sites or companies that have high profitability (profitability attained by another variable such as positive market conditions) may use some of that profit and invest it in a greater commitment to health and safety. In this scenario, in a time frame the profit is made first, thus it is the IV; investment and a commitment to health and safety follows second, as the DV. The variables 'melt' into one another since the high profitability that led to higher commitment in health and safety then leads to more profitability which leads to higher commitment still etc. A circular effect then ensues, as illustrated in figure 4.3.

You may need to write in your document to demonstrate that you understand that for some studies variables 'melt' into each other. For clarity, it is useful to label your variables as IV or DV. Which variable is the IV and which the DV depends on the context of your study, as defined by

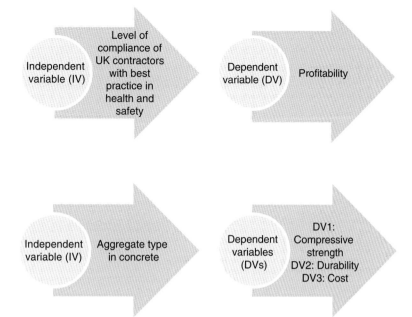

Figure 4.2 A time frame: the relationship between the independent variable (IV) and the dependent variable (DV).

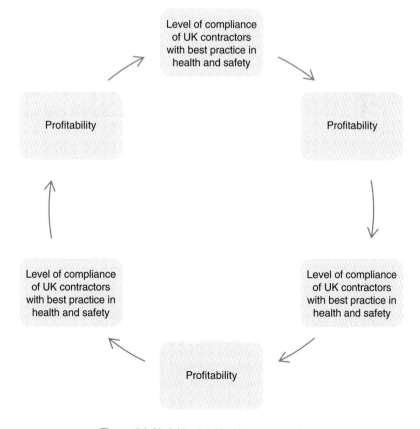

Figure 4.3 Variables 'melting' into one another.

your description of the problem and your objectives. Being clear about which is which helps to focus the whole structure of the research, but do not get hung up about it. For your own study, know your IV, know your DV and recognise that in another study they could be classified the other way round. It may be more accurate to say that there is a relationship or consistent association between the variables; a movement of one variable may result in the movement of the other variable; something is going on, there is a link between the variables that needs an explanation. Perhaps the issue of which is the IV and which the DV is for the discussion chapter.

On each occasion that you need to write about the two variables, perhaps when you write RQs, OBs and Hs, it is better that the IV be written first followed by the DV. For example, in the context of health and safety, let us label commitment to health and safety as the IV, and profitability as the DV, thus:

H1: The level of compliance of UK contractors with best practice in health and safety (IV) influences profitability (DV).

Is better than:

H1: Profitability (DV) is influenced by the level of compliance of UK contractors with best practice in health and safety (IV).

4.7.2 Manipulation or observation variables in research?

In laboratory or fieldwork type studies, the researcher may manipulate the IV as research proceeds, and simultaneously measure the impact of the movement of this IV on the DV. For example, in concrete science you may manipulate by volume or type, the constituent parts of a mix design (course aggregate, fine aggregate, cement and water). The mix design is the IV; it may have two or more values – two or more mixes. You may then measure the impact of this manipulation on several possible DVs: compressive strength, durability, water absorption.

In non-technical research, the measurement process may be to observe variables in practice; you cannot manipulate them as a researcher, for example an IV 'commitment in companies to best practice in health and safety', and a DV 'profitability'. To enable you to get different values of the IV, you may observe or collect data from a number of construction sites or a number of companies. As a part-time student you may also have access to the DV profit figures for these sites. If it is found that those sites or companies that have 'lower commitment to health and safety' have 'lower profits', and those with a 'higher commitment to health and safety' have 'higher profits', there can be a recommendation to those where commitment is not so high as to improve that commitment; what will hopefully follow, as indicated by the results of the research, will be higher profitability. Manipulation of the IV will occur, though not by you as a researcher.

4.7.3 A relationship, not a cause: strength of relationships

If a relationship is found between two variables, it does not necessarily mean cause. It may be stated that there appears to be an association between two variables, and that association warrants further investigation. Managers have control of IVs that they can manipulate in the hope of changing a DV related to business performance. There may appear to be an association between unhealthy lifestyles (IV) (e.g. smoking or too little exercise or poor diet) and illness in people (DV), but the first is not necessarily the cause of the second. Therefore try to use the connecting words between the IV and the DV carefully. You may consider association, influence, impact, relationship, correlation, link, effect or cause. To assert that an IV has a causal effect on a DV may need more work, including some careful preliminary speculation around the 'why' issues.

In science-based research, if a relationship is found, judgements can be made about the strength of that relationship. Will it need a large move in the IV to trigger a small move in the DV? Or will only a small move in the IV instigate a large move in the DV? In social science research, judgements can also be made about the strength of that relationship by plotting scatter diagrams and calculating correlation coefficients (see chapter 9.).

4.8 Writing the hypothesis: nulls and tails – a matter of semantics

When writing hypotheses, there are two concepts to be grasped:

- Is the hypothesis written as the null hypothesis or the alternative hypothesis?
- Is the hypothesis written as a one-tailed or a two-tailed hypothesis?

The null or the alternative hypothesis?

The null hypothesis is the hypothesis of no association, no difference or no relationship. Dictionary definitions of null are 'not binding', 'non-existent', 'amounting to nothing', 'no elements' – all the n's. The IV does not influence the DV:

Null hypothesis: 'The level of compliance of UK contractors with best practice in health and safety (IV) *does not* influence profitability (DV).'

The alternative hypothesis is written in a style such that it is suggested that the IV *does* influence the DV:

Alternative hypothesis: 'The level of compliance of UK contractors with best practice in health and safety (IV) influences profitability (DV).'

In statistics, tests of hypotheses are executed against the null. Upon execution of a statistical test, one of two findings may be expected:

- 'The level of compliance of UK contractors with best practice in health and safety (IV) *does not* influence profitability (DV).' In this case, the terminology adopted is that the null hypothesis cannot be rejected (this may be interpreted as saying the null hypothesis is accepted but by convention this latter terminology is not used);
- 'The level of compliance of UK contractors with best practice in health and safety (IV) *does* influence profitability (DV)'– in this case the null hypothesis is rejected – reject the null – reject the notion that there is no relationship.

A hypothesis intended to be a provisional supposition would seem more appropriately written as the alternative hypothesis. Indeed, it can seem odd to commence a study with a statement of the hypothesis written in the null. You may be excited to assert and prove your alternative hypothesis as being definitively correct. The issue of writing hypotheses as the null or alternative is only a matter of semantics. It is only words, and not to worry about. It can be argued that the writing style of the alternative hypothesis is more apt to the 'provisional supposition' definition of a hypothesis. If you write your hypotheses as the alternative, put a statement in your document something like 'hypotheses are written as alternative hypotheses, although it is recognised that statistical testing is undertaken against the null'. Chapter 9 gives quantitative examples of how to test null hypotheses.

The one or two tailed hypothesis?

The concept of whether the hypothesis is one-tailed or two-tailed is an issue of direction – direction of movement in the variables. The hypothesis 'the level of compliance of UK contractors with best practice in health and safety (IV) influences profitability (DV)' does not predict the direction of the movement. Will an increase in compliance lead to an increase or decrease in profitability? The writing style for a one-tailed could be 'high levels of compliance of UK contractors with best practice in health and safety (IV) leads to higher profitability (DV)' or 'high levels of compliance of UK contractors with best practice in health and safety (IV) leads to lower profitability (DV)'.

The difference between the two-tailed and one-tailed hypothesis is more important than whether the hypothesis is written as the null or the alternative. The rule is that the two-tailed hypothesis should always be used unless there is strong established evidence to predict the direction of movement of variables (Hays, 1988, pp. 276–277). In many research projects however, evidence to prove a relationship is not available, and in this case the hypothesis tested is often the two-tailed hypothesis. In the case of best practice in health and safety and profitability, firm evidence does not exist and it should be two-tailed. Going into your research you may try and anticipate that, if a relationship is found, what direction it will be in. You should still resist the temptation to write one-tailed hypotheses, unless the evidence is already there. Statistically there is twice the chance of rejecting a two-tailed hypothesis than rejecting a one-tailed hypothesis; more of that in chapter 9.

Integrating the null and the tails

There are thus four permutations for hypotheses: (a) the null, two-tailed hypothesis, (b) the null, one-tailed hypothesis, (c) the alternative, two-tailed hypothesis, (d) the alternative, one-tailed hypothesis.

(a) Null hypothesis, two-tailed: 'The level of compliance of UK contractors with best practice in health and safety (IV) *does not* influence profitability (DV)',
(b) Null hypothesis, one-tailed: 'The level of compliance of UK contractors with best practice in health and safety (IV) *does not* improve profitability (DV).'
(c) Alternative hypothesis, two-tailed: 'The level of compliance of UK contractors with best practice in health and safety (IV) *does* influence profitability (DV).'
(d) Alternative hypothesis, one-tailed: 'High levels of compliance of UK contractors with best practice in health and safety (IV) leads to higher profitability (DV).'

It is not to complicate the situation. Only one is necessary, but you must be clear in your own mind when you write your hypothesis which one of the above it is. In the above, option (c), the alternative two-tailed hypothesis may be preferred.

If some of your study objectives are being met by the literature review, a hypothesis may not be appropriate. Therefore give questions and objectives, and clearly state that there are no hypotheses for those objectives.

4.9 'Lots' of variables at large, intervening variables

In life, there are 'lots' of variables at large. In social sciences particularly there will often be many IVs impacting on one DV. For example, there are many variables (IVs) that affect profitability, not just the IV 'commitment to health and safety'. In hard sciences, there are many variables that impact upon the behaviour of structures, soils, fluids and materials. A research project may wish to seek out as many IVs as possible, and determine the strength of their influence upon the DV. Some IVs will have a strong influence, others less strong. Some of the more sophisticated statistical techniques are able to test for the influence of more than one IV upon one DV. It is adequate at undergraduate level, in any one test, to look for the effect of one IV on one DV. If a relationship is found, it is not to draw conclusions that are too strong, given that the strength of a relationship may only be modest, and there may be other more important variables that warrant attention. A modest relationship may indicate that significant manipulation of the IV is required to instigate less significant movement in the DV.

A study may initially want to determine whether there is a relationship between an IV and a DV, but it may discover that there are lots of intermediate relationships or intervening variables (IntVs). Again using safety as an example, and the proposal to investigate if there is a relationship between a commitment to health and safety and profitability. After examining the literature, it is apparent that another variable, 'the quality of risk and method statements', sits somewhere in this equation. Therefore, datasets are collected about three variables, labelled initially:

IV: commitment to health and safety
IntV: the quality of risk and method statements
DV: profitability

Analysis may show that relationships between the IV and the IntV and between the IntV and the DV are stronger than the one between IV and DV. That may be because there are more variables than just the IV impacting on the IntV. An outcome of the study could be a recommendation to seek out what other variables are impacting on the IntV to thus improve the DV. Could there be other variables between the IntV and DV? That could be another study. Figure 4.4 illustrates potential relationships.

Other examples on intervening variables are given in table 4.3.

4.10 Ancillary or subject variables

In your findings you may wish to assert that an IV influences a DV. Repeating from section 4.9, there are 'lots' of variables at large. It may be the case that there are variables, other than intervening variables, impacting on your DV. Bryman and Cramer (1997) call these 'subject

Figure 4.4 Intervening variables.

Table 4.3 Examples of Intervening variables.

Independent variable	Intervening variable	Dependent variable
Leadership style of bosses	Employee satisfaction	Productivity
Office environmental conditions	Absenteeism rates	Profitability
Method of paying craftspeople	Motivation of workers	Time predictability of completing projects
Volume of safety training	Knowledge of construction workers	Frequency of wearing personal protective equipment (PPE)
Implementation of best practice on partnering projects	Levels of trust between parties	Successful projects
Use of cost control systems	Psychological approach to managing resources	Lower costs
Supply chain management systems	Snag list applications	Defects at handover
Recruitment policy	Training and CPD for employees	Client satisfaction

variables' since in psychological research they are characteristics of the subject or person taking part in the research. Perhaps the term 'ancillary variables' is useful, since in the built environment, it is not only the person or subject we are concerned with; we are also are interested in projects, methods of working etc. The ancillary variables (AncV) may be a characteristic of the way that projects are procured, or their size or type. You cannot measure all the variables that are at large; to measure one variable can be a whole study. It is not good to attempt two studies in one. If another variable is very important in helping to understand a problem, let that be a separate study, but there will often be some further data that can be collected without putting unreasonable demands on your time or the time of others. If you are collecting data from archives, there will often be a whole host of information available. In surveys you can ask general questions, to get data about other variables. Typical general questions are about participants' personal data (sometimes called demographic data), such as age, gender, job title, ethnic background or qualifications. Also easily collected data is items such as type of project, value and size, method of procurement or size of company.

Consider again the objective 'to determine if the level of commitment of UK contractors to best practice in health and safety influences profitability'. There are lots of ways to measure the level of commitment. Perhaps if you were a part-time student and able to take data from your own sites, you could design a measurement instrument based on observation of site conditions, talking to professionals and inspecting site documents. You may also be able to easily

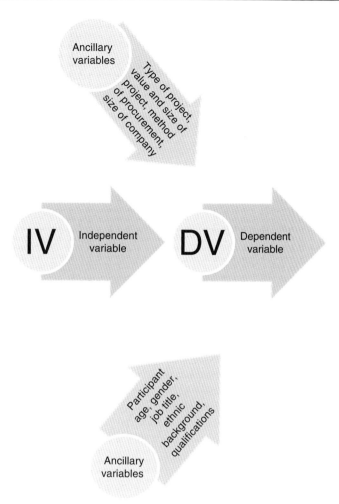

Figure 4.5 Ancillary or subject variables also simultaneously acting on a DV.

ascertain personal details of site managers on each project viz: age, gender, job title, ethnic background and qualifications. These are ancillary or subject variables that may impact on your measures. As a part-time student you may get access to the actual profit figures on your sites. Ancillary subject variables that are easily collected, and that might affect profit are for example method of procurement, type of contract, size of project or type of project. It may also be interesting to determine if these variables impact upon measures of the IV. Figure 4.5 illustrates how ancillary or subject variables may impact upon IVs and DVs.

While variables such as age and gender may be often classified as ancillary variables, if a problem is articulated on the basis that age or gender etc. is one of the issues to be investigated, they may then be placed as IVs or DVs in your research.

In your analysis you will determine whether or not the ancillary variables influence the IV and the DV. Assuming the IV is found to influence the DV, if the ancillary variables are also found to influence the DV, it would be difficult to draw strong conclusions about the relationship between the IV and the DV. A relationship between the IV and the DV may be spurious; the variable that really impacts on the DV is the ancillary variable; all material for discussion in the closing parts of your document.

However, if again the IV is found to influence the DV, but the ancillary variables are not found to influence the DV, that may lead you to draw strong conclusions about the relationship between the IV and the DV. You are sure it is the IV and not other variables, impacting on the DV.

If you find a relationship between an ancillary variable and a DV, it is not to throw away the finding about the relationship between the IV and the DV; you may wish to express some caveats in your findings, or it may be possible to do some more exploration and analysis of the data. If you go into your research with the hope of finding a relationship between the IV and the DV (though you should not necessarily be doing that), your Eureka moment will be when you get significant results, that is a relationship between the IV and the DV. Another Eureka moment should be if you do an 'eyeball' analysis around ancillary variables, and that shows only small differences; that is, there is no relationship between the AncVs and the DV. That would give you greater confidence that any relationship between the IV and the DV is genuine; it is not spurious. It also indicates that the dataset that you have is homogeneous; the matrix of numbers that you have hangs together well around an assertion that the IV influences the DV.

Collecting data about ancillary variables and making observations helps to improve the validity of your study. You could be asked, perhaps in a viva, 'How do you know that it is the IV impacting on the DV, and not some other variable?' Your answer is clear, 'I have collected data about as many other variables as I reasonably could, and (if this is the case) I have found no other relationships.' If you do find relationships between ancillary variables and the DV, do not be disappointed. You will get marks for doing the analysis; but you should also write in your conclusion chapter about how this is a limitation on the validity of your work. If you realise and understand limitations, you can get marks for that too, but if your work is limited and you do not realise that it is, then marks could be deducted. You may find it appropriate to compare scores of ancillary variables against the IV as well as the DV, especially if it is the case that the issue of which is the IV and which is the DV is open to interpretation. The statistical tests in the above examples to determine whether relationships are significant are Spearman's correlation (for testing between the IV and the DV) and the Mann–Whitney test (for testing between the ancillary variables and the DV). Chapter 8 will illustrate how these tests and probability can be used to determine whether relationships are significant.

Tables 4.4 and 4.5 detail some hypothetical data about the impact of the ancillary variable 'age of site managers' on the DV profitability. The main hypothesis is to test if 'method of procurement influences profitability'. Just ten projects are listed for illustrative purposes; this is a very low number for n; chapter 9 advises you should aim for n to be as high as possible. The statistical tool used to eyeball the data to look for differences or relationships is 'means'. Chapter 9 illustrates statistical tests to determine whether these differences and relationships are considered 'significant'. The independent variable (IV) 'method of procurement' is in just two groups or has two values: projects procured in competition with other companies and projects procured by negotiation. It could be possible to have three or more groups or values

Table 4.4 The impact of the ancillary variable 'age of site managers' on the DV profitability: the IV **does** influence the DV, but ancillary variables **do not** influence the DV.

	A	B	C	D	E	F	G	H	I
1		Independent variable; categorical level	Dependent variable: profitability %. Interval level			Ancillary variable: age of site managers. Ordinal level	Dependent variable: profitability %. Interval level		
2	Project number	Method of procurement: competition = A, negotiation = B	All projects	Projects procured by competition	Projects procured by negotiation	<35 years = A, 35 to 50 years = B, >50 years = C	For site managers <35 years; A	For site managers 35-50 years; B	For site managers >50 years; C
3	1	A	4.80	4.80		A	4.80		
4	2	B	5.60		5.60	C			5.60
5	3	B	6.50		6.50	B		6.50	
6	4	A	2.10	2.10		C			2.10
7	5	A	3.50	3.50		A	3.50		
8	6	A	2.20	2.20		B		2.20	
9	7	A	1.10	1.10		A	1.10		
10	8	B	4.40		4.40	B		4.40	
11	9	A	5.20	5.20		C			5.20
12	10	B	8.20		8.20	A	8.20		
13	Means; profitability %		4.36	3.15	6.18	–	4.40	4.37	4.30

Table 4.5 The impact of the ancillary variable 'age of site managers' on the DV profitability; the IV **does not** influence the DV, but ancillary variables **do** influence the DV.

	A	B	C	D	E	F	G	H	I
1		Independent variable; categorical level	Dependent variable: profitability %. Interval level			Ancillary variable: age of site managers. Ordinal level	Dependent variable: profitability %. Interval level		
2	Project number	Method of procurement: competition = A, negotiation = B	All projects	Projects procured by competition	Projects procured by negotiation	<35 years = A, 35 to 50 years = B, >50 years = C	For site managers <35 years; A	For site managers 35-50 years; B	For site managers >50 years; C
3	1	B	4.80		4.80	B		4.80	
4	2	B	5.60		5.60	A	5.60		
5	3	A	6.50	6.50		A	6.50		
6	4	A	2.10	2.10		C			2.10
7	5	A	3.50	3.50		C			3.50
8	6	B	2.20		2.20	C			2.20
9	7	A	1.10	1.10		C			1.10
10	8	A	4.40	4.40		B		4.40	
11	9	B	5.20		5.20	B		5.20	
12	10	A	8.20	8.20		A	8.20		
13	Means; profitability %		4.36	4.30	4.45	–	6.77	4.80	2.23

for this variable. The ancillary variable 'age' is arranged in three groups or has three values. Young site managers are classified as being under 35 years of age; middle-aged site managers 35 to 50 years and older site managers over 50 years. It could also be possible to have two, or four or more groups or values for this variable.

In table 4.4, mean profit values in cells 13D and E based on 'type of procurement' are different at 3.15% and 6.18%. The profit values in cells 13G, H and I based on age of site managers appear similar at 4.40%, 4.37% and 4.30%. It appears that the IV method of procurement influences the DV profitability; negotiated projects give much better profit levels. But the age of site managers as an ancillary variable, is not impacting upon profitability.

In table 4.5 the hypothetical dataset is rearranged. Mean profit values in cells 13D and E based on 'type of procurement' are very similar at 4.30% and 4.45%. The profit values in cells 13G, H and I based on age of site managers appear different at 6.77%, 4.80% and 2.23%. Given this set of data, it appears that the IV method of procurement does not influence the DV profitability: negotiated and competitive projects have similar profit levels. But the age of site managers as an ancillary variable is impacting upon profitability; younger site managers attain best profits.

Using authoritative classifications to measure variables

You will need to prejudge the possible values for your variables; if so, it is not for you to design new criteria. For example, if you wanted to classify projects by type as a variable, the Office of Nationals Statistics (ONS, 2014) in its publication Construction Statistics No. 15, 2014 edition, classifies private work into four major categories or four variable labels: (a) new housing, (b) infrastructure, (c) industrial and (d) commercial. Industrial is subdivided into three categories and commercial into eight. Use these categories or some other similar authoritative source in your research. Similarly if you were to measure sustainability ratings, the appropriate values are 'outstanding', 'excellent', 'very good', 'good', 'pass' and 'unclassified', which come from the Building Research Establishment Energy Assessment Method (BREEAM, 2015).

4.11 No relationship between the IV and the DV

There may be a temptation to hope that, at the end of a study, a relationship is found between an IV and a DV. However, it is possible that in quantitative work, your hypothesis will not be proven; no relationship is found between variables. You must not view that as though your study has failed and you must not be disappointed. It cannot be that every piece of research finds relationships; that would be preposterous. The nature of research is that provisional suppositions are put forward as hypotheses for testing, but a positive outcome is not assured. If no relationship is found, that is fine; in the next study it is just to back track a little and put forward new hypotheses for testing – 'Well if it is not A that influences B, I wonder if it is variable C?' Whether or not you find a relationship is not a factor in judging the success of your study. The idea is not to force relationships, so that after analysis you can assert 'I told you so'. The purpose of research is to determine whether there are relationships between variables and if so to tease them out gently. If relationships are found, they should be re-tested in subsequent research using different methodologies to ensure validity. Society needs to understand these relationships to build its knowledge. We need to know whether your IV influences the DV; or is it another IV? Society needs to establish the links from variable A to variable Z, and it can only make those links in very small, often minute steps over long time periods. The process is like going through a maze. We want to get from start to the finish, and that journey involves linking a chain and a web of variables. It is inevitable that some research will take the wrong route in the maze; it may explore possible relationships and find none. You should mark the

Why we need journals with negative result

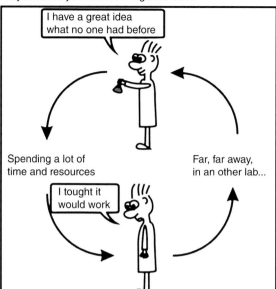

Figure 4.6 Why we need journals with negative results. Source: Science for all.

wrong road that you have taken, and turn back so that you or someone else can try a different route. The wrong road is marked by publishing your work, and telling other researchers that you have tried this route and it is a dead end. Following researchers can then devote their attentions to other methodologies or other variables.

Studies that do find relationships between variables may have conclusions written with vigour, championing the study as though a victory was won. If you do not find a relationship, you must not let your conclusion suffer. You should still write it with vigour, perhaps with recommendations to suggest slightly different methodologies for retesting, or suggesting slightly different variables. It is often the case that the studies that grab media headlines are those that find relationships between variables. So be it; you are not after media headlines from your research. Figure 4.6 illustrates the concept of someone else, somewhere else in the world, wasting time discovering the same failures that you have discovered; but they did not know it would fail, because journals are most interested in publishing positive outcomes.

4.12 Designing measurement instruments; use authoritative tools and adapt the work of others

There is a view that your dissertation or project is about measuring something really well. You should not try to measure too many things. To measure a few things well should attract good marks. To measure many or complex things superficially will not attract good marks. For example, it would not be possible to measure in one piece of research the success of companies around the cost, time and quality triangle. Better to just measure one, and indeed just one small part of one of these variables.

Philosophical or qualitative analysts may argue that they are not about measuring things. Their approach is more about gaining insights into the world and about how other people see things. Qualitative analysts collect rich data. Certainly an outcome of qualitative research is not to score things on scales. If research is about the study of constants, that is correct in an absolute sense. More likely, most qualitative research is about things in life that are changing and are therefore variables. Qualitative research collects evidence and makes judgements one way or the other about issues. An outcome of research may be that something is allocated the qualitative label of 'good' or 'bad'. This outcome could have gone one of two ways; it is variable.

Only as a last resort should you use your own measurement tool. You should be seeking to build from a platform created by others, not reinventing the wheel; it is not for you to design new things and new ways. It is far better to use established authoritative methods to help you. In the context of measuring your variables, delve deeply into the literature to see if a measurement tool has been used before. Using the work of others to provide a platform for your work improves its validity, and should give you more marks. However, where you do use or adapt the work of others, make sure that you cite the original source: 'This survey is adapted from the work of Arewa (2013)'. Do not just use any measurement tool that you stumble across. Browse them all; mix and pick the most authoritative sources that are relevant to the context of your study. Better than just using the internet, also look at what may have been used in academic journals. Make sure that you do not breach any copyright.

In scientific work you should measure things in the same way that they are measured in British Standards, such as 'BS EN 772-3:1998. Methods of test for masonry units. Determination of net volume and percentage of voids of clay masonry units by hydrostatic weighing' (BS, 1998). If you wanted to measure or test any of these properties of bricks, you would not design your measures or your own way of doing it; follow the British Standard.

In social sciences, you may wish to measure leadership style. To design your own measurement tool would be foolhardy. On a continuum, authoritative research anchors leadership style at the extremities by terms such as 'task orientation' and 'people orientation'. Put simply, task orientated managers speak to employees only about the work in hand, while people orientated managers take some time out of work to talk to employees socially. People orientated managers have been found to get better results in the workplace. Managers may be judged to be at one end or the other, or somewhere in between. Web-based searches will show that there are lots of questionnaires available, using key word phrases such as 'leadership style questionnaire', e.g. Sage Publications (2015).

While the most valid measurement tools are those already in the public domain, do not let that deter you from thinking innovatively about your own designs. If you do this, the argument that measures in the public domain are proved to be more valid by regular use can be compensated by robust piloting of your questions, and by doing tests of internal reliability (see chapter 9).

4.12.1 Variable values with high or low numbers as best?

In scientific work, whether a high or low number is best is decided by what is being measured e.g. in concrete, high compressive strength is better than low, while low water absorption properties are better than high. Constructing Excellence measures the number of accidents per 100,000 man-hours worked; this is an objective measure, and clearly low numbers are best.

In social sciences, rating systems often start with qualitative words or descriptors which are then converted to numbers. You may attribute low or high scores to the best; that is your choice. There are many rating systems for hotels across the world. One popular system is by stars, where 1* hotels are thought to be not so good, while 5* are excellent. The best hotels get the highest scores.

In the UK, the Office for Standards in Education (OFSTED, 2015) is responsible to government for inspecting the quality of teaching in schools. It has grade descriptors, with the best score being lowest, thus: 1 = outstanding, 2 = good, 3 = requires improvement and 4 = inadequate.

There may be some scales where it is arguable whether there is a best or worst at each end of the spectrum, for example leadership style. In such a case, it may not matter whether task orientation and people orientation are scored with high numbers or low.

On multiple-item scales, if there is clearly a better end to the spectrum, it may be better to allocate the high score to the best end. Chapter 5 provides more details about multiple-item scales used to measure concepts.

4.12.2 Measurement scales of 0–10 and 0–100; multiple-item scales

Many modern key performance indicators (KPIs) are designed around 0–10 scales. If you do design your own measurement scale or adapt a scale from the literature, it is logical to do the same or possibly expand it to the range 0–100 like an examination scale. Note that the scale should be anchored at 0 and not 1. If for example your initial measurement scale has a maximum score of say 40, you may arithmetically convert that to a percentage whereby 30 out of 40 = 75%.

Example

A foreman is required by 'his' employer to rate the quality of work of bricklayers. Rather than just award each bricklayer an anecdotal subjective number out of ten for quality of work, it is decided to try to improve the quality of this subjective measure by using a multiple-item scale of eight criteria to make an overall judgement. A frequency scale of 'how often' is used. Initially the foreman allocates values where the best score is a low number, such that:

1 = always
2 = usually
3 = sometimes
4 = rarely
5 = never

The eight criteria are that bricklayers:

C1. ensure tools have been cleaned after use
C2. leave enough time to point or joint work
C3. only lay to heights that can be comfortably reached
C4. ensure that work is covered to protect it from weather if necessary
C5. ensure that work is carried out from a firm level platform
C6. are especially careful when setting out, building corners and in plumbing work
C7. ensure that cavities are clear of mortar droppings
C8. are careful to ensure that ties and DPCs are installed correctly.

Table 4.6 Scoring judgements for ten bricklayers based on eight criteria; scoring range 1 to 5 with 1 as the best score.

	A	B	C	D	E	F	G	H	I	J	K
		Criterion (C) 1	C2	C3	C4	C5	C6	C7	C8	Total: best possible score = 8, worst possible score 40.	Mean
1											
2	Bricklayer (B) 1	1	1	3	2	3	2	3	2	17	2.13
3	B2	1	2	2	3	3	2	4	3	20	2.50
4	B3	1	2	3	2	4	3	4	1	20	2.50
5	B4	2	3	3	2	4	4	4	3	25	3.13
6	B5	1	2	1	2	4	2	5	3	20	2.50
7	B6	1	3	2	2	3	2	5	2	20	2.50
8	B7	1	1	2	2	3	2	4	2	17	2.13
9	B8	2	3	2	3	3	3	3	3	22	2.75
10	B9	5	5	5	5	5	5	5	5	40	5.00
11	B10	1	1	1	1	1	1	1	1	8	1.00
12	Totals	16	23	24	24	33	26	38	25	209	26.13
13	Mean	1.6	2.3	2.4	2.4	3.3	2.6	3.8	2.5	**20.9**	2.61

The results for ten bricklayers are presented in Microsoft Excel and shown in table 4.6. Each bricklayer obtains a personal score derived from adding columns B to I; the answer is in column J. Each criterion obtains a score derived from adding rows 2 to 11; the answer is in row 12.

The results show, reading horizontally that bricklayer 10 is judged the best bricklayer with a score of 8 (mean of 1.00) and bricklayer 9 the worst with a score of 40 (mean of 5.00). Reading vertically, bricklayers are best at criterion C1 with a score of 16 or a mean of 1.6, and worst at criterion C7 with a score of 38 or a mean of 3.8.

Using this scoring scale, a typical question is, 'How good is bricklayer B4 with a score of 25 given a range of scores of 8 to 40?' Hmm. Need to think about that? What about 'How good are these bricklayers as a group overall with a score of 207 and a range of possible scores from 80 (all ten bricklayers scoring 1 on all eight questions) to 400 (all ten bricklayers scoring 5 on all eight questions)?' Need to think about that too?

Now consider table 4.7 with the same results, but with two changes made to the scoring scale, thus: (a) the highest number is allocated to the best performance, and (b) numbers are allocated in the range of 0 to 4 rather than 1 to 5 (anchoring the scale at 0 not 1), thus:

4 = always
3 = usually
2 = sometimes
1 = rarely
0 = never

With the highest score allocated to best performance, and very importantly with the scale anchored at zero, the opportunity arises to have a 0–10 or 0–100 percentage scale that is perhaps easier to interpret.

Table 4.7 Scoring judgements for ten bricklayers based on eight criteria; scoring range 0 to 4 with 4 as the best score.

	Criterion (C) 1	C2	C3	C4	C5	C6	C7	C8	Total: best possible score = 32, worst possible score 40.	Percentage score
Bricklayer (B) 1	4	4	2	3	2	3	2	3	23	71.88
B2	4	3	3	2	2	3	1	2	20	62.50
B3	4	3	2	3	1	2	1	4	20	62.50
B4	3	2	2	3	1	1	1	2	15	46.88
B5	4	3	4	3	1	3	0	2	20	62.50
B6	4	2	3	3	2	3	0	3	20	62.50
B7	4	4	3	3	2	3	1	3	23	71.88
B8	3	2	2	2	2	2	2	2	17	53.13
B9	0	0	0	0	0	0	0	0	0	0.00
B10	4	4	4	4	4	4	4	4	32	100.00
Totals	34	27	25	26	17	24	12	25	190	
Mean	3.4	2.7	2.5	2.6	1.7	2.4	1.2	2.5	2.38	
Percentage	85.00	67.50	62.50	65.00	42.50	60.00	30.00	62.50		**59.38**

Now interpret the results again. Percentage scores are calculated, for example reading horizontally Bricklayer 1 has 23 out of 32 = 23/32 × 100 = 71.88%. Reading vertically, criterion 1 scores 35 out of a maximum of 40 = 35/40 × 100 = 85%. Bricklayer 10 is judged the best bricklayer with a score of 100% and bricklayer 9 the worst with a score of 0%. Reading vertically, bricklayers are best at criterion C1 with a score of 85%, and worst at criterion C7 with a score of 30%.

A typical question is, 'How good is bricklayer B4?' With this scoring scale a clear answer is possible: 46.88%. Is it also possible to determine how good these bricklayers are as a group? Answer = 59.38, say 60%. In the discussion chapter, it will be possible to write about this score of 60% compared to what might be reasonably expected; as a foreman you may expect a score of 85% plus?

In the context of multiple-item scales described in chapter 5, you may reflect on the questionnaire in the discussion; since the word 'and' appears, should C6 really have been three separate questions (are especially careful when setting out, building corners and in plumbing work), and C8 two questions (are careful to ensure that ties and DPCs are installed correctly)? Also, notably, speed of working is not one criterion.

4.13 Levels of measurement

On an everyday basis we collect data to measure things. Variables can be measured at different levels: (a) categorical (nominal), (b) ordinal, (c) interval, (d) ratio. The sequence is important: the richness of the data and the power of statistical tests that are possible increases as the scale moves from (a) to (d). Categorical variables are also called 'nominal variables' in many texts; you need to be able to use the two descriptors interchangeably. The sequence of the variables may be remembered by the mnemonic COIR (the fibre from the outer husk of the coconut) or if you prefer to use the term nominal, NOIR (French for black).

Table 4.8 Exemplar of categorical, ordinal and interval data based on student assessment scores.

	A	B	C	D	E	F	G	H
1		1	2	3	4	5	6	7
2		Student name	Interval examination percentage score	Observations about the interval score	Categorical / nominal score	Observations about the categorical / nominal score	Ordinal score; criteria 0-39 = F, 40-49 = D, 50-59 = C, 60-69 = B and 70 - 100 = A.	Observations about the ordinal score
3		Anne	0		Fail		F	
4		Barry	39		Fail	Barry has scored significantly better than Anne, but has the same outcome	F	Barry has scored significantly better than Anne, but has the same outcome
5		Clare	40		Pass	Clare has scored marginally better than Barry, but has a significantly different outcome	D	Clare has scored marginally better than Barry, but has a significantly different outcome
6		Dave	51		Pass		C	
7		Emma	69	Differences between all scores completely transparent	Pass		B	Emma has scored significantly better than Dave, but has an outcome only marginally different. Also Emma has a score that is minutely different to Fred, but the 'marginal' outcome is 'relatively' large.
8		Fred	70		Pass		A	
9		Gary	100		Pass	Gary has scored significantly better than Clare, but has the same outcome	A	Gary has scored significantly better than Fred, but has the same outcome

Table 4.8 illustrates the difference between measurement levels with an example based upon an examination percentage scale. The pass mark is 40%. Seven students take an assessment. The work is initially marked at the interval level in column 2; it can be seen that one student has scored 0%, one 100% and the others range from 39 to 70%.

On some academic programmes only a pass or grade mark is awarded; that is at the categorical level. Students are not told the percentage mark they achieved in an assessment; only whether they passed or failed. Suppose for this assessment, a pass or fail judgement is given to students. Reading column 4, two students are in the fail category, and five in the pass category. But what is being hidden? As a student, if you received your result at the categorical level, you may be pleased to pass, but will wonder how well you passed. Did you do as well as Gary or only as well as Clare?

Alternatively, suppose students are not told the percentage mark achieved in an assessment; only the grade: A, B, C, D, E or F. What is being hidden now? Reading from column 6, if you achieved a grade B like Emma, perhaps you would wonder how close you were to a grade A. If she had access to her interval score, since she scored 69%, she would perhaps be unhappy that she did not make a grade A. With 51%, Dave might wonder how close he was to a grade B; he was actually a long way short, since grade B requires a minimum of 60%.

By awarding marks at the categorical or ordinal level, some of the raw data is being masked. Examining boards that award ordinal marks are masking the richness of the data they hold.

In some instances, collapsing data from the interval to the ordinal, and even categorical level, may be justified, and indeed necessary. Interval data may be too much to digest in a world of information overload. While government holds information about percentage marks for students at the interval level, the headline data for GSCE results is published at the categorical level; historically that has been the percentage of students who gain five passes at grade C or above, including English and Maths. There are two values or categories: either a student gets the five passes or they do not.

When you collect data for some variables you may have a choice of whether to collect it at the categorical, ordinal or interval level. Get the data at the richest level possible. Getting data at the richest possible level can improve the validity of your research. However, there may be good reasons for not seeking information at the highest levels. Participants may be reluctant to give details at the interval level about age (precise age in years) or salary (precise earnings in £). Also ethically you may consider it inappropriate to intrude on information that participants are reasonably entitled to keep private. You may, however, tease out details at the ordinal level (age 25–37 or 38–55 etc., salary £20–30k etc.). In materials science, you may have a concrete mix designed to 28 N/mm^2. If you were to test some cubes, you could just collect data at the categorical level; each cube at destruction would go in the fail category (destruction below 28 N/mm^2) or pass (destruction 28 N/mm^2 or above). This would seem to be nonsense. You would obviously record at the interval level, the compressive strength of each failure, for example cube A failed at 32.80 N/mm^2, cube B failed at 36.92 N/mm^2. In non-technical work, if you were able to collect profit data based upon actual monetary figures or percentages, it would be a nonsense only to record profit in two potential categories of 'as or more than antici-pated' and 'less than anticipated'. Take the data at the richest level possible; report the percentage profit figures.

4.14 Examples of categorical or nominal data in construction

For a categorical or nominal variable, the 'values' do not have a relationship between each other that can be measured; they are just different and there is no distance between them. The values of the variable are put into categories or boxes; they are not on a continuum or an order. For example, if the variable name or label is 'type of material', in the proposed research it may have three values: steel, concrete and timber. These three materials cannot be placed in order. Steel is no closer to concrete than it is to timber.

Some variables by their very nature are only measurable at the categorical level, e.g. gender, which usually has only two values, male and female. It might be the case that you wish to do a comparative study. Your variable label may be 'type of procurement', and you have two categorical values, traditional and design/build. Or your variable label may be 'methods of pouring concrete', and you have two categorical values, crane and pump. You may also choose the categorical level of measurement, but have three or more values of the one variable: again using materials as an example, variable label 'type of material' with three values. steel, concrete and timber. Be careful though, since examination of three materials in this case may give you too much work. Similarly for three or more types of procurement or pouring concrete.

Students often ask participants to make judgements or give opinions, with the only possible categorical responses being 'yes' or 'no'. This is often not appropriate since you can do better and obtain richer ordinal data. Often people cannot answer 'yes' or no' because they wish to give ordinal responses somewhere in middle: 'definitely yes', 'probably yes', 'unsure', 'probably no' or 'definitely no'. It could alternatively be the student making a judgement, perhaps about the aesthetic appeal of civil engineering structures, where a five or six point scale is better than just two or three.

4.15 Examples of ordinal data in construction

Ordinal variables are most often applicable to subjective and non-technical subjective work. Table 4.9 illustrates this concept with the marks of Fred (70%), Emma (69%) and Dave (51%), classified in column 6 as A, B and C. Ordinal variables are placed in sequence, but the distance between ordered categories is not measurable. The distance between 'A' and 'B' may not be the

Table 4.9 Variables to measure in construction at the categorical/nominal, ordinal and interval level.

	A	B	C	D	E	F	G	H
1	Category.	Variable name	Perhaps measured by	Objective or subjective data	Primary or secondary data	Option 1; measure at the categorical level	Option2; measure at the ordinal level	Option 3; measure at the interval / ratio level
2		Tensile strength of steel	N/mm^2					
3		Deflection of timber beams	mm		Potential to derive your own primary data in laboratories / fieldwork or take secondary data from the literature			
4		Speed of fluid flow	m/hour					
5		Bearing capacity of soils	N/mm^2					Potential to measure all at the interval or ratio level
6	Technical	Traffic pollution; air quality	Index 0 to 10	Objective				
7		Heat flow through external walls	W/m^2K					
8		Lighting levels in libraries	Lumens/m2					
9		Sound reduction through party walls	Db					
10		Profit of sites	Internal cost reports	Objective	Primary			Percentages
11		Time predictability of completing projects	Percentage early or late completion or on time	Objective	Primary			Percentages
12		Aesthetic appeal of architectural styles	Public judgements	Subjective	Secondary		Six groups: outstanding, excellent, very good, good, poor, very poor	
13		Method of procurement	Site records	Objective	Primary	Two groups; traditional and design / build		
14		Quality of workmanship	Number of defects at handover per £100k	Objective	Secondary			Frequency counts
15		Frequency of wearing personal protective equipment (PPE)	Clip board observations	Objective	Secondary			Frequency counts
16		Method of payment to craftspeople	Site records	Objective	Secondary	Two groups: fixed pay schemes, incentive schemes		
17	Non-technical	Method of bills of quantities production	Site records	Objective	Secondary	Two groups: manual, computer		
18		Motivation level of maintenance workers	Questionnaire	Subjective	Primary	Two groups		
19		Accessibility of historic buildings	Clip board observations	Subjective	Primary	Two groups: good, not good	Six groups: outstanding, excellent, very good, good, poor, very poor	Eleven rating scores: 0 to 10 (or 0-100)
20		Cost of buildings	£	Objective	Primary	Two groups: high, low		£/m2
21		Quality of service	Questionnaire	Subjective	Secondary			
22		Compliance with best practice	Questionnaire	Subjective	Primary	Two groups: good, not good	Six groups: outstanding, excellent, very good, good, poor, very poor	Eleven rating scores: 0 to 10 (or 0-100)
23		Client satisfaction	Questionnaire	Subjective	Primary			
24		Environmental performance	Energy use	Objective	Secondary			
25		Systems of pouring concrete	Observation	Objective	Primary	Two groups: hoisting by cranes, pumping	–	–

same as the distance between 'B' and 'C'. The sequence of A, B, C is important; it is in order: A is better than B, which is better than C.

Consider also the ratings 'excellent', 'very good' and 'good' of three products which are clearly in 'order', but a participant may reflect that the 'very good' product was quite close to the 'excellent' product, while the good product, although good, was a lot lower quality than the very good product. Those latter judgements are lost in the data analysis; it is just in the order of A, B and C.

Quantitative analysts will allocate numbers to these qualitative judgements, such that in the context of five labels, excellent will be coded 4, very good 3, good 2, satisfactory 1 and poor zero. Multiple-item scales (see 4.12.2 and sections 5.3.2 and 3) are then used to add numbers together to produce overall scores. Fink (1995, p. 50) states that ordinal data is habitually treated as though it were numeric data, but some purists may argue that allocating real numbers to ordinal data is inappropriate. However, there is a need to simplify data; it serves to convert some qualitative judgements or statements by participants to the quantitative. Therefore perhaps the leap is justified, providing that it is acknowledged by analysts.

Lots of subjective judgements are made about products and services on the ordinal scale. Just briefly here frequency judgements (always, sometimes, rarely, never), importance judgements, (very important … not important), and agreement (strongly agree … strongly disagree). There are more examples in section 5.4.5.

4.16 Examples of interval and ratio data in construction

The values for interval and ratio level measurements are placed in a sequence, and the differences between the values are assumed to be definitively apparent and measurable. The difference between 2 and 4 is the same as the distance between 4 and 6. Table 4.8 illustrates this concept with the marks of Fred (70%), Emma (69%) and Dave (51%). Emma's mark is nearly as good as Fred's and a lot better than Dave's. Ratio scales have a definitive zero point. A subjective measure of perhaps job satisfaction on a 0–10 point scale is interval not ratio; a person may score zero on this scale, but that leaves the possibility that there is someone else who is less satisfied still.

In section 4.15, the qualitative judgements 'excellent', 'very good' and 'good' were described as being ordinal since they were in 'order', but with the distance between 'excellent' and 'very good' not necessarily being the same as the distance between 'very good' and 'good'. Therefore these words are not interval data. But it becomes a fuzzy issue when the qualitative words are replaced by numbers thus: excellent 4, very good 3, good 2, satisfactory 1 and poor 0. Now the distance between 4 and 3 is clearly the same as the distance between 3 and 2; so can we argue that excellent, very good etc. is actually interval data? Perhaps it is difficult to justify that, but many subjective measures are based on multiple-item scales. The example in section 4.12.2 illustrates eight criteria used to measure the quality of bricklayers' work. There are more multiple-item scales in chapter 5. It is hard to argue that the range of 0 to 4 is interval data when the numbers originated from qualitative judgements, but much easier to substantiate that it is interval when the eight criteria added together have a range of 0 to 32. While debatable, such multiple-item scales are routinely treated as interval data by many statisticians when they select appropriate tests – see chapter 9. More powerful tests can be executed on interval data.

Many technical and numerical datasets may be considered ratio. Technical data may comprise large or small measurements in the SI units electric current, mass, length, time, temperature, substance or luminous intensity. They may be used as singular measures or be used in calculations of velocity, pressure, density, heat transmission etc. Similarly for sound and angular measurements. In your research you may derive these measures from fieldwork, laboratory work or from the literature. In the case of concrete mix design, there may be only one adjustment made to the mix design, such as quantity of cement. So if the cement content increased by say $1 \, kg/m^3$ and there were ten mixes (ten values of the variable 'cement content'), this is considered ratio. Statistical data is often at the ratio level. Examples are: number of vehicles on UK roads, number of people employed, number of apprentices, number of companies in business, number of accidents, number of sites, turnover of the UK construction industry in £, indices, age.

For most statistical tests, data at the interval level is treated similarly to data at the ratio level. A key principle to ascertain the difference between interval and ratio is that the zero point on the interval scale is arbitrary, but on the ratio scale it has real meaning. This difference is illustrated in the literature by comparison of the Celsius and Kelvin temperature scales. On the Celsius scale, the freezing point of water has an arbitrary value of zero, and it is thus considered interval. However, the Kelvin temperature scale has a non-arbitrary zero point of absolute zero, which is denoted 0K and is equal to −273.15 degrees Celsius. The 0K point is non-arbitrary, as the particles that comprise matter at this temperature have zero kinetic energy. There are many technical measures that are ratio with definitive zero points, such as mass, velocity; and non-technical such as age, income, work absenteeism.

Table 4.9 shows some other variables that you may wish to measure in the context of problems in management, finance, measurement, maintenance, design, structures, geotechnics, hydraulics and drainage, materials science, geomatics or land surveying and transportation.

4.17 Types of data

Chapter 1.6 described the difference between qualitative and quantitative. Consideration also needs to be given to whether your data is primary or secondary, and whether it is objective or subjective. There are eight the possible permutations, which figure 4.7 helps to illustrate:

qualitative, primary, objective
qualitative, primary, subjective
qualitative, secondary, objective
qualitative, primary, subjective

quantitative, primary, objective
quantitative, primary, subjective
quantitative, secondary, objective
quantitative, primary, subjective

You should clearly state which categories your data falls under. You may measure several variables, perhaps in different ways, but each variable should be classified.

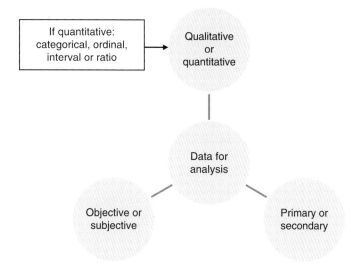

Figure 4.7 Permutations for types of data; given the eight possible permutations, what kind of data do you have?

4.17.1 Primary or secondary data

When setting objectives, it is reasonable that you are mindful of limitations in getting data. You cannot be too ambitious nor aim for objectives that can only be met by data that is likely to be confidential or that involves concepts that are too difficult to measure.

Data may be loosely classified as primary or secondary. The boundaries between the two can be blurred. Primary data can be defined as new data generated by the efforts of the researcher; that is, the words or numbers that you use in your analysis created specifically for your research. Examples of primary data are physical surveys, surveys of people, individual interviews, focus groups, laboratory experiments, fieldwork, action research or participation.

Secondary data is already existing data. If you use secondary data you will try to create some originality by analysing the data in a different way, or perhaps bringing together data from two different sources for analysis. The data may be recent, or historic and perhaps in archives. Often it is in the public domain, either paper based or electronically. The possibilities for collecting secondary data are endless. The European Union publishes a whole host of statistical data (Eurostat, 2015). Data is available on economic performance, cost, environmental pollution, traffic flows and safety. Other sources include the government's Office for National Statistics (ONS), government quangos, the Health and Safety Executive, the Construction Industry Training Board, local authorities, professional institutions, charities, company reports and price books. Data may be taken from the published work of other researchers, provided that it is acknowledged and cited.

Part-time students have many opportunities to obtain secondary data from their workplace. It could be items such as production data, minutes of meetings, diaries, photographs, material receipts, labour and plant timesheets, correspondence files, costs, safety statistics, personnel records, snag or defect sheets, contract terms, drawings or specifications. Again, the possibilities are endless. If you take this route you must ensure that permissions are obtained and anonymity is preserved; follow the ethical guidelines.

Students often prefer the primary route as a quick and easy exercise in data grab; it should not be that way. The secondary data route requires perhaps a little more thought, insight and ingenuity. A secondary approach may involve collecting data about lots of variables within a given problem area, and then 'poking around' that data to look for relationships or causes and effects. If you are measuring two variables, it may be appropriate to use primary data for one and secondary for the other; or of course primary for both or secondary for both. Whether you use primary or secondary data in your research should be driven by your study objectives, mindful of the limitations that you face.

4.17.2 Objective or subjective data: hard or soft

The Oxford Dictionary (Stephenson, 2010) defines objective data as 'actually existing, real …
facts uncoloured by feelings or opinions', and subjective data are defined as 'a person's views …
proceeding from personal … individuality'. Do not confuse 'objective' in the context of a
'study objective' with 'objective data'. It may help to think of objective data as 'hard' data, and
subjective data as 'soft'. Lots of quantitative or statistical data is objective or hard. Scientific
measures around heat, light, sound, strengths of materials, velocity and densities are also
hard. Opinion surveys can be classified as subjective or soft data. On the one hand, it can be
argued that just because people are of the opinion that product A (perhaps a car or a structure) is better than product B, it does not mean that this is the case in a technical sense.
However, peoples' opinions influence their behaviour and purchasing decisions; these types
of decisions can be drivers in the economy, and it is important that we measure them.

The boundaries between subjectivity and objectivity can be blurred. The soft opinions
expressed qualitatively by participants in closed questions are converted to numbers and treated
as though they were hard data. In your document, you need to make a judgement about where
your data lies on the continuum between the two. This judgement will underpin assertions that
you will also make about the validity of your research. Oppenheim (1992, p. 179) distinguishes
between opinions, attitudes and values. Opinions may be fairly loose and people may be per-
suaded to change them. Attitudes are more deep-rooted; it is more difficult to change attitudes
than to change opinions. Values are more deep-rooted still and are only likely to change over a
long time span. It may be argued that data at the values end of the continuum are best measured
qualitatively. The world of politics may be used to illustrate this: clearly some people change their
opinion from one UK election to another when they cast their vote. Others would argue that for
generations their families have voted for one party or another, based upon the values that they
hold, and they absolutely never change just because … Until …? Values can change too.

Businesses need to know and measure how customers, employees and other stakeholders
are judging them subjectively. If your study objectives are related to judgements about issues
such as job or product or systems satisfaction, or aesthetic appeal, you should use subjective
measures. Measuring the knowledge of professionals, perhaps in the area of sustainability, is
arguably subjective.

Some research may use multiple-item scales or a basket of measures, comprising subjective
and objective items. Subjective measures will be used where appropriate, but to improve the

validity of the measures, wherever possible, objective data will be used. Part of a piloting process may involve trying to move subjective measures to objective. For example, quality of life is a concept measured universally. It is an important issue for society, but it also has its applications to construction. There are independent comparative measures of quality of life in different countries. The UK government, through the former Audit Commission (2007 cited in Higgins et al., 2014), measured quality of life in local communities. Its published initiative was called 'Local quality of life indicators – supporting local communities to become sustainable'. The areas measured are: people and place, community cohesion and involvement, community safety, culture and leisure, economic wellbeing, education and life-long learning, environment, health and social wellbeing, housing, transport and access. One measure of culture and leisure is subjective; 'the percentage of residents who think that for their local area, over the past three years, has got better or stayed the same'. One measure of environment is objective; 'the volume of household waste collected and the proportion recycled'.

You may wish to measure some part of quality of life in organisations. You may be able to base your measure on a subjective opinion survey, and justify that on the basis that quality of lives is 'in the eyes of the beholder'. Whether you use objective or subjective 'data in your research should again be driven by your study objectives, mindful of the limitations that you face.

4.17.3 *Using subjective data since objective data is too difficult to secure*

You may make the judgement that your study objective is best met by hard data. However, given that you are limited by your time and your ability to get access to data (some of which may be confidential), you may be able to justify in your methodology using soft data. The issue of justification is important, and you should write an acknowledgement in the 'limitations and criticisms of the study' in your conclusion chapter, about how using soft measures has limited the validity of the study. If you cover these latter two items well, there is no reason why your research should attract lower marks than research that has been able to use objective data.

Consider a study involving waste, where you may wish to measure the amount of material waste on a number of sites. It may be limited to (the population) work carried out directly by main contractors or alternatively by specialists. Your definition of waste may be relatively broad; it draws on authoritative definitions, but includes waste caused by design mistakes, variations, cutting, theft, over-ordering, breakage before use and damage after fixing. It does not include other elements of waste, such as time wasted by designers, managers, craftspeople or operatives. You do not want to measure waste by the number of skips used on site, since that measure includes all trades. You decide that the objective measure is to determine from material invoices, the quantities of materials paid for. This will be compared with quantities measured from drawings. You will need to accurately re-measure work, taking account of variations, similar to those often used in final accounts. Part-time students may be able to get at this objective data in their own organisations for a reasonably large sample size; if it is possible, do it.

However, full-time students may find it impossible. There is the possibility that some element of material waste could be measured by a clip board survey, walking around sites and making observations; perhaps taking photographs (with permission). However, this too may be difficult for full-time students. Alternatively therefore, full-time students may design a subjective questionnaire. The example below shows a multiple-item scale or basket of questions administered to site foremen, with possible ordinal responses:

How often do you check material deliveries for quality?
Never/not very often/sometimes/often/always;

How often do you check material deliveries for correct quantity delivered?
Never/not very often/sometimes/often/always;

How often do you check materials while they are in storage?
Never/not very often/sometimes/often/always;

How important do you think it is to give tradespeople 'designed cutting lists' to minimise cutting waste?
Very important/important/possibly important/probably not important/definitely not important;

How important do you think is it to protect vulnerable materials on site after installation?
Very important/important/possibly important/probably not important/definitely not important.

On the one hand, you may wish to argue that these questions give you a 'measure' of material waste. However, such an argument is tenuous; these questions are not a valid measure of waste. Therefore, it may be better to change your objective, such that you do not seek to measure waste in its hard or objective form. Also better that you change your definition of waste. Perhaps the five questions are subjectively measuring 'care in controlling waste'. You should articulate in your introduction or methodology that your intent was to objectively measure waste, but recognising your constraints, you had to change this to the subjective measure 'care in controlling waste'. Since there are five possible responses to each question, the most positive answer will be coded 4 and the most negative as 0; similar to the scoring scale in the example in section 4.12.2 and table 4.7. You will be able to calculate an overall percentage score for 'care in controlling waste'. You would also need to be mindful that foremen may give you answers to show themselves in a good light; this will be another study limitation to consider in your conclusion.

If your objective has a variable that can reasonably be measured using hard data, you should design your methodology around that measure.

4.18 Money and CO_2 as variables

Whatever your construction discipline, money cannot be ignored. It is not expected that money will necessarily be a key strand of your work, but it is arguably remiss of you if you did not give money at least a peripheral mention. Also repeating the citation of Morrell from section 1.4 'we're going to need to start counting carbon as rigorously as we count money, and accepting that a building is not of value if the pound signs look okay, but the carbon count does not' (Richardson, 2009). Designers may consider the technical or aesthetic merits of methods of construction. Surveyors may appraise methods of maintaining facilities. Technical work may recommend the choice of new materials or products; that may be a futile recommendation if they are far more expensive than other choices. A finding in favour of one method or another should at least lead to a recommendation to investigate money in another study.

It may be that money is a key variable in your work. If that is the case, how is money defined and measured in the context of your study? It could be on the basis of capital costs, running costs, life cycle costs, budgets, profits, cash flow, retention, value, turnover, return on capital employed, liquidity, gearing or price earnings ratio. Measures could be compared as £s, percentages, ratios, unit rates. You may wish to study currency exchange rates and their impact on material prices or delivery lead periods. What do monetary measures include and exclude? Perhaps they include site overheads and do not include head office overheads, or perhaps they include neither or both. Costs may or may not include interest charges, discounts, professional fees, cost of land and value added tax. Profits may be before or after tax or exceptional costs.

Lowest cost to one party in the supply chain may not be the same as cost to another. Consider a specialist plastering contractor A whose labour, plant, material, preliminary and head office costs for a square metre of plaster are £10.00/m^2. The specialist adds 10% for profit to give a gross rate of £11.00/m^2. Specialist B may have total costs of £10.20/m^2 but adds only 5% for profit to give a gross rate of £10.71/m^2. Specialist A' has the lower costs, but clients who employ specialist A will have higher costs than those who employ specialist B.

There is a lot of cost information in the public domain through price books and the Building Cost Information Service (BCIS, 2015); while this service is by subscription, many universities do have licences. If you are to study the financial performance of other companies, you will only be able to get information that is available to the public. The construction and national financial press summarise the performance of the companies in the construction sector. Those companies that are public liability companies (PLCs) are required to publish annual reports detailing financial performance. Companies that are privately owned (private limited liability, limited liability partnerships or partnerships) are required to submit returns to Companies House for taxation purposes. Information for the larger, privately owned companies finds its way easily into the public domain. If you want details about the financial performance of smaller companies, you may have to go to Companies House yourself and pay a small fee.

Companies are very unlikely to give you access to figures that arise from internal management accounting systems. You will not get details about profit figures, valuations or final accounts amounts. However, as a part-time student, perhaps the accounts of your company are open to you. You may be able to get valid data to use for analysis, provided you ensure that the source of the data cannot be traced and there is anonymity.

Where in your study lies CO_2?

4.19 Three objectives, each with an IV and DV: four variables to measure

The introduction states that three objectives are often doable in a dissertation or project, plus perhaps one or two objectives to be met by the literature review. It is arguably best if each objective contains two variables, an IV and a DV. To avoid measuring six variables, each objective may contain the same IV or the same DV:

OB1 = the influence of IV1 on DV1
OB2 = ditto of IV1 on DV2
OB3 = ditto of IV1 on DV3

Or

OB1 = the influence of IV1 on DV1
OB2 = ditto of IV2 on DV1
OB3 = ditto of IV3 on DV1

4.20 Summarising research goals; variables and their definition

The middle part of your document will be where you will describe the design of your data collection instrument. This instrument will measure in detail your variables. Your variables need to be unpicked in detail by the measurement instrument. An early part of that unpicking should be in your introduction chapter, and should comprise a definition of your variables within a paragraph or so of text. If your variable is 'compliance with best practice in health and safety', at first reading that may mean one thing to you as the writer, but something completely different to the readers. At the end of the introduction, your readers should share with you a common understanding of what your variables mean.

Example: research goals based upon the work of Arewa (Appendix D6)

Title: Health and safety; ways of implementing best practice to achieve performance improvement.

Aim: To investigate processes for improvement in UK construction industry health and safety performance.

RQ1: What is best practice in health and safety?
OB1: To determine what is best practice in health and safety
H1: There is no hypothesis for testing related to RQ and OB1
RQ2: Does the level of compliance of UK contractors with best practice in health and safety (IV1) influence site profitability (DV1)?
OB2: To determine if the level of compliance of UK contractors with best practice in health and safety influences site profitability
H2: The level of compliance of UK contractors with best practice in health and safety influences site profitability
RQ3: Does the level of compliance of UK contractors with best practice in health and safety (IV1) influence company profitability (DV3)?
OB3: To determine if the level of compliance of UK contractors with best practice in health and safety influences company profitability
H3: The level of compliance of UK contractors with best practice in health and safety influences company profitability
RQ4: Does the level of compliance of UK contractors with best practice in health and safety (IV1) influence turnover (DV3)?
OB4: 'To determine if the level of compliance of UK contractors with best practice in health and safety influences turnover
H4: The level of compliance of UK contractors with best practice in health and safety influences turnover

Definition of the variables

IV1: Level of compliance of UK contractors with best practice in health and safety: average score allocated by the company's health and safety officer in monthly visits on a range of 0–10, scores allocated against standard company health and safety criteria.

DV1: Site profitability: percentage profit or loss on individual projects based on comparison with net cost before overheads and profit.

DV2: Company profitability: based on head office figures, comprising typically 20 to 30 sites over a five-year period

DV3: Company turnover: based on head office figures, comprising typically 20 to 30 sites over a five-year period

Ancillary variables

Ancillary variables: the research will also examine the impact of the ancillary variables age, gender, job title, ethnic background, qualifications of participants. Also method of procurement, type of contract, size of project, and type of project. Sub-hypotheses for testing based on these ancillary variables, will be developed in later chapters.

Summary of this chapter

The study aim is supported by research questions, objectives and hypotheses. Questions, objectives and hypotheses should all imitate each other. It may be preferable to use 'objective' as the key term around which research revolves. Objectives should be clear and words within them should not change as the document progresses. All parts of the study should hang around the objectives, which may be written as strap lines. In your introduction chapter you should identify the variables in objectives and write a narrative to define them in the context of your study. If you have two variables in an objective, identify one as the IV and the other as the DV. Whether your study is qualitative or quantitative, you should focus on measuring something. You may gain more marks if you are able to observe the impact of ancillary variables on your study. On some occasions it is self-explicit how variables are measured, e.g. interest rates, but for some variables you may need to design your own measurement scales. Use authoritative scales wherever possible. Quantitative studies should be clear about the level at which the dataset is measured: categorical, ordinal or interval/ratio. You should not be disappointed if you do not find a relationship between your variables. If money is a variable in your study, be mindful that it is not always best measured simply by cost or profit.

References

Arewa, A. (2013) A review of construction SMEs' safety performance; where are we and why? *IPGRC*. University of Salford.

BCIS (2015) Online products. www.rics.org/uk/knowledge/bcis/?gclid=Cj0KEQjwvJqvBRCL77m2-uKczsIBEiQAkx8VjDSxLwFC1Qg5Vxqch6o5Y42WRGLnDejX1_el06eEEp8aAt3p8P8HAQ Accessed 03.09.15.

BREEAM (2015) The world's lead design and assessment method for sustainable buildings. BRE. Building Research Establishment. www.breeam.org/ Accessed 27.08.15.

Bryman, A. and Cramer, D. (1994) Quantitative Data Analysis for Social Scientists. Routledge, London.

BS (1998) 'BS EN 772-3:1998. *Methods of test for masonry units. Determination of net volume and percentage of voids of clay masonry units by hydrostatic weighing*'. http://shop.bsigroup.com/ProductDetail/?pid=000000000001532230 Accessed 03.09.15.

Eurostat (2015) Your key to European Statistics. http://ec.europa.eu/eurostat. Accessed 30.07.15.

Fellows, R. and Liu, A. (2008) *Research Methods for Construction.* 3rd edition. Blackwell Science: Oxford, UK.

Fink, A. (1995) The Survey Handbook. *The Survey Kit.* Volume 1. Sage, London.

Glenigan (2015) *KPI zone.* UK Industry Performance Report. https://www.glenigan.com/construction-market-analysis/news/2015-construction-kpis Accessed 23.09.15.

Hays, W.L. (1988) Statistics. 4th edition. Holt, Rinehart and Winston Inc. London.

Higgins, P., Campanera, J. and Nobajas, A. (2014) *Quality of life and spatial inequality in London.* European Urban and Regional Studies. http://eur.sagepub.com/content/21/1/42.full.pdf+html

Holt, G. (1998) *A guide to successful dissertation study for students of the Built Environment.* 2nd edition. The University of Wolverhampton, UK.

Kinnear, P.R. and Gray, C.D. (1994) *SPSS for windows made simple.*L. Erlbaum Associates Ltd, East Sussex

Marks and Spencer (2010) *Doing the right thing. Our plan A commitments.* http://corporate.marksandspencer.com/plan-a/85488c3c608e4f468d4a403f4ebbd628 Accessed 03.08.15.

NPL (2015) SI based units. www.npl.co.uk/reference/measurement-units/si-base-units/ Accessed 03.09.15.

Ofsted (2015) *Ofsted grade descriptors – Quality of teaching in the school.* Office for Standards in Education. www.clerktogovernors.co.uk/ofsted/ofsted-grade-descriptors-quality-of-teaching-in-the-school/ Accessed 29.07.15.

ONS (2014) *Construction Statistics No. 15*, 2014 edition, www.ons.gov.uk/ons/rel/construction/construction-statistics/index.html Accessed 04.09.15.

Oppenheim, A.N. (1992) Questionnaire Design, Interviewing and Attitude Measurement. 2nd Ed. Cassell, London.

Richardson, S. (2009) Morrell takes the stage with a warning to government. *Building* 27.11.09. pp. 10–11 issue 47.

Sage Publications (2015) *Leadership styles questionnaire.* www.lboro.ac.uk/service/std/ilm/Becoming_an_effective_leader/89527_03q.pdf Accessed 29.07.15.

Stephenson, A. (2010) *Oxford Dictionary of English.* 3rd edition. Oxford University Press, Oxford. Available online.

5 The Methodology chapter; analysis, results and findings

The titles and objectives of the sections of this chapter are the following:

5.1 Introduction; to provide context for research design
5.2 Approaches to collecting data; to classify approaches and give examples
5.3 Data measuring and collection; to describe a variety of data collection tools
    5.3.1 Populations and samples
    5.3.2 Single and multiple-item scales
    5.3.3 How many response points on each Likert item? Words or numbers or both? Weighting.
5.4 Issues mostly relevant to just questionnaires; to emphasise the necessity for multiple-item scales and ordinal scales
    5.4.1 Introduction
    5.4.2 Administration
    5.4.3 Piloting the questionnaire
    5.4.4 Statements or questions
    5.4.5 Alternating poles on statements and questions
    5.4.6 Things to do and not to do in questionnaires
5.5 Ranking studies; to describe this type of study and provide an example
5.6 Other analytical tools; to provide ideas for analysis that do not fall into the traditional qualitative and quantitative schools
5.7 Incorporating reliability and validity; to explain these concepts and describe how they can be achieved
5.8 Analysis, results and findings; to distinguish between them as a platform for writing this chapter in the dissertation or project

5.1 Introduction

Before you start your methodology chapter, you should be clear about your objectives. In this chapter of your document, you will design your method to meet your objectives; how can you design your method if you are not clear about your objectives? You should be clear about the names or labels of variables you wish to measure. Also be clear about the range of potential values for each variable; will the values be measured at the categorical, ordinal, interval or ratio level? If you are not clear about these issues, stop and take advice. Since they are all important, the point is emphasised in figure 5.1.

Writing Built Environment Dissertations and Projects: Practical Guidance and Examples, Second Edition.
Peter Farrell, Fred Sherratt and Alan Richardson.
© 2017 John Wiley & Sons, Ltd. Published 2017 by John Wiley & Sons, Ltd.
Companion Website: www.wiley.com/go/Farrell/Built_Environment_Dissertations_and_Projects

Figure 5.1 Are you clear about your objectives, your variables and the potential values of each variable? If not, stop and seek advice.

Students often race into the data collection exercise without clarity about their objectives. Lots of data is collected which, on reflection, is random and uncoordinated. The data collection exercise is a large part of your research; you will not have time to do it twice, so it needs to be correct first time. Speak with your supervisor and get informal approval before you begin experiments or start to use a data collection tool. If your data collection is poor, that will limit your ability to conclude against your objectives; most likely a poor mark will follow. It is a significant milestone in your research when you start to collect data.

Your methodology chapter should thoroughly describe the way you have conducted your study, such that it may be replicated by others; you should also justify the way you have done it. You need to answer the questions 'Why, how and what?' Often students do not write at length in this chapter. It is as though they think it is 'obvious' what they have done, but while it may be obvious to the student and supervisor since they have lived the research together, it also has to be clear to other readers of the document. While your research is about you becoming expert in your subject area, it is also about you having insight and a generic appreciation of the role of research in society. It is in the methodology chapter that you can demonstrate that you have these attributes. The necessity to justify your methodology means that you must understand several approaches, so as to dismiss those that are not appropriate. This does not mean that you must be able to use and apply all research methodologies, but you do need to have some depth to your knowledge.

Methodology is about why you designed your research at a strategic level. It is about the generic principles underpinning the process you followed and about your choices. The method itself is at an operational level, and describes the mechanics of what you have done. When designing your methodology, you must remain focused on your objectives.

Mindful of resource constraints, the methodology must be the best way of meeting those objectives. Methodology is not just about the way you collect your data. It is about the whole research process from beginning to end, including how you founded the research questions and objectives from the problem, how you conducted the literature review, pilot studies, data collection, analytical methods and the process of developing findings and conclusions. There is a lot of potential for weaknesses in the methodology of your research which could invalidate it and make it fundamentally flawed. Studies are invalidated for example, if populations are not selected correctly, or inappropriate methods of analysis are used, or if there is poor design of surveys. You need to detail what steps you have taken to maximise the reliability and validity of your study.

You must describe in detail each step along the way. Your findings will not be 'believable' if you do not describe your method. You may be familiar with describing methods used in scientific experiments in laboratories. The principles of description are similar in social sciences research. Based on findings from research, there may be the possibility of seeking funding for investment and application in industry. As one piece of research is published, someone else may wish to devote substantial resources to developing that work, perhaps £100k or more. It may be that research contains mistakes, or that unreasonable assumptions were made. By reading your method, subsequent researchers may be able to make initial judgements as to whether what you have done is valid. If their preliminary judgements are positive, they may wish to replicate the research 'in the field or laboratory', to truly validate it. Retesting, repeating, replication will take place before business people will commit to spend.

Your justification must be robust. It is not just to describe the way you have done it, and that is that. Why have you done it that way? What alternatives did you consider? The design you use should be based on authoritative methodologies in the literature. When you read about the subject matter of other people's research, you should also digest their methods. You should read specialist methodology textbooks. Cite all your sources. Your methodology chapter may include statements something like 'the method was adapted from the work of Holt (1997) and Breach (2009). The approach advocated by Bell (2005) was considered but rejected because …'. Cite this book if you wish to substantiate your method, but citing just one book does not demonstrate breadth in your reading. You cannot read from front to back lots of texts, but you can browse parts or chapters relevant to your work, and cite those parts. Read texts relevant to the built environment e.g. Holt (1997), Knight and Ruddock (2008); Borden and Ray (2006); Breach (2009), Naoum (2013), Fellows and Liu (2015) and Lucas (2016). Also those relevant to research in all disciplines such as Bell (2005) (available as a free pdf download) and Creswell (2009).

If you have performed experiments in laboratories, they may be based on British Standards, so cite those standards, and describe if and why you may have deviated from them. Describe the equipment used. If you have observed a production process, describe the process itself and the precise circumstances of your observation.

If you have asked questions in interviews or surveys, define/quantify your population and describe how you picked your sample. What was the size of your sample and what response rate did you get? What alternative sampling methodologies were considered? Give each question and justify why you asked it. Which objective is each question related to? Have questions been adapted or taken verbatim from another source? If so, that is fine, but cite your source. If interviews were undertaken, describe how they were conducted. How were introductions handled, how long did the interviews take, were they formal/informal, how was the data recorded? If questions were administered other than face to face, how was that undertaken? Possibly by hand delivery or electronic web surveys.

Whether you have gone down the qualitative or quantitative route, the data analysis process must be described and you must substantiate why a particular tool was selected. There is a variety of qualitative approaches suggested by the literature, so why have you used the approach of one author and not another? If you used a quantitative approach, did you limit yourself to only descriptive analysis? If you were able to undertake inferential analysis, did you use non-parametric or parametric tests?

You should describe what steps you have taken to ensure that you comply with ethical codes, and if appropriate undertaken risk assessments. Cite your university ethical guidelines; that is compulsory – deduct ten marks if you do not. Include in an appendix research ethics, health and safety and health templates; see appendix B. In the better studies, methodology chapters will write at length about what steps were taken to ensure that the study is as reliable and valid as is reasonably possible, and will demonstrate genuine understanding of these concepts.

5.2 Approaches to collecting data

Bell (2005) suggests seven approaches to research: (a) action research, (b) case studies, (c) surveys, (d) experiments, (e) ethnography, (f) grounded theory, (g) narrative enquiries. The definitions vary between authors, and the boundaries are therefore somewhat blurred. They may be called approaches to collecting data, and are applicable if the data to be collected are qualitative or quantitative. These classifications are not closed boxes. You may collect some data that could easily fit into both categories; you may choose not to place your collection method emphatically into one box. You need to be a realist in whichever approach you use; some data may just be too difficult to get at. But you should also be innovative and proactive and not just follow the usual methods, e.g. surveys, and do not be passive. Section 4.17.1 suggests a variety of ways that part-time students can collect secondary data from their workplace.

Action research is a possibility for part-time students in construction. You may examine a production process, perhaps shuttering methods for insitu concrete. You may make take measurements around productivity or safety and take action to introduce new methods to hopefully secure improvement. Measurements for the new method would also be taken to determine if the hoped for improvements have occurred.

Case studies themselves can be undertaken as an insider, or you may be given access to a case by another organisation. There is also the increasing possibility that there may be sufficient data available in the public domain about a past or current case. This gives full-time students the opportunity to do applied work that they may feel is mostly only possible for part-time colleagues. You may be able to get some data through Freedom of Information Act requests, but if you go down this route, do it sensibly, not vexatiously. Case studies can be related to a single project, such as a flagship development, a framework agreement or infrastructure for one-off sporting events, or alternatively a single event at some point in time, or a one-off financial crisis, or liquidation of a single company, or a landmark legal dispute. Alternatively, you may investigate the impact of amendments to standard forms of building contract on the success of projects, again either qualitatively or quantitatively. Part-time students may be able to get access to archived files from completed projects and evaluation undertaken against a chosen variable. Interviews could be held with professional staff who worked on the projects to enhance the validity of judgements about success of the same projects. Proverbs and Gameson (in Knight and Ruddock, 2008) detail an exemplar case study applicable to construction.

A word of caution when doing quantitative analysis based on data from action research or case studies: it is possible when taking information from one source, perhaps your own company if you are a part-time student, to get large datasets that present themselves as suitable for quantitative data analysis. It is perfectly acceptable that you do the quantitative statistical work but with the proviso that you do not infer that your findings would be replicated in the wider world; only surveys that use a representative sample from the whole of the population allow you to make inferences.

Surveys embrace a wide variety of approaches to collecting data. They are often associated with questionnaires of some kind, usually but not always involving people. They may be surveys of buildings. Surveys may imply intent to collect data from a wider audience. In construction, surveys of single plots of land or single buildings perhaps fall more easily into the case study category. However, they can become surveys of a population, if there are larger numbers involved, for example, accessibility surveys of public buildings. Similarly focus groups may fall between being a case study or, if used often, a survey. A study that mostly analyses secondary data from archives or the internet becomes a survey of relevant literature.

Construction presents many opportunities to undertake laboratory or fieldwork experiments. Land surveying, materials technology and science, geotechnics, hydraulics, transportation and environmental science are parts of the built environment curriculum in which you can use experiments as a way of collecting data. Various concrete mix designs can be trialled to test properties of concrete such as compressive strength, density and water absorption. The design mixes may vary by type and volume of aggregates, cement, water content and plasticiser. Experiments on people are often undertaken in other disciplines, perhaps with a control group (that carries on as normal) and an experimental group (that has some variable imposed upon it). A construction company may trial a new piece of technology on some sites. Some parallel sites continue using old systems as the control group, and the trialled sites become the experimental groups; some degree of success is then measured across both groups to see if there is improvement.

An ethnography involves you immersing yourself in the life of participants to attain a cultural understanding of their environment. You may wish to share in their experiences and, detaching yourself from inherent bias, see things from their perspective. Construction managers may seek cultural understandings of craft workers, particularly in the context of health and safety. You may for example take time out from normal everyday site activities to get closer to craftspeople than would normally be the case.

Grounded theory lends itself to collection of qualitative data through unstructured or semi-structured interviews. There are no preconceived ideas; the term 'grounded' is used to denote new theories that may start to emerge from the data. The process is inductive (bottom up, from data collection to the theory) rather than deductive (top down, from the theory to the data collection). These theories may be developed into hypotheses from the data for subsequent testing in other studies.

Narrative enquiries also lend themselves to collection of data by interviews, though the intent is that participants will tell of their stories, rather than using a semi-structured or structured format. The data collection process may need to be over a longer period of time than that normally associated with interviews. You would need to be respectful of the time of interviewees. Most often, individuals will not have time to sit in several interviews over an hour or two each, but you may find situations where this is possible. For example, you may find willing professionals between projects, or serving notice at one employer before moving to another, or towards the end of their careers.

5.3 Data measuring and collection

This section is written in the context of all data collection. Many but not all parts are relevant to questionnaires. Section 5.4 is specifically related to questionnaires. There are many instances where you may use data measuring instruments, for example you may observe and rate the aesthetic appeal of modern housing or the safety practices of construction operatives, or accessibility of historic buildings or deterioration in concrete etc. In these cases it may be you taking the measurements or making the judgements, not the participants. Both civil engineering and building students need to be mindful of the need to describe population and sampling techniques.

Most often these instruments will gather subjective data qualitatively, for example you or your participants may rate products as poor, satisfactory, good, very good, or excellent. These responses will be converted to numbers 0 to 4, such that they become quantitative. Section 4.12.2 includes a multiple-item rating scale for bricklayers. There are more examples in this chapter.

Ideally your research will contain at least one IV and one DV. Perhaps also some ancillary variables. Since you are creating this data afresh with a measurement tool (it is not data already in files) it is primary data. The objective of your study may best be met by using the same measurement tool to score all variables. Alternatively, the measurement tool may only seek to secure data for one of the IV or the DV, plus perhaps ancillary variables. The other variable may be measured using data obtained from elsewhere, perhaps secondary data from files.

Section 4.12 suggests that only as a last resort should you use your own measuring tool. It is likely for many concepts that you wish to measure, a measurement instrument has already been designed.

5.3.1 Populations and samples

In your methodology chapter there should be a subheading labelled 'Populations and samples'. They are related concepts, but different. Students often use the words interchangeably, and as though they were exactly the same, so be careful to use the correct word in the correct context. While populations and samples are often thought to be mostly relevant to people, sampling is frequently undertaken in laboratory and fieldwork, and also if you wish to study types of buildings, or types of contracts, or companies, or construction plant, or materials waste etc. On many occasions media headlines report on what the UK general public think about a particular issue as though everyone had been asked. Clearly not everyone has been asked, just a sample of people, and it is inferred that what the sample has said, you and I would say.

Populations

The population is the whole category that is being studied. For example, in the UK 2015 general election, that would be all those people who are eligible to vote. The Electoral Commission (2015) specify that:

To vote in a UK general election a person must be registered to vote and also:

be 18 or over;

be a British citizen; a qualifying Commonwealth citizen or a citizen of the Republic of Ireland;

not be subject to any legal incapacity to vote.

Additionally, the following cannot vote in a UK general election:

members of the House of Lords (although they can vote at elections to local authorities, devolved legislatures and the European Parliament);

EU citizens resident in the UK (although they can vote at elections to local authorities, devolved legislatures and the European Parliament);

anyone other than British, Irish and qualifying Commonwealth citizens;

convicted persons detained in pursuance of their sentences (though remand prisoners, unconvicted prisoners and civil prisoners can vote if they are on the electoral register);

anyone found guilty within the previous five years of corrupt or illegal practices in connection with an election.

This definition gives 'inclusion criteria' and 'exclusion criteria': what you must be and what you must not be. In your study you must describe your population, give inclusion and exclusion criteria and estimate the population size. The population size for the 2015 UK election was 46 million and the response rate was 66% (BBC, 2015).

You must select the population that best enables you to meet your objective. If you want to determine the environmental awareness of civil engineers, would it be best to ask the engineers themselves or people they work with? You decide. You must estimate the size of the civil engineering population. Your first point of contact may be the Institution of Civil Engineers website. How many members does it have? Does your population include those who are retired, those who are part qualified, students? If your population involves testing materials, how many bricks are manufactured in the UK each year and what types? For concrete, how many m^3 of concrete have been poured? How much do these sectors contribute in £ to the UK economy? If your research is about private finance initiative (PFI) or PF2 (HM Treasury, 2012) projects, how many are there in the UK and what is the value in £ of such work?

Samples

A sample is drawn from the population. It is a smaller part taken for testing purposes. It is not possible to test whole populations because of time and cost. If you wanted to determine the likely compressive strength of bricks drawn from a kiln, you would clearly not test them all to destruction; just a sample carefully selected from different parts of the batch. If six bricks were selected and found to have similar strengths, you will infer that all the bricks from that kiln have the same compressive strength. Indeed, you may not test all batches from the kiln, and you may infer other batches have the same compressive strength. There are some occasions when a survey would be taken of the whole of the population e.g. democratic elections for people into government other important positions in society. National referenda or trade union ballots are other examples. It is for example illegal in the UK, to take strike action based on a survey of a sample.

In quantitative work, when sampling there is an inference that:

> if the whole of the population had been surveyed, the results from the sample would be same as the results that would have been obtained if the whole of the population had been surveyed.

This inference is made using statistical tools as described in chapter 9. The higher the sample size, the more confidence you can have in this assertion. In surveys of people, perhaps a minimum number of replies is around 25 for undergraduate work; better nearer to 50. For masters or PhD work higher samples are required. You will need to consider the response rate in making judgements about how many people to ask. A response rate of 50% may be considered good and would require you to contact 100 people to achieve 50 returns. Report your response rate in your methodology chapter. Sample size in technical work is also important; if tests are relatively quick and inexpensive – perhaps testing concrete cubes to destruction – your sample size for each mix design may be six cubes or more. You may, however, be building one prototype model of part of a structure to test to destruction; most likely just one model would suffice. In qualitative work, inference to the population is not the goal (see chapter 7). Your sample will comprise small numbers of people to gain insights into how they see the world.

If your sample is buildings that you need to visit and survey, perhaps historic churches, it would be good to select your sample from all churches across the UK so that your findings could be inferred to exist in them all. However, your university will not expect you to spend unreasonable amounts of time and money on travelling. You can describe the population in your methodology chapter for the whole of the UK, but then write that because of resource limitations the data collection is restricted to your own locality, say a survey of historic churches in Lancashire. In your conclusions, you may then consider how likely your findings are to be replicated across the UK.

There are lots of terms used to describe sampling methods. In non-technical work, often but not always involving people, they all come under the umbrella names of probability or non-probability sampling.

Probability sampling

In all sampling the inference is that if the whole of the population had been surveyed the same results would have been obtained as if the whole of the population had been surveyed. This is relied upon most heavily in probability sampling. Sometimes sampling can get it wrong. Using again as an example in the 2015 UK general election there were many samples taken (polls) to predict in advance the likely outcome of the election. While all polls seemed to predict a coalition, the outcome of a majority government was not predicted by taking electorate opinions in samples.

Probability sampling requires that the sample is selected carefully to avoid bias. It should be picked 'randomly' from the entire population, and statistically speaking, the characteristics of the sample should match closely the characteristics of the sample. If 90% of the population is male and 10% female, a randomly selected sample should have approximately the same proportion. Similarly for other criteria such as ethnicity, age, qualifications and employment status. There are lots of examples of populations available to you in the public domain, especially if you write criteria that limit their size, for example a survey of listed buildings or highest-spending clients in construction, or consultants, or the top 100 contractors, or local authorities, or brick manufacturers. You can get details from the weekly construction press or trade associations.

Random sampling is truly that. If all members of the population are arranged in a list, Excel will generate a list of random numbers to allow you to pick from the list. Assume that you have a population list of 500 and require a sample of 100; place the cursor in cell 1A of Excel, and then type into the function box f_x '=RANDBETWEEN(1,500)'. Copy and paste the formula from cell 1A into cells 2A to 100A. Excel will generate 100 random numbers between 1 and 500. Use these numbers to pick your sample from the list.

Non-probability sampling; a sample of convenience or purposive sampling

In your undergraduate research, it may often be the case that you cannot get a list of the entire population. If you do not have access to the population, you cannot draw a random sample. In surveys involving people, personal details may not be in the public domain because of an obligation placed on holders of such information to comply with the Data Protection Act.

Samples of convenience are often used by students since populations are sometimes difficult to identify. Write clearly in your research if you have used a sample of convenience. You should not be penalised, provided you write to justify why you have done it and about the limitations that it may have on your findings and conclusions.

Part-time students may wish to use data from their own workplace. They may use as samples people they work with or projects within their company. It is also possible to access directly parts of the supply chain: clients, consultants, contractors, specialists, suppliers or manufacturers. On the one hand, you cannot infer that what you have found in your own company would be replicated across a wider population in the UK, but you can write in that manner and then in a section under 'Limitations and criticisms of the study' you can acknowledge that your sample was one of convenience, potentially biased and taken in recognition of the difficulty of getting data from elsewhere. If your sampling technique is biased, and you do not write to demonstrate that you understand that, you may have marks deducted. Alternatively, you could write up your research as a case study, that is, it is a survey of people or projects etc. in your own company, and there is no attempt to infer that the findings may be replicated elsewhere.

Purposive sampling may be required in construction research when there may be a limited number of people who have expertise in the area of your study. Relatively few construction professionals may claim to be expert in, for example, the private finance initiative, construction adjudication, masonry arch structures, epoxy coated reinforcement or whatever. Also you may not be able to compile a list of all those people who are expert: the population. In such cases you may seek out names of authoritative people, rather than circulating a survey widely to construction professionals generally. You may write your objective in the terms that your survey is one of expert people, but you must write to recognise that since you did not take a random sample, there are limitations on inferring that your findings may apply to the whole population.

5.3.2 Single and multiple-item scales

When you measure your variables, you need to measure them really well. It is better to measure a few things well than many things superficially. Some variables are easily measured and take little of your time; those variables may be IVs, DVs or perhaps ancillary variables, such as to determine the method of procurement on projects (possibly just two values: traditional or design and build) or to determine the ancillary variables age or gender of participants. If one of your IV or DV is easily measured, it is likely that the other variable will involve you in more work.

Clearly, method of procurement, age or gender etc. are measured on a single-item scale. Technical results in fieldwork or a laboratory can be just one measure, though if you were for example testing materials to destruction, you would perhaps perform tests several times to be sure of your results, that is to improve validity.

Rensis Likert (1903–1981), an American psychologist, first used Likert multiple-items and scales. Items and scales are based on providing statements with which participants will agree or disagree. While Likert first used the scale with statements to which participants would agree or disagree, the principle was then adapted to be used as a multiple-item scale for questions. It was also extended to the measurement of many variables that do not involve participants. There is a difference between a 'Likert item' and a 'Likert scale':

Likert item: one statement

Likert scale: multiple 'items' or statements which are taken together to be a measure of one variable

To attain an overall measure of the variable, numbers are allocated to answers, and an arithmetic total determined. Section 5.3.3 gives an example of a multiple-item scale based on questions. Section 4.12.2 and table 4.7 are examples of a multiple-item scale based upon observations or judgements by a bricklaying foreman.

Subjective measures lend themselves to multiple-item scales. If you wanted to measure for example 'the effectiveness of quality systems in organisations', you would not base that judgement on just one criterion, perhaps 'number of defects'. That would be nonsense. You would want to take many sub-measurements of this variable and add them all together to come to a final judgement. If you were making judgements about the quality of products, or in technical work the durability of materials, you should use multiple criteria to make those judgements.

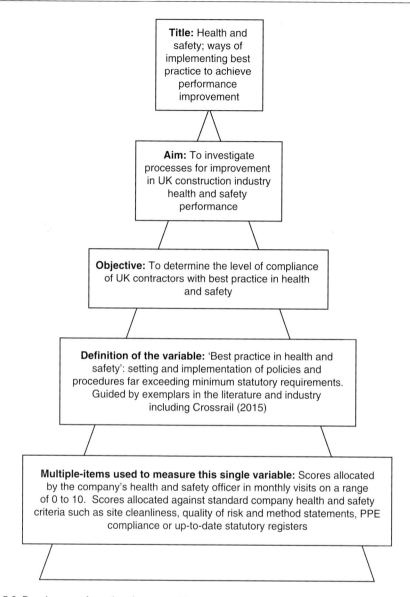

Figure 5.2 Development from the aim to an objective with a variable; subsequently definition of the variable and a measurement tool that focuses upon the definition (Based upon the work of Arewa, 2013, Appendix D6).

In section 4.20, it is recommended that you are clear about your variables, and that you write a paragraph to define them. When you design your measurement tool, it is merely a case of opening up that definition, so that you have detailed criterion that are measureable. The multiple-item or Likert type scale is the way to open up the definition of your variable. Figure 5.2 illustrates how a study may start with an aim, and as it opens up, an objective and variable are identified, defined and then finally measured using multiple-items; all are brought together with a title.

5.3.3 *How many response points on each Likert item?*
Words, numbers or both? Weighting

The number of response points or range of possible answers for each individual Likert item should be given careful thought. Students often rate – or ask other people to rate – things on a mere two point yes/no scale. In the context that you wish to attain really good measurements and maximise the potential for robust analysis, it is not appropriate to limit answers so narrowly. Life is often more complex than merely yes or no. You need to tease out and measure intermediate possibilities such as definitely yes, probably yes, unsure, probably no, definitely no. At the other extreme, why not a ten point scale (1 to 10), or eleven point (0 to 10), or even wider?

Whether to include an odd or even number of points should also be given careful thought. An odd number allows participants to take the easy middle option and be neutral in their answers: to sit on the fence. An even number forces participants onto one side or the other. Fink (1995, p. 53) argues that current thinking suggests that 5–7 point scales are adequate for most surveys that generate ordinal data, and there is no conclusive evidence supporting odd or even scales; it depends on the survey's needs. The even six point responses of 'very poor, poor, satisfactory, good, very good and excellent', can be increased to an odd seven point response by adding 'outstanding'. If you are rating yourself or asking participants to rate, it is usual to have a qualitative descriptor, word or statement to tick, without a number, such as only the qualitative descriptors excellent, very good etc. Numbers are not visible to participants, and are only allocated by the researcher later.

However, on some occasions numbers are used, perhaps with a ten or eleven point range, with no intermediate qualitative guidance. For example, 'On a 0–10 scale with 0 as very poor and 10 excellent, how do you rate the quality of product A?' It is arguably difficult for you or participants to rate on this scale, since what for instance 8 may mean to one person in terms of quality, is not what it may mean to another. Take for example 7 out of 10. On an undergraduate degree programme, 7 out of 10 is an 'excellent' score at first class honours standard. However, 'excellent' in Constructing Excellence KPIs is thought by some to be only achieved at 9 or 10 out of 10. To improve validity of scales, numbers should also have a descriptor. To write descriptors for all eleven points on a 0 to 10 scale can be difficult.

Table 5.1 is a Likert type scale and illustrates a compromise position, based on observations taken by a researcher to rate the disabled access provision in five public buildings. Qualitative labels are written to cover a narrow range of numbers. If you wish to rate buildings in such a way, ensure that you follow your university's ethical guidelines. For example seek permissions, take your student identity card with you and undertake a risk assessment. When there is a need to distinguish between say 6 and 7 'good', you may have in mind a judgement at the lower or upper end of 'good'. School teachers routinely use upper and lower scores, e.g. B+ is clearly grade B, and not quite so good as –A, and –A is clearly an A; or is it? The criteria used to make judgements are from authoritative sources. One criterion of ten in table 5.1 is the quality of disabled toilet provision, and the sub-criteria used to make judgements is based on the work of Spooner for Local Authority Building Control (LABC, n.d.) which details good practice for compliance with the Building Regulations.

Weightings are given to each of the ten criteria. The most important items are weighted 1.00, and items of less importance less than 1.00. Whatever the weightings are, they should not be attributed arbitrarily. Evidence should be drawn from the literature to support judgements about weightings, and it could be a large part of the research doing perhaps some qualitative work with participants to get their judgements about appropriate weightings.

Table 5.1 Eleven point (0–10) points for the rating of disabled access provision; ten criteria weighted.

Building number ↓	Quality of disabled toilet provision: 1.00	Level access entrances: 1.00	Provision and proximity of disabled car parks: 1.00	Staff support: 1.00	Allocated seating areas: 0.90	Circulation space area: 0.90	Door widths: 0.90	Braille signage: 0.70	Tactile surfaces: 0.70	Communication aids: 0.50	Total score out of 100	Total weighted score out of 86	Percentage score M/86*100
A	8	9	6	10	6	8	5	0	0	10	62	55.10	64.07
B	7	8	5	10	8	8	8	5	2	0	61	56.50	65.70
C	3	2	6	3	6	9	3	8	1	2	43	37.50	43.60
D	10	10	9	6	8	9	6	8	4	6	76	67.10	78.02
E	6	6	5	2	5	8	0	0	10	7	49	41.20	47.91
Total score out of 50	34	35	31	31	33	42	22	21	17	25			percentage score for all buildings
Total weighted total	34	35	31	31	29.7	37.8	19.8	14.7	11.9	12.5			59.86
Percentage score for each criteria	68.00	70.00	62.00	62.00	66.00	84.00	44.00	42.00	34.00	50.00			

Criteria and scoring matrix for quality of disabled toilet provision	
0 to 1	Very poor
2 to 3	Poor
4 to 5	Satisfactory
6 to 7	Good
8 to 9	Very good
10	Excellent

Criteria based on Building Regulations in Practice: Accessible toilets (LABC, n.d.). Height of WC pan; position of the basin; position of the flush handle; provision of toilet tissue not toilet paper; position of taps, provision and positioning of mirrors; friction hinge on drop down rails; access to cubicle; manoeuvring space; outward opening door; door furniture operable with closed fist; position and type of taps et al.

In figure 5.2, measuring the variable 'best practice in health and safety', perhaps site cleanliness could be weighted 1.00, quality of risk and method statements 0.90, PPE compliance 0.90, up-to-date statutory registers 0.80?

In table 5.1 the four criteria in columns B, C, D and E are weighted at 1.00. In column F, the criterion 'allocated seating areas' is considered less important and scored at 0.90. Columns G to K are also weighted. There are ten criteria all scored on an eleven point scale 0–10. Therefore the maximum score each building in column L is 100, but when the weightings are taken into account it becomes 86 (10+10+10+10+9+9+9+7+7+5 = 86). In column N a percentage score for each building is given, such that 43 out of 86 would equal 50%. The best is building D with a score of 78.02% in cell 5N; the overall mean score for all buildings is 59.86% in cell 8N. Is this mean score better or worse than we might reasonably expect? That is for your discussion chapter. The criterion that scores best is 'circulation space area' with a score of 84.00% in cell 9G.

When making judgements about each individual score for each criterion in each building, the principle of weighting can also be applied. For example, in a disabled toilet, manoeuvrability may be considered more important than the provision of toilet tissue rather than toilet paper. LABC (n.d.) states 'Some people can only use one hand, which can make the job of tearing sheets of paper off of a traditional toilet roll quite difficult.'

Figure 5.3 contains an example of two Likert scales used in questionnaires. Appendix D4 concludes with a Likert scale questionnaire to measure commitment to best practice in health and safety.

5.4 Issues mostly relevant to just questionnaires

5.4.1 Introduction

If you propose to administer a questionnaire, you should also read section 5.3. Some people argue that questionnaires are used too frequently by students. They are a convenient way to 'grab' some data. If students are busy, questionnaires are quick to administer and they allow students to take an easy passive approach to data collection rather than the more difficult proactive approach. Some people become extremely cynical about questionnaires, to the extent that they are the subject of severe criticism. Industry can become weary of receiving questionnaires from students in local universities; there is a need to respect time constraints on professional people and others.

A judgement may need to be made about whether questionnaires are the best way to meet the study objectives. If the answer is 'no', then do not use one. If the answer is 'yes', then you should use one, but be prepared to robustly justify its use in your methodology chapter. When external examiners review dissertations and projects at the end of the academic year, they should reasonably expect to see a range of data collection methods used by students. There is something wrong if most students use a questionnaire, but also it is unlikely that none do. To reiterate, if the best way to meet the study objectives is by questionnaire, do it.

Leadership style of bosses and motivation of workers

*Required

What is your gender? *

○ Male

○ Female

○ Prefer not to say

What is your age? *

○ 18-25 years

○ 26-37 years

○ 38-55 years

○ over 55 years

○ Prefer not to say

Do you have the following qualifications *

	Yes	No	Prefer not to say
A levels or equivalent	○	○	○
First degree or equivalent	○	○	○
Post graduate degree or equivalent	○	○	○
Member of a professional body	○	○	○

How often is do you experience the following at work?: *

	Always	Mostly	Often	Occasionally	Rarely	Never
My boss asks me politely to do things, gives me reasons why, and invites my suggestions	○	○	○	○	○	○
I am encouraged to learn skills outside of my immediate area of responsibility	○	○	○	○	○	○
I am left to work without interference from my boss, but help is available if I want it	○	○	○	○	○	○
I am given credit and praise when I do good work or put in extra effort	○	○	○	○	○	○
I am incentivised to work hard and well	○	○	○	○	○	○
If I want extra responsibility, my boss will find a way to give it to me	○	○	○	○	○	○

Figure 5.3 Questions constructed in Google Drive questionnaire software.

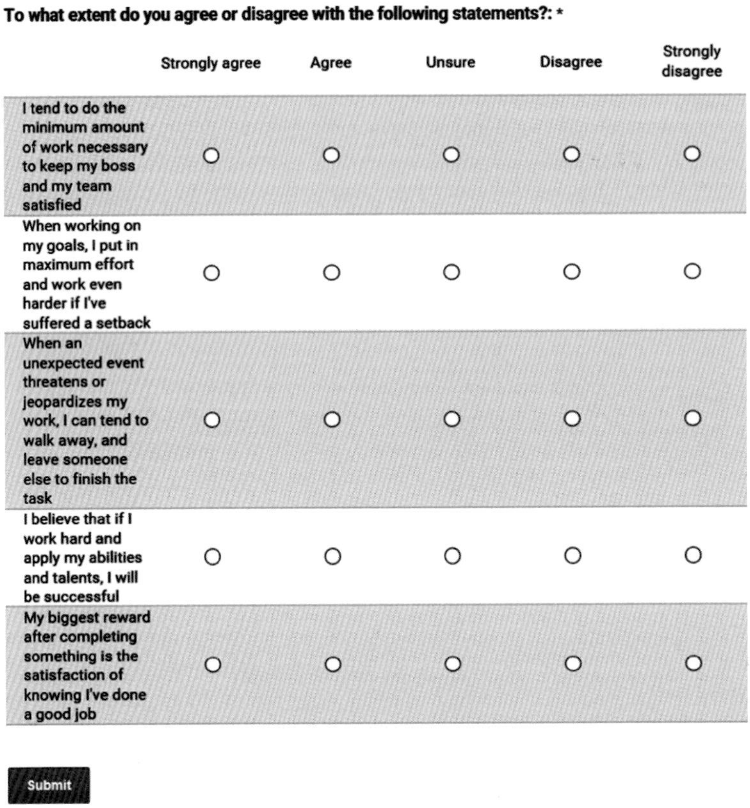

To what extent do you agree or disagree with the following statements?: *

	Strongly agree	Agree	Unsure	Disagree	Strongly disagree
I tend to do the minimum amount of work necessary to keep my boss and my team satisfied	O	O	O	O	O
When working on my goals, I put in maximum effort and work even harder if I've suffered a setback	O	O	O	O	O
When an unexpected event threatens or jeopardizes my work, I can tend to walk away, and leave someone else to finish the task	O	O	O	O	O
I believe that if I work hard and apply my abilities and talents, I will be successful	O	O	O	O	O
My biggest reward after completing something is the satisfaction of knowing I've done a good job	O	O	O	O	O

Submit

Figure 5.3 (Continued)

There are many legitimate concerns about the validity of questionnaires as a method of collecting data. Validity is defined here as 'how well a measure really measures what it purports to measure'. Biased data may result from low response rates, e.g. 30 of 100 questionnaires returned: what would the other 70 participants have said if they had completed the questionnaire? Therefore even if the numbers in statistical analysis indicate significant results, they may be of dubious value. Consider two examples: (a) A questionnaire survey to social housing tenants to measure the quality of service provide by landlords in dealing with maintenance issues. There is a possibility that the tenants who will have the greatest propensity to reply are those who are dissatisfied with the service, and therefore scores may be poor; (b) A questionnaire survey to measure commitment to health and safety will most likely be responded to by companies that are highly committed, and therefore scores will be good. In both cases, the results do not reflect the total population; they are biased.

A questionnaire is used to 'measure' something within a defined population. Careful selection of the sample must take place. It is very useful to keep focused on the concept of 'measurement'. These measures are often about people's opinions, values, knowledge or behaviours. Questionnaires that are not administered face to face usually have closed questions with a choice of preselected answers. There may be a few open questions that require participants to express their own words qualitatively, but be mindful that you are not asking participants to write long answers as though for an examination. If you really want essay type answers from

> Students sometimes mistakenly use questionnaires to seek answers to interesting questions not directly related to study objectives. There seems to be a 'scatter-gun' approach to collecting all sorts of irrelevant data. Do not stray from the study objectives, and the variables in those objectives that you want to measure. Measure variables using multiple-item scales.

participants, the questionnaire is not the appropriate method; you should revert to qualitative data collection through interviews.

There may be four sections in a questionnaire: perhaps section A to measure ancillary variables or demographic data, section B to measure the IV and section C measuring the DV. It may be the case that you propose to measure one of the IV or DV by another means e.g. secondary data, and you will not need both sections B and C. The closing section D may thank participants, and ask for contact details if they wish to see results from the study. Also you should include a final question 'Please add any comments you may have below.' Such qualitative responses may give you new insights that you can use to flavour your discussion and conclusions. If your questionnaire is interesting, topical and has reached the correct people or population, you should get lots of comments to help you. Some participants may 'pour their hearts out to you' about how passionately they feel about the topic area of your study. You may get some comments that are quite profound and that give you new insights. Type such comments up verbatim, and include them in an appendix.

5.4.2 Administration

Most often, questionnaires are administered electronically. However, paper based questionnaires may be distributed by hand, if that is appropriate to your population and/or sampling methodology. Some examples of the need to use paper are: surveys of homeowners of newly built properties, questions to members of the public that you may wish to ask as they are shopping, surveys to old people or craftspeople who may not use electronic systems. Your population may be such that you stay with people as they complete your survey, giving prompts as you may need to, or you read the questions to participants and complete the form yourself. You decide the best approach based upon the circumstances of your work, but be careful to follow your university ethical guidelines.

There are many commercial applications for distributing your survey electronically. Use an application that is free; your university does not expect you to pay. Some applications are free for only a limited number of questions; some are free for an unlimited number. Appropriate key words in web search engines will locate alternative software. Universities may provide free access to Bristol-online. You are likely to get a low response rate if you send a Word document to participants, which requires them to save it to their own computer, and email it back to you as an attachment; that would take them too much time. Electronic surveys have easy-to-read presentation styles and are quick to complete. You can give participants a choice of methods to respond; perhaps the two most popular are 'multiple choice' and 'check boxes. Participants should be able to go through the questionnaire at a fairly high speed, merely ticking appropriate answers. There is the option to ask participants to write in text of their own choosing, but it is best perhaps to restrict the use of this facility to the final question: 'Please add any comments you may have below.' Questionnaires are best for participants to select answers from a given choice, that you will later convert to numbers to do some quantitative analysis.

To distribute an electronic questionnaire, send an email to potential participants to request their contribution, with a web link for them to open the survey. Follow the ethical guidelines with details in your email similar to those in appendix B2. Select your words carefully since you are trying to entice participants. Potential participants will make one of two instant judgements, having read your email invitation, either to click onto to the web link to open the questionnaire, or to delete it.

On completion of the survey they will merely press the 'submit' button. As the administrator for the account, you will be able to access the results. One difficulty with the electronic questionnaire is to get the email addresses of your sample. If you wish to contact people at their place of work, some organisations will place email addresses of employees in the public domain, but most will not. That may lead you to a 'purposive' sampling technique as described in section 5.3 of this chapter, where you target individuals that you have been able to locate for their expertise. Figure 5.3 illustrates the layout of the software in the Google Drive application; it is free to use. If you wish, you can type into your browser the following web link:

```
https://docs.google.com/forms/d/1nbJu9aundYNsoDV6Fa3QdTpGjkm0_xBkMR8_
WIizWzM/viewform?c=0&w=1
```

It is a survey that you can complete, just so that you have some idea of what it is like to be a participant. It should take you no more than five minutes.

The survey measures three ancillary variables: gender, age and qualifications of participants. The IV is leadership style based on the work of Douglas McGregor (1906–1964) who used a multiple-item scale to measure management style (Business, 2010). Theory X was an authoritarian management style; theory Y was a participative style. He suggested that enlightened managers are theory Y and get better results. Measures were based upon assumptions of managers about workers. Workers scored their managers. The multiple-item scale is a better measure of leadership style than the single question 'What leadership style does your boss have?' The original work contained 15 questions; for brevity in figure 5.3 just six are shown. The DV that is measured is motivation of workers. The questions are sourced from Mindtools (2015). The objective is: To determine if leadership style of bosses (IV) influences motivation of workers (DV). Hypothetical responses to the DV, and illustrations of how this multiple-item scale is used to calculate descriptive statistics are presented in section 8.2 and tables 8.1, 8.2 and 8.3. Responses to the DV are also included in section 8.6, and table 8.7. The IV and DV are brought together in figure 8.9.

There is also the possibility of using business and social networking sites. If you use these methods, keep in mind that your sample, depending your objectives, may not be representative of the population. Therefore in your conclusion chapter, you should write under a subheading of 'Limitations and criticisms of the study' about how this has influenced the validity of your work.

5.4.3 Piloting the questionnaire

You should pilot any measurement tool. It is particularly important where you have participants involved. Piloting a questionnaire helps to improve its validity. A questionnaire that you have designed may be perfectly clear to you but not clear to readers. In the first draft of a questionnaire there is the possibility of mistakes. Consider a situation where a researcher is given £100k to administer a questionnaire, and spends all this money in administering it quickly. When responses are received, it is realised there were mistakes – but realised too late. Better to spend £10k in piloting, and then £90k after any potential mistakes have been rectified. When you send the questionnaire out, that is an important milestone in your study – no turning back. If the questionnaire contains a fundamental error or is poor even after piloting, the rest of the study will most likely be poor. You will not have time to re-administer a poor questionnaire. Work with your supervisor on several iterations. When you are both happy, start the piloting process. Stage 1 of piloting is to select three or four people; for part-time students perhaps colleagues at work, or full-time students may ask part-time students. Sit with them while they complete the questionnaire. Remain silent, since you will not be able to speak to participants when the questionnaire goes live. Observe them while they complete the document, and time how long it takes. After they have completed it, have an informal feedback session:

How did they find it?
Were the instructions clear?
Were the questions clear?
If not, which ones?
If they paused over any questions, why?
What do they think of the scales?
Did they object to answering any questions?
Was the layout attractive?
Do they think the questions are good measures of the variables?
Do they have any other comments or suggestions for improvement?

Stage 2 of the piloting process is to administer the questionnaire to say 10% of the sample. There is potentially a huge constraint on you here: time. To send out a questionnaire, get the data back, do some analysis and make changes before administering it again to the whole sample is potentially very time consuming. Make sure that you allow for it in your programme. Not all students do so. You may 'target' the questionnaire or take a purposeful sample to help get best response rate, recognising that the data is arguably biased. That is no matter though, on the basis that it will not be used in the main study analysis. Tell participants it is part of a piloting process, and at the end of the questionnaire, ask the same questions you have asked the people who gave you feedback face to face. You should tabulate results, and proceed to analyse the data using appropriate analytical methods. Ideally go on also to formulate interim findings, conclusions and recommendations. Section 9.10 details some possible statistical checks on internal reliability of questions. Check to see if results cluster around the central point. Any faults in the research questionnaire should be corrected before it is issued to the remainder of the sample. Explain clearly in your method chapter your piloting process, and justify any changes made to your questionnaire between the pilot and main study. You should be awarded 'extra' marks for a thorough piloting process.

5.4.4 Statements or questions

Section 5.3.2 introduces Likert items and Likert scales and explains that the agree/disagree points that Likert originally used have been adapted to questions and other measurement tools.

Students often mix statements and questions. There may be a question, clearly ending with a question mark, and participants are asked to agree or disagree with the question. This is not appropriate. In the context of your study, decide whether statements or questions are appropriate, as follows.

One statement in a multiple-item scale

Please indicate the extent to which you agree or disagree with the following statement:
I tend to do the minimum amount of work necessary to keep my boss and my team satisfied. (Note there is no question mark.)
Possible responses: strongly agree, agree, unsure, disagree, strongly disagree.

OR

One question in a multiple-item scale

Please answer the following question:
How often do you tend to do the minimum amount of work necessary to keep your boss and your team satisfied? (Note the question mark.)
Possible responses: always, frequently, sometimes, rarely, never.

But not like this:

Inappropriately mixing a statement and a question

I tend to do the minimum amount of work necessary to keep my boss and my team satisfied:
Possible responses: always, frequently, sometimes, rarely, never'.

You must be clear which style you will use, and preferably use the same style where possible throughout your questionnaire. If you are using statements, the five points can be widened to seven to include very strongly agree and very strongly disagree at the extremities. The word used in the central position is important; can it be left out to give just an even number of responses e.g. strongly agree, agree, disagree, strongly disagree? If you do this, you need to be sure in the context of your study that participants will be able to pick one possible answer, so perhaps include a 'don't know' option. If you want a central point, in the context of your study, what is the correct word in the middle? 'Neither agree nor disagree', 'neutral', 'unsure', 'uncertain'? This will influence your scoring system, since numbers are allocated thus: 'strongly agree = 4, agree = 3, unsure = 2, disagree = 1, strongly disagree = 0'. It is arguable that 'unsure' or 'uncertain' is really a 'don't know' answer, and if a participant does not know, the response is not mid-range and should not be included in the analysis at all. The scoring scale should therefore be: agree = 3, agree = 2, unsure = – (dash), disagree = 1, strongly disagree = 0'.

However, responses of 'neither agree nor disagree' or 'neutral' do indicate knowledge or opinion, and these should be scored 2.

Whether asking for agreement or disagreement to statements, or asking questions, there are lots of possible responses. Four, five and six point response scales are illustrated:

Satisfaction: completely satisfied / very satisfied / somewhat satisfied / somewhat dissatisfied / very dissatisfied / completely dissatisfied (six point range)

Satisfaction: completely satisfied / somewhat satisfied / somewhat dissatisfied / completely dissatisfied (four point range)

Agreement: definitely true / true / don't know / false / definitely false (five point range)

Agreement: definitely yes / probably yes / probably no / definitely no / uncertain or no opinion (five point range)

Importance: definitely unimportant / probably unimportant / probably important / definitely important / no opinion / don't know (six point range)

Frequency: always / very often / fairly often / sometimes / almost never / never (six point range);

Frequency: always / frequently / sometimes / rarely / never (five point range)

Intensity: none / very mild / mild / moderate / severe (five point range)

Intensity: very much / much / a fair amount / a little / not at all (five point range)

Influence: big problem / moderate problem / small problem / very small problem / no problem (five point range)

Comparison: much more than others / somewhat more than others / about the same as others / somewhat less than others / much less than others (five point range)

5.4.5 *Alternating poles on statements and questions*

When using multiple-item scales to measure a variable, the 'poles' should be switched to avoid the consequence of 'yea-sayers' or 'nay-sayers'. Figure 5.3 illustrates five questions to measure the 'motivation of workers'. Multiple-item scales in your work should ideally be longer than five, perhaps a minimum of ten. For questions 1 and 3, a highly motivated worker would select the response 'strongly disagree'. The poles of questions 2, 4 and 5 are opposite, so that a highly motivated worker would select the response 'strongly agree'.

As another example, the Sunday Times conducts an annual survey of the best UK companies to work for. Construction companies are rightly proud if they achieve a ranking in the top 100. The survey measures one key variable 'How good is the company to work for?' If it were just to ask the single question 'How good is your company to work for?', that would be a poor measure. To measure this one variable, the survey uses a multiple-item scale comprising 70 statements in eight categories thus (Sunday Times, 2015):

Leadership: how employees feel about the head of the company and its senior managers

Wellbeing: how staff feel about the stress, pressure and the balance between their work and home duties

Giving something back: how much companies are thought by their staff to put back into society generally and the local community

Personal growth: to what extent staff feel they are stretched and challenged by their job

My manager: how staff feel towards their immediate boss and day-to-day managers

My company: feelings about the company people work for as opposed to the people they
   work with
My team: how staff feel about their immediate colleagues
Fair deal: how happy the workforce is with their pay and benefits

Employees can indicate agreement or disagreement over a seven point scale (strongly
disagree, disagree, slightly disagree, neither agree nor disagree, slightly agree, agree,
strongly agree). Each answer is given a number 0 to 6 (note, not 1 to 7), with a high score
indicating a good company to work for. The minimum score for a company, though
unlikely, is zero and the maximum score is 420 (70 statements $\times$ 6). For simplicity, total
scores can be expressed as a percentage, such that 210 points out of 420 = 50%. Each
employee will have his or her own mean score; an overall company score is taken as a mean
of all employee scores. The 'poles' of the 70 statements are switched. For example in the
category 'leadership', one question is 'the leader of this company is full of positive energy';
on the seven wide scale strongly disagree is scored 0 and strongly agree is 6. Another question is 'I'm under so much pressure at work I can't concentrate.'; strongly disagree is 6 and
strongly agree is 0.

If questions were not switched, participants would quickly grasp that ticks for being highly
motivated are on one side, and unmotivated on the other. They may become flippant or
bored in answering questions and merely tick statements in the same vertical column without giving too much thought to their responses. In long surveys, it cannot be expected that
participants will labour long and hard over each response, but it can be expected that they
pause momentarily and digest each question. If questions are switched, that slows participants down a little, and subconsciously forces them to think a little harder before answering.
There should not be a structure to the switching, e.g. in long multiple-item scales, five questions or statements in one direction and then five in the next; it should be thoughtful but
fairly random. If you do change the poles on questions or statements, you must be careful to
change the codes too. For example one item of the scale may be scored thus: 0 = strongly
agree, 1 = agree, 2 = unsure, 3 = disagree and 4 = strongly disagree. For an item on the scale
with a reversed pole, the scores must be 4 = strongly agree, 3 = agree, 2 = unsure, 1 = disagree
and 0 = strongly disagree.

5.4.6 Things to do and not to do in questionnaires

Some things to avoid in questionnaires:

(a) Loaded questions, such as, 'Do you think method A is better than method B?' – you
     must use a neutral format.
(b) Two questions in one since the word 'and' is included, such as 'Do you have responsibil-
     ity and authority in your workplace?' Perhaps a participant has responsibility but no
     authority, so this should be written as two separate questions.
(c) Sensitive personal questions: people may not wish to answer questions about income,
     religious beliefs, alcohol consumption or precise age.
(d) Vague questions: 'How would you describe your health?' is better written as 'How would
     you describe your health in the last three months?'

(e)   Limiting closed responses, e.g. age bands, such that the top range for professionals is 55 to 65; better to say 'over 55'.
(f)   Questions that require participants to reach into files for data; if that is the case non-response rates will be very high – most questionnaires should allow participants to complete them reasonably quickly, such that it is: tick, tick, tick, some qualitative comments, finished and submit.
(g)   Time-consuming questionnaires to busy business people – perhaps ten minutes maximum.
(h)   Skip patterns: a skip pattern anticipates that some questions will not apply to all participants, e.g. if answered 'yes' to this question, please go to question ...

Do not attempt to ask for sensitive business information, such as safety statistics, cost data or anything else that businesses may consider confidential.

Some things you should do in questionnaires:

(a)   Be conversational in your tone.
(b)   Consider the target audience, e.g. questions to young people should be different from those to executives.
(c)   Make sure that any closed responses to questions are exhaustive. This may mean including options such as, 'don't know', 'unsure', 'uncertain', 'neither agree nor disagree'.
(d)   Use tick boxes or table structures where possible to facilitate speedy completion.
(e)   Assure participants that data provided will be kept confidential – and make sure you comply with that.
(f)   Assure participants that their anonymity will be protected.
(g)   Offer participants a summary of your results. You will need to ask for contact details if they indicate that they would like them.
(h)   Always finish with a 'thank you for completing the questionnaire'.

You will need to predict a response rate to your questionnaire to ensure that you have a sufficient number of responses for analysis. Postal survey response rates, though not often used these days, may be as low as 25%, so if you need 30 responses, 120 should be distributed. A more optimistic response rate may be predicted for electronic surveys, but the actual response rate depends upon many variables, such as relevance of the topic area to the sample selected, attractiveness of design and time for participants to complete.

5.5 Ranking studies

The objective of a study may be to ascertain ranking judgements of participants about issues. A list of factors may be given, and you ask participants to rank them in terms of importance, quality, frequency of occurrence, frequency of use or such like. The example in table 5.2 is a study that seeks to rank participant opinion about factors that keep sites safe. There are three generic areas: bespoke risk assessments, supervision and training operatives. Other areas could be selected too as part of a study, such as 'Use of personal protective equipment', 'Training and education for managers' and 'Extent to which productivity incentive schemes are used'. Rankings are sought for five items in each generic area, and then mean rankings taken. Table 5.2 shows a small hypothetical dataset; note that a sample size of five is too small for a

Table 5.2 Ranking of data; factors that keep sites safe.

Participant ↓	Q1: When writing risk assessments, the people involved should be:					Q2: Supervisors should be:					Q3: Operatives and craftspeople become most safety aware by:				
	a. The operative or craftsperson	b. Trade foreman	c. Site manager	d. Contracts manager	e. Safety officer	a. Experienced in construction	b. Experienced in the type of project under construction	c. Ex-craftspeople	d. Graduates	e. Sufficient in number on site	a. Site experience	b. Tool box talks	c. Off-site short courses	d. Threat of disciplinary action	e. Safety incentive schemes
A	1	2	4	5	3	1	2	5	4	3	1	2	3	4	5
B	2	3	1	4	5	1	2	5	4	3	2	3	5	4	1
C	1	3	2	4	5	1	2	5	4	3	3	4	1	5	2
D	2	4	1	5	3	1	2	5	4	3	4	3	2	1	5
E	1	5	2	3	4	1	2	5	4	3	5	3	4	1	2
Total of ranks	7	17	10	21	20	5	10	25	20	15	15	15	15	15	15
Mean rank	1.4	3.4	2	4.2	4	1	2	5	4	3	3	3	3	3	3

dissertation or project. The instruction to participants is 'Please rank each of the following factors in order of their importance in keeping sites safe, ranking the most important as 1 and the least important as 5'.

For question 1, the lead rank item is 1a, while the worst ranked items are 1d and 1e. These mean ranks should promote discussion, including identifying items that should be targeted for action. The dataset shown for Q2 is unlikely to occur in practice; there is perfect agreement between the judges: item 2a is unanimously the lead rank and item 2c the worst ranked. The dataset shown for Q3 is again unlikely to occur in practice; there is no agreement between the judges, since the mean ranks for all items 3a to 3e are identical. In practice, whether your mean ranks are close to Q1, Q2 or Q3 should drive the type of discussion.

If you wish to undertake inferential tests for ranking data, the appropriate test is Kendall's coefficient of concordance as explained by either Siegel and Castellan (1988) or Cohen and Holliday (1996). This test will tell you whether differences in rankings between participants are significant.

Ranking studies lend themselves to items such as causes of: delay in projects, material waste, defects, plant breakdown etc. When providing a list of items to rank, provide enough items in the list to give you the opportunity to get rich data, but not a list of items that is too long, since meaningful ranking may be difficult for participants. Perhaps ten items to rank is too many and difficult for participants.

5.6 Other analytical tools

The two analytical tools most often described are qualitative and quantitative, but the quantitative approach does not have to involve conventional descriptive or inferential statistics. Other tools can be used too, which fit only loosely into the qualitative and quantitative domains. Providing they help to improve knowledge and understanding in construction, they are fine. Here are some examples:

- drawings or sketches used to predict how materials may behave when placed together in composite structures, or how moisture may penetrate into buildings
- comparison of alternative designs to yield technical or environmentally friendly solutions
- photographic surveys used to compare and contrast material defects in existing properties; use with computer images to enhance visualisation
- building surveys: comparisons across a range of properties or areas
- critical path programmes or bar charts to illustrate the impact on time of choice of a particular method of construction, perhaps timber frame versus traditional housing or concrete versus steel multi-storey structures
- quantity surveyors may wish to take off parts of a building using SMM7, and/or the New Rules of Measurement and/or International Rules of Measurement; comparisons can be made about speed of take-off
- comparison of price estimates undertaken based on elemental cost plans, approximate quantities and composite rates or on different methods of construction
- land or GPS surveys to compare the accuracy of instrumentation
- time and motion studies to appraise new methods of working

- laboratory experiments, such as materials testing, soils testing, behaviour of fluids, heat, light and sound
- mathematical computations; use of formulas, such as those used in U-values, energy calculations or surveying
- comparison of site investigation reports
- risk analysis: considering alternative projects in 'what if' models to assess their financial viability, e.g. what if interest rates change, ditto inflation?
- decision trees: comparing established theoretical models to a case study situation
- software programs: designing a measuring tool, and using it to compare two competing products on the market
- brainstorming alternative building designs to yield best value solution
- feasibility studies
- cost/benefit analysis techniques
- criterion scale judgements; writing qualitative criteria and making judgements against about where actual performance sits on a scale

5.7 Incorporating reliability and validity

Reliability and validity are very important and related concepts in research; if you cannot demonstrate that your research is reliable and valid, it is worthless. You should work very hard to maximise reliability and validity. They are not considered in isolated parts of research: assuring validity is not a single act, it is the whole process. You should write at length in the methodology chapter about how you have striven to achieve them. Reliability is part of validity, indeed you cannot have validity without reliability. Reliability is consistency over time. Validity is 'how far a measure really measures the concept that it purports to measure' (Bryman and Cramer, 2005, p. 80). Therefore it follows that the validity of the whole study is whether it achieves what it purports to achieve; to what extent does it truly meet its objectives? Also be mindful that your objectives should be set to accomplish something worthwhile: modest objectives, but not just to achieve the obvious or something too simplistic.

Reliability is required, for example, with laboratory or surveying equipment. A weighing scale must consistently weigh a 10 kg weight at the same weight, not 10 kg one day, 9.9 kg the next day and 10.1 kg the next day. As part of your study you should report in appendices the manufacturers' recommendations for maintenance of equipment and records of service histories that provide evidence that recommendations are being followed. An automatic survey level must consistently give the same difference in height between two fixed points; the difference in height must not change each time the instrument is reset and new readings are taken. You should carry out tests and retests on surveying equipment before collecting data for your research, and report the results of those tests.

A reliable questionnaire would generate the same answers if the survey were repeated. That might suggest that you ask participants to complete a survey, and then a few weeks later repeat it again with the same participants. You may then judge the survey reliable if you get the same answers. In theory, this is correct, and you would be able to assert that your questions are reliable. But in practice, of course, you cannot usually do this. Reliability may be asserted if the survey has been proved to be successful in many studies elsewhere. If your survey is not well established or if you are designing your own questions, multiple-item scales are used to improve the internal reliability of surveys. Section 9.10 illustrates how scatter graphs and correlation coefficients can be used to test internal reliability. The piloting process should be used to detect any questions that are unreliable. If questions are found to be unreliable in the pilot, they should be changed or removed when the main study questionnaire is administered.

The relationship between reliability and validity is often illustrated by drawing an analogy to a clock:

- A clock that measures 61 minutes for one hour, then 60 minutes and then 58 minutes is not reliable; nor is it valid. It cannot be relied upon to measure time.
- A clock that measures 61 minutes for one hour, and does that every hour, to the exact second, is highly reliable. We can be sure that for the next hour it will again measure it as though 61 minutes. After the clock has run for 61 minutes, we can definitively say that one hour has just passed. But, it is not valid, since it does not do what it purports to do, that is, measure one hour as 60 minutes.
- A clock that measures 60 minutes for one hour, and does that every hour, to the exact second, is highly reliable and valid. We can be sure that for the next hour, it will again measure it as 60 minutes, and that is what it purports to do.

In the example of the weighing scale used above for reliability, a scale that consistently weighs a 10 kg weight at 10.1 kg is reliable, but not valid. We can see from these examples that reliability comes before validity, and we can have reliability without validity. But we cannot have validity without reliability.

You should work to assure the validity of the whole of your study and each individual part, which is done by careful design and pragmatic action. You may also consider the validity of say the measurement part, or the validity of the sampling strategy. The validity of your research holds together like a chain: if one link in the chain is weak, the whole study becomes invalid. Figure 5.4 illustrates a weak data collection instrument. You should seek to demonstrate validity in your methodology chapter. Validity first recognises that you are the lead player, and you will self-scrutinise all facets of your study. In the introduction chapter, you must ensure that the problem is well founded, has a theoretical base and

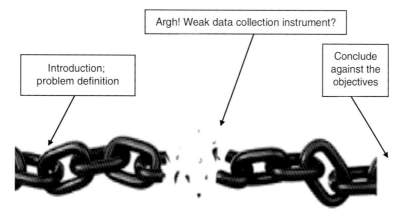

Figure 5.4 Do you have a weak link in your research? Validity of the whole study broken?

that the variables in objectives are well defined. Take parts of your study to other people who are knowledgeable in the field and ask them to give you feedback. That can be by formal or informal interviews. In the review of the literature you must ensure that you are critical and do not miss previous important work. In your methodology you must select the best method to meet the objective, including the most appropriate population and sample. Large sample size improves validity. Pilot parts of your study, perhaps the data capture and analysis part. Ideally, use a measurement tool that is considered to be the gold standard in your discipline; that is, use something that has been developed elsewhere by authoritative sources (be sure that you cite them), and that has been tried and tested by years of experience in respected studies. The analytical method must be robust, rather than using lightweight analytical tools. Towards the end of the document, the discussion and conclusions should be informed by insight, and best developed through dialogue or interviews with experts.

If your study held together reasonably well but, for example, you were careless in how you picked your sample (one weak link in the chain), or your sample was biased, the whole study would be flawed. The consequence of this weak link is that your sample may not reflect the population, and therefore any inference that the results from the sample reflect results if the whole of the population had been surveyed is spurious. The analysis that follows will use spurious data, and the discussion and conclusions are formed around spurious results.

If possible, try to use objective means to measure your variables. There are lots of validity issues around subjective data, questionnaires and collecting data by observation. When answering questions, do participants tell you what they think you want to know, do they try to show themselves off in good light, do they really answer honestly, do they feel restricted in what they want to say by their employers? If you observe people to determine their behaviour, do they change their behaviour because they know they are being observed? You need to design your methodology to minimise the effects of some of these issues on the validity of your work.

At the end of your study you should self-appraise the reliability and validity of your research holistically. It may be useful to consider both concepts on a scale of 0–10, as illustrated in figure 5.5. Zero could be anchored with a qualitative label of 'little value' or even 'worthless' – that is zero reliability and validity. Ten could be anchored with a qualitative label of 'first proof, beyond doubt, in a significant area of work'. Some intermediate labels may help in making judgements. How far can you expect to get on this hypothetical scale? In your research you cannot be expected to score ten. While your dissertation or project is a really significant piece of work for you, and while you will get some help, it is mostly limited to what you can do personally. You do not have unlimited time, unlimited money and a team of researchers to help. It is not to score your work objectively on this 0–10 scale; the scale is suggested as a way to think the issues through in your own mind.

Consider the following three practical examples of validity.

Example 1

There is an objective to measure stress in site managers; it could alternatively be any professional, technician or tradesman in the supply chain. That is, only to consider reliability and validity of the measurement of stress, not reliability and validity of the whole study. This objective may be founded in a problem summarised by Langford (1988) 'there is awareness amongst older site managers that retirement is all too often followed by a sudden death'. One part of the methodol-

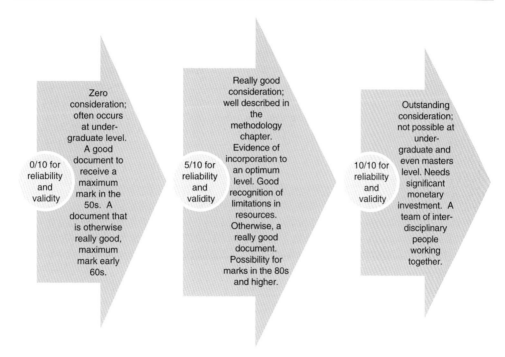

Figure 5.5 How reliable and valid is your work on a 0–10 scale?

ogy might be to measure stress by observation of managers in the workplace, or by interviews, or by some kind of survey. The criteria used to make judgements about stress levels and/or the questions asked must be based on the literature. There are lots of studies and templates in the healthcare professions. You would be foolhardy to design your own measures of stress; that would constitute zero validity. If you design a measurement system adapted from the literature, that is a good foundation for asserting good reliability and validity. If you use a measurement tool that is authoritative, you may judge the reliability and validity of your measure as say 5 out of 10. In the context of an undergraduate research, that can be fine. However, better measures of stress are possible if site managers are subjected to a range of medical tests to measure heart rate, blood pressure and cholesterol levels. There was a study (BBC, 2002) where football managers were wired up to heart rate monitors during matches. At one point, the heart rate of one manager reached 160 beats per minute, four times his normal resting pulse. The study methodology tried to get a 10 out of 10 score for reliability and validity in measuring stress. While it may be possible as part of university/industry funded research to get site managers wired up to heart rate monitors, you cannot reasonably do it as part of your research. However, what you need to do is to research and write about what the authoritative sources suggest is the most valid way to measure stress; the gold standards use heart rate monitors, blood pressure measures and take cholesterol levels. Then, compare what you have been able to do, given your limited resources and access to site managers, with gold standard measurement methods.

While you cannot be marked down for not 'scoring' 10 out of 10, you will get marks for reaching as high up the scale as you reasonably can. Also, you will get more marks for recognising where on the scale your work is. This would mean that you have good insight into the concepts of reliability and validity and that you understand the limits of what you have done. You will lose marks if you do not understand the limits of the reliability and validity of your work. You may be able to assert that some parts of your study are highly valid, but recognise that other parts are less so. You may start the process of recognising where your work is not as valid as you would like in your methodology chapter. You may expand on that in the conclusion chapter, with a subsection titled 'Limitations and criticisms of the study'. Section 9.10 illustrates examples of how to use correlation coefficients to measure reliability and validity.

Example 2

Some insitu concrete columns are poured on a site. One ready mix delivery is used of 6 m³, 28 N/mm² mix design. A sample of concrete is taken from the mix and six 150 mm cubes are made and cured. Two cubes are crushed at seven days; they should have attained 65% of their design strength at this time or 18.2 N/mm². The results are 15.5 and 17.1 N/mm². Two more cubes are crushed at 14 days; 90% design strength should have been attained or 25.2 N/mm². The results are 23.9 and 24.1 N/mm². The last two cubes crushed at 28 days; 99% design strength should have been attained. The results are 24.2 and 24.4 N/mm². In this 28 day period a suspended concrete floor has been cast on the columns. Questions; should the floor and columns be demolished? Are the cubes' results a valid measure of the compressive strength of the concrete that has been poured into the column formwork? The concrete in the columns could be defective; quality control systems may have failed, and the 6 m³ may have been an isolated bad mix.

However, reluctant to unnecessarily break out good concrete, the site team examine reasons why the cubes may have failed even though the mix was actually OK. Who made the cubes? Were they made in accordance with the British Standard? Were the cubes cured? How quickly were the moulds stripped and placed in the tank? Was the water in the tank at the correct temperature? Assume some failings were identified in the making of the cubes; Schmidt hammer tests are performed and it is found that the concrete in the columns actually has a compressive strength of 35 N/mm². If necessary 150 mm cores may have taken by diamond tipped drill bits. All is fine, but the cube-making process needs to be examined. The compressive strength of the cubes was not a valid measure of the compressive strength of the concrete in the columns.

There could have been a bigger problem if the cubes were fine, but the compressive strength of the concrete in the columns did not achieve the strength indicated by the cubes. Concrete cubes are often made and cured in excellent laboratory conditions and may attain better results than concrete poured on site in bad weather and with perhaps inadequate supervision. Site concrete may not be compacted properly, may have water added by operatives while supervisors' backs were turned or may not have been cured adequately. In a large pour, there may be one rogue delivery of ready mix concrete that has not had cubes taken from it. The same question again: are the cubes' results a valid measure of the compressive strength of the concrete that has been poured into the column formwork? Perhaps not, and if there are no obvious visual problems with concrete that has been placed, what are the potential consequences?

Example 3

In the recruitment process, interviews are used universally to appoint people to jobs. The question is 'Are interviews a valid way of predicting future likely performance in the workplace?' It is possible to imagine people who interview poorly, perhaps they are nervous, but perform really well at work. Similarly, there will be people who talk a really good job, but do not perform. Some of course will interview well and perform well. Because the simple interview is known to be a poor predictor of job performance, companies seek to introduce other measures into the selection process, such as references (although these are notoriously invalid measures, especially when interviewees name the referees), panel and two-stage interviews, intelligence tests and personality tests.

Sections 9.9 and 9.10 explain about correlation coefficients. Dattner (2010) reports that the validity of unstructured interview (IV), a predictor of future likely job performance (DV), is 0.20, and structured interviews 0.50; figures much closer to 1.00 would be expected if interviews were a valid measure of future likely job performance.

5.8 Analysis, results and findings

The methodology chapter provides a lead-in to the analysis, and the subsequent discussion and conclusions chapters. The terms 'results, findings and conclusions' are often used interchangeably in the literature and the media. It may be appropriate to consider the definitions in the context of your study. You need to be clear in your own mind how you apply these terms to your research. The way you use the terms may be driven by whether your analysis is quantitative or qualitative. For quantitative work the terms 'analysis, results and findings' may be presented in the same chapter and in that order. They are key elements in helping you to meet study objectives; they are all focused on the objectives.

The analytical process is one whereby all raw data is brought together and sorted systematically. That may involve some numerical coding of qualitative labels. It may involve summarising the data in spreadsheets. Data that is available numerically can be subjected to a variety of academic analytical tools. It may be put into simple or complex formulas. There may be some adding, multiplication, calculations of means, medians, modes or standard deviations. There may be some comparison tables or simple frequency counts. More complex calculations may be performed using computer software. The strength of relationships between variables and significance levels may be determined. U-value or sound insulation calculations may be executed. If you have few calculations being performed by hand, these may be written out in full in this chapter. If there are many calculations, they may be best placed in an appendix. The raw data goes into the analytical process unstructured; it comes out structured.

After the analysis, come the results. Calculations terminate with an answer. The answer is the result. It may be a figure, such as $2.0\,W/m^2K$, or $28\,N/mm^2$, or 50%, or £50.00/m^2, or $p \leq 0.05$. As a figure, it is short, it is clear, it requires few words. If you are testing hypotheses, you should state the hypothesis in full before each result. A contingency table used in chi-square calculations (see chapter 9) can summarise frequency counts, or scatter diagrams can be used to show relationships between variables. Test results from laboratory work may be illustrated in graphs or line diagrams. If there are many results, they are likely to be presented in tables.

After results, there can be statements of findings that stem from results. If your quantitative study has only one variable, the finding mirrors the result, e.g. the study finds that: the U-value of this composite structure is 2.0 W/m²K, or this concrete cube failed at 28 N/mm², or the mean quality score was 50%, or the cost of floor finishes was £50.00/m². If your quantitative study has two variables and you have tested a hypothesis, you should state as part of your findings that you either 'reject the null hypothesis' or 'cannot reject the null hypothesis'. You may also be able to assert findings such as the IV influences the DV, or method or group A is better or lower cost than B. Alternatively, the IV is not found to influence the DV, or method or group A is not found to be better or lower cost than B.

The analysis, results and findings chapter will also summarise any demographic data that you may have collected. That may include ages, qualifications and job positions of people involved in surveys. It may be most simply summarised in tables, rather than too many pie charts or histograms. If your work involves lots of analysis, it may be useful to summarise the results in a table towards the end of the chapter.

In qualitative studies, the raw data comprises words, possibly verbatim transcripts of interviews. The analytical process comprises ordering, labelling and sorting paragraphs of text. Transcripts and labelling are likely to be in appendices. There may be some frequency counts of key words and tables to integrate with the literature. A narrative will be derived from the tables. The tables and the narrative may be in the document's main body. The boundary between analysis and results, and results and findings is not so clear. Labelling, sorting and frequency counting are part of the analysis. The process of writing the narrative may be thought of as part of the analytical process, but it is also the outcome (the result) and it asserts what is found.

Summary of this chapter

You should design your study carefully. In your methodology chapter describe in detail what you have done so that it can be replicated by others, and justify why you have done it that way. Seven approaches to collecting data are identified. You should be proactive and innovative in collecting data; only use questionnaires if they are the best way to meet your objectives. Primary datasets are generated by the researcher and secondary datasets are existing, compiled by others. Students arguably too infrequently use secondary data. Objective datasets are facts uncoloured by people's opinions, while subjective datasets are people's views. For some studies, only subjective data is appropriate. For other studies, objective or hard datasets may be preferred, but if it is too difficult to secure that data, alternatively subjective measurements may be used. In such cases you should recognise the limitations this has on your study. If you use questionnaires, multiple-item scales should be used to measure variables. Only use your questionnaire to measure your variables, not to ask other interesting questions. If possible, use a questionnaire from an authoritative source, adapt if necessary to your study, and pilot it. Score and code responses so that they can be summarised in a spreadsheet. Invite participants to add their comments, and write those comments up verbatim in an appendix. There are many analytical tools that can be used to analyse data, which draw on the qualitative and quantitative techniques. If you demonstrate understanding of reliability and validity, and incorporate these concepts into your study, that gives you the possibility of achieving high marks. Reliability is consistency over time, and validity is whether a study achieves what it purports to achieve. Analysis is followed by results, which is followed by findings. Be careful to distinguish between the three and provide all of them in a chapter under this or a similar heading.

References

BBC (2002) The heart of the matter. http://news.bbc.co.uk/sport1/hi/football/1758132.stm Accessed 06.08.15.

BBC (2015) Election 2015. Available at; www.bbc.co.uk/news/election/2015/results Accessed 06.08.15.

Bell, J. (2005) *Doing your own research project*. 4th Edition. Open University Press. McGraw-Hill. Available as free full text download.

Borden, I. and Ray, K.R. (2006) *The dissertation; an architecture student's handbook*. Architectural, London.

Breach, M. (2009) *Dissertation writing for engineers and scientists*. Pearson, Harlow, UK.

Bryman, A. and Cramer, D. (2005) *Quantitative data analysis with SPSS 12 and 13: a guide for social scientists*. Routledge, London.

Business (2010) Business Balls. www.businessballs.com/freepdfmaterials/X-Y_Theory_Questionnaire_2pages.pdf Accessed 04.09.15.

Crossrail (2015) Sharing health and safety best practices. Crossrail Health and Safety – Industry Best Practice Guides. www.crossrail.co.uk/sustainability/health-and-safety/industry-best-practice-guides Accessed 11.09.15.

Cohen, M. and Holliday, L. (1996) *Practical Statistics for Students*. Chapman Publishing, London.

Creswell, J. (2009) *Research design: qualitative, quantitative, and mixed method approaches*. 3rd Ed. Sage, London.

Dattner (2010) Improving Interviews. Dattner Consulting. www.dattnerconsulting.com/presentations/interviews.pdf accessed 13.09.15.

Electoral Commission (2015) Who is eligible to note at a general election? www.electoralcommission.org.uk/faq/voting-and-registration/who-is-eligible-to-vote-at-a-uk-general-election? Accessed 26.09.15.

Fellows, R. and Liu, A. (2015) *Research Methods in Construction*. 4th Edition. Wiley-Blackwell, London.

Fink, A. (1995) The Survey Handbook. *The Survey Kit*. Volume 1. Sage, London.

HM Treasury (2012) A new approach to public private partnerships. https://www.gov.uk/government/uploads/system/uploads/attachment_data/file/205112/pf2_infrastructure_new_approach_to_public_private_parnerships_051212.pdf Accessed 20.09.15.

Holt, G.D. (1997) *A guide to successful dissertation study for students of the built environment*. The Built Environment Research Unit. Wolverhampton, UK.

Knight, A. and Ruddock, L. (2008) *Advanced research methods in the built environment*. Wiley. Chichester, UK.

LABC (n.d.) Building Regulations in Practice. Accessible toilets. Local Authority Building Control. https://www.charnwood.gov.uk/files/documents/accessible_toilet_diagram_and_guidance/Accessible%20Toilet%20Diagram%20and%20Advice.pdf Accessed 27.09.15.

Langford, V. (1988) Stress, satisfaction and managers in the construction industry. *Occupational Psychologist*. 6, 30–32.

Lucas, R. (2016) Research Methods for Architecture. London: Laurence King Publishing Ltd.

Mindtools (2015) How self-motivated are you? www.mindtools.com/pages/article/newLDR_57.htm Accessed 27.09.15.

Naoum, S.G. (2013) *Dissertation research and Writing for Construction students*. 3rd edition. Butterworth-Heinemann, Oxford.

Siegel, S. and Castellan, N.J. (1988) *Nonparametric Statistics for the Behavioural Sciences*. 2nd Ed. McGraw Hill, London.

Sunday Times (2015) The Sunday Times 100 best companies. www.thesundaytimes.co.uk/sto/public/article1523544.ece Accessed 03.09.15.

6 Laboratory experiments

The titles and objectives of the sections of this chapter are the following:

6.1 Introduction – to set the scene for technical dissertation or project production
6.2 Methodology – outline a possible structure for the methodology
6.3 Sourcing test materials
6.4 Reliability of findings
6.5 Sample size – determine relevance within practical parameters
6.6 Laboratory recording procedures
6.7 Writing a laboratory report
  6.7.1 Conclusion
  6.7.2 Literature References and Citations
  6.7.3 Appendices:
6.8 Health and safety in the lab; COSHH and risk assessments.
  6.8.1 Health and Safety Induction
  6.8.2 Housekeeping
  6.8.3 Confidential Health Questionnaire for students and staff participating in Laboratory Work
  6.8.4 Example of Risk assessment for Laboratory work
6.9 Role of the supervisor
6.10 Possible research topics for technical dissertations or projects; construction and civil engineering
6.11 Examples of research proposals – to provide the researcher with examples of topics to investigate and the salient findings.
6.12 Research objectives and sample findings by the author
  6.12.1 Comparative performance of fibre reinforced polymer (FRP) and steel rebar
  6.12.2 Equating steel and synthetic fibre concrete post crack performance to BS EN 14651: 2005+A1:2007
  6.12.3 Concrete with crushed, graded and washed recycled construction demolition waste as a coarse aggregate replacement.
  6.12.4 Surface coating of traditional construction materials using microbially induced calcite precipitation
  6.12.5 Pull-out performance of chemical anchor bolts in fibre concrete

Writing Built Environment Dissertations and Projects: Practical Guidance and Examples, Second Edition.
Peter Farrell, Fred Sherratt and Alan Richardson.
© 2017 John Wiley & Sons, Ltd. Published 2017 by John Wiley & Sons, Ltd.
Companion Website: www.wiley.com/go/Farrell/Built_Environment_Dissertations_and_Projects

6.1 Introduction

Chapter 6 examines laboratory based investigations and is aimed at built environment and civil engineering students who will develop their study skills and look at areas associated with their interests. Civil engineering students would be expected to cope with mathematical analysis and design processes with relative ease while gaining the opportunity to extend their management skills associated with project planning, organising people and dealing with health and safety issues. Built environment students should have the management skills to permit a full investigation and they will benefit from the opportunity of expanding their numerical and analytical skills. A dissertation or project defines your degree, and an interest in a technical area such as sustainable materials makes undergraduates very employable when they demonstrate the same interest and show their ability in this area.

In essence, laboratory based projects and softer management style investigations are very similar in that it is the logical process that is being examined. The skill set required for an empirical based quantitative project is fundamentally no different from that of qualitative project. From the author's experience, the easier of the two forms of analysis is quantitative research. The student should consider what it says on the title of their programme of study. If you are studying a BSc then surely should there not be some science in the document? A BSc in construction management need not necessarily be completely science or technology based, but if it is not, then the softer management style of investigation should at least be analysed with a reasonable degree of mathematical rigour to achieve a satisfactory pass grade. When a student embarks upon a BEng, the need for carrying out a technical investigation is much greater. The term project for a BEng and dissertation for a BSc can be interchanged reasonably easily as the final product will be much the same.

An extract from a final year civil engineering project module descriptor sums up what is required of a BEng student.

> The individual project provides a vehicle for the student to undertake a substantive piece of self-directed work which will focus on a chosen aspect of civil engineering, studied elsewhere in the programme. This technical investigation will be characterised by the determination of an area of research, that establishes aims and objectives to be realised in pursuit of the research question/problem statement where contemporary literature on the subject area is presented and appraised. On determining and selecting a method for investigation, contextualised those that already exist, the student will pursue the collection and analysis of data to realise the project aims. The project will be examined by a presentation verbally and also in writing, a critical analysis, justifiable conclusions and a reflection on the study undertaken, through the production of a technical thesis and via a verbal voce.

To defend the research through the medium of a viva requires a thorough understanding of the subject studied and confidence in the results as presented.

All dissertations or projects must identify a worthwhile topic worthy of investigation, a means of carrying out the investigation, justifying the choices in terms of logic and selecting appropriate statistical analysis to present results that hopefully will remove the need for investigator judgement as to whether or not the results are significant. Industry partners will help identify areas in need of laboratory research. Networking at professional body events will prove to be beneficial to the investigator.

A well-designed laboratory investigation will provide reliable data from which deductions can be safely made, free from interpretation. To achieve this reliability within the obtained results, a standard test procedure would normally be adopted. Alternatively, a suitable bespoke test may be carefully designed. The use of standard or non-standard test procedures should be justified and informed by substantial background reading.

Selecting a suitable test procedure is not as easy as it sounds. One view would be that it is simply following a predetermined method without any operator input and that this will provide suitable results. This could not be further from the truth.

For example, a freeze/thaw test that examines the durability of concrete can be carried out to ASTM standards over a variety of freeze/thaw cycles, such as 300 cycles as detailed in ASTM 666 or 50 cycles as per ASTM 672. The BS recommends 56 cycles, and all standards have different measurement criteria from which conclusions can be drawn. Which one is best? How long do you have to complete a test? This may be the deciding factor. Can we accelerate the effect of the test by using conditions that are extremely aggressive? Has the laboratory sufficient equipment to carry out the test? Are suitably qualified staff available to assist and is there laboratory time available for all test proposals? Careful consideration of all of these factors should be made to determine a suitable test programme.

One of the main parameters for a short duration investigation such as a dissertation or project, is the length of time available to design and manufacture test materials. If concrete is chosen as a test material, we must not forget the time required to cure the test materials and the interim and final test time required to provide sound, defect-free data.

This chapter mainly focuses upon the design, batching and testing of concrete. It has been written because concrete being a conglomerate has many variables and is a perfect foil for this work. The testing of any other material may be subject to very similar constraints and conditions.

6.2 Test methodology

A test methodology should initially consider the following issues:

- What are the primary variables?
- What controls or limits need to be set?
- What ranges of measurement are needed?
- How much data should be gathered?
- What are the acceptable uncertainty tolerances?
- Is the problem static or dynamic?
- Can the variables be measured?
- Are there any safety issues?
- Are there any financial restrictions?
- Are there any ethical issues?

The test methodology is informed by substantial background reading. It cannot be emphasised too strongly that exposure to the work of other researchers through the reading of journal and conference papers forms the foundation upon which a successful technical project can be achieved. The methodology should inform the reader of what is proposed to be carried out, linked to the aims and objectives of the project. The researcher should then clearly explain what procedures are to be adopted during the course of the research. The level of detail should be sufficiently complete as to allow another researcher to repeat this investigation and precisely compare the results. If your intended audience is within your own country and the reader is conversant with current standards for materials testing, then it is satisfactory to refer to the more common standards such as compressive strength testing of cubes. However, if the standard is not commonly used, a full explanation of the test procedure should be included within the work. If the audience is international, then include every detail of the standards used because an international reader may not be conversant with the standards being used. Formulas used should always be included and ensure the formulas are explained in the text as to what is to be achieved from the calculations. A flow chart or visual representation of the work to be carried out allows the reader to see at a glance what is proposed and how the materials are to be tested. An example of this is shown in Figure 6.1.

6.3 Sourcing test materials

Having a research idea based upon substantive background reading is generally considered to be a safe way to develop material knowledge incrementally. It is strongly recommended that the student reads around the subject area to inform the possible avenues that are left to be explored.

As the research will be laboratory based, there is a need for material to test. To purchase these materials can be expensive and therefore it is important to contact material manufacturers with a good case as to why they need you to test their products. If a strong enough case is put to the supplier, the materials are normally shipped free of charge with the proviso that the supplier obtains a copy of the results. Publication of the results is a bonus for the student and the supplier. The student receives a degree with a publication attached to it. The supplier receives free product development.

It is strongly recommended to forge links with potential material suppliers well before there is a need for material to test. A lot of suppliers encourage students to use their products as it places them in a future marketplace. After all, that is what the student population are: tomorrow's workforce. Suppliers can be very supportive in terms of material supply, but also with material performance expertise. Why waste years of knowledge of product development by ignoring a potential free source of consultancy?

6.4 Reliability and validity of findings

Reliability is consistency or repeatability, but the question must be asked, 'How valid are the results being presented'. For results to be valid, the measurements must be well founded and must accurately represent what is actually being observed and recorded, free from interpretation. A valid result will truly measure the hypothesis being tested. Reliability of results falls into three main areas:

(1) reliability and current calibration of the test equipment
(2) design of a test procedure that reduces the number of potential variables
(3) operator competence when manufacturing materials and operator competence when using the test equipment

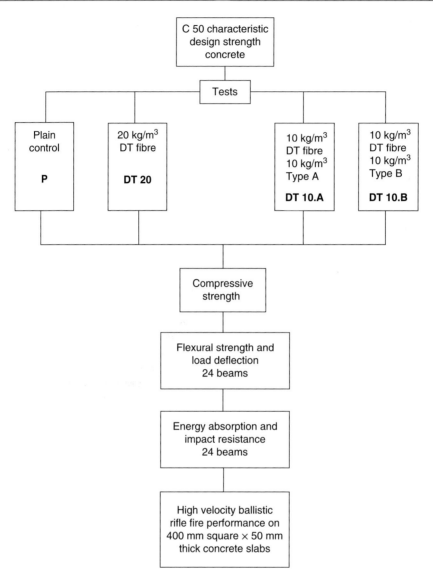

Figure 6.1 Test programme.

To obtain reliable data, the environment in which the readings are taken can play a part in achieving accuracy; for example, high temperatures can cause misreading errors when made close to the ground and viewed through air. Severe cold can cause loss of sensation in the finger tips when operating measuring apparatus. That is without considering the effect on the motivation of the operator who is uncomfortably cold.

Before embarking upon any test, it is imperative that current test certificates for equipment are in place, and in your methodology chapter you should state that you have checked this. This verifies the state of the calibration of the equipment to be used. The satisfactory design of the test is dependent upon control of variables. For instance, if 30 concrete beams and 30 concrete cubes are required, it would be beneficial to batch the material in one batch, thus reducing the number of variables. This may require a ready mixed concrete supplier delivering to the laboratory.

Thorough training should be provided to the operator of test equipment to ensure that rigorous health and safety standards are achieved, and operators fully understand the processes and are competent to carry out such work. This is where trained technicians are invaluable.

6.5 Sample size

The physical sample sizes will be determined by the objectives and aims of the research and the facilities available in the laboratory. If these research parameters are not achievable then it may be possible to outsource to another university or to a local material testing specialist. It is always worth asking, as the answer is generally a simple yes or no.

For most undergraduate projects samples sizes may be determined by the laboratory capability. For instance, if 50 students wish to carry out laboratory based projects then the available time per student would be very much less than if five students wished to use the laboratory facilities.

As a general rule, try to test 30 samples of each variable, as this is a number of samples where you may achieve a statistically representative sample. Comparing plain beams with fibre concrete beams may provide an opportunity for two sample sizes of 30, but if four aggregate types were replaced in a concrete mix then the sample size may be 6–15 to avoid overloading the student and laboratory space.

Also as a general rule try not to have less than six samples to test. From six samples, a standard deviation can be obtained and there is the possibility of removing outliers and still having sufficient samples to make a statistically significant statement as to the results. An outlier is a result obtained that for some reason (perhaps an operator error) is substantially different from the rest of its group. This may sound like very few samples to test, but if the test is recording the strength development over 3, 7, 14, 21, 28 and 56 days, this would equate to 36 samples for each variable to be tested. It is not an insignificant amount of work to batch and to test. Not everyone is physically fit enough to handle large volumes of heavy materials. In terms of risk management, it is always better to be self-reliant than to be waiting for the services of others who do not have such a high stake in the research.

6.6 Laboratory recording procedures

When embarking upon a laboratory test programme, the methodology must be prepared prior to the test; however, a log book to record the process and variations is a very useful tool. Note that the log book is a precursor to the obtained experiment data and forms the basis of information to allow students to write a professional report. It must be durable as the laboratory conditions may be aggressive. This can normally be achieved by laminating the front and back cover of a book. Pencil is the preferred method of recording in the log book as the records will not be affected if the book gets wet. Pen is subject to damage when in the presence of water. Guidelines for a log book are as follows:

(1) You need a log book where one of the sides is a graph-formatted page while the other is a normal page with horizontal lines.
(2) The log book should be clearly handwritten, for you to be able to reproduce the results and information for your report.
(3) What is the purpose of a log book?
     (a) The log book is used for you to record your experiment procedures and results.
(4) What you should write in a log book?
     (a) Record extra steps that are missing from the laboratory sheets.
     (b) Record troubleshooting procedures if any.
     (c) Record all the necessary details to reproduce the results.
     (d) Record your results.
     (e) Sketch details.
     (f) Sketch graphs.
(5) What about the format?
     (a) Date of experiment
     (b) Objectives
     (c) Extra procedures
     (d) Equipment and parameters used
     (e) Results
     (f) Observations.

The log book should be for the students' reference only. It is unlikely it that will be neat enough for inclusion in an appendix within the document, but the observations made and notes taken will inform the main body of the work. The use of photographs will supplement the log book to record and convey events to the reader. Photographic plates need to be used sparingly within the main body of the work, using representative images only, otherwise the document may suffer from image overload. However, these images may well form a solid support for statements made within the main body of text, if included in the appendix.

6.7 Dissertation/project writing (introduction, methodology and results)

Writing is an essential skill for any person involved in construction and civil engineering (built environment). The dissertation or project is the primary medium by which students are able to tell the outside world about their research or experiment. If a person cannot communicate results effectively to others, the research or experiment becomes less useful or maybe even worthless or a liability – do not forget that patent lawsuits can cost companies billions of pounds, unless research can be backed up by clearly understandable reports and notes. This means that student reports should be clear, concise, consistent and comprehensible to the audience for which they are intended.

From little acorns, great oak trees grow. An isolated dissertation or project may not be thought to be a very valuable piece of research until it is needed as evidence to add to the body of knowledge required to make a legal judgement. If the work is using material covered by intellectual property rights, ensure that the owner of these rights is clearly identified in the document. If the work is protected by a patent, ensure that all ethical procedures have been adopted to ensure the security of the findings until such time as the owner of the patent wishes to release the results. A confidentiality agreement should be considered, and this must never be breached, and the student should consider taking out intellectual property rights or a full patent. There is a cost to both of these options and the cost can be substantial. It would be worth considering a funding option prior to the start of the test to have it in place if the results prove to be promising.

The 'audience' for the researchers/students will be examiners, who might not necessarily be familiar with the experiments. There should be a logical thread linking one section to the next, and each section should summarise the findings and inform the reader of what is to come next. The experiment should be described in detail, especially the *objective* – what is intended to get out of your experiment? – and the method itself – how does the experiment work? The student must define all symbols used, and reference any statement, equation or graphic that has been quoted, which have not been derived personally. It is usual to place a glossary of symbols at the start of a document to inform the reader of what is to be read within. References are very important for two major reasons:

(1)   They allow you to justify your use of the experiment and your application of the theory to your problem by referring to other people's work – otherwise you would have to derive everything you state from first principles.
(2)   They allow readers to look up other articles and texts which are directly relevant to understanding your work.

The way the results are obtained should be carefully described, including sample calculations. The *discussion* should be as consistent as possible and aided by a clear *presentation* of these results, as needed. Try to look behind the experiment, to see the significance of your results as measured by that experiment. This is very important. You must describe what your results mean, why they are important and how reliable they are. In construction and civil engineering the reliability of a measurement is of utmost importance. Concrete for instance has a default standard deviation of 8 when no reliable data is present to evaluate the characteristic strength. This indicates that a large variance is expected with concrete. Steel having fewer component parts and strict manufacturing processes would be expected to have a much lower standard deviation with regard to manufacturing. Timber possesses multiple options for a variation in strength and should be accurately classified using the variable parameters of the timber, such as knot size, grain direction, moisture content, density, growth rate and, if it is reused, nail holes or notches. The material under examination must be viewed with these possibilities in mind.

The error is a measure of how far the result can be trusted, and so conversely a result quoted without an estimate of its error cannot be trusted at all, however accurate it might actually be. You must therefore try to estimate the error in your readings and thus the error in your results, preferably by using some statistical tools if available, such as standard

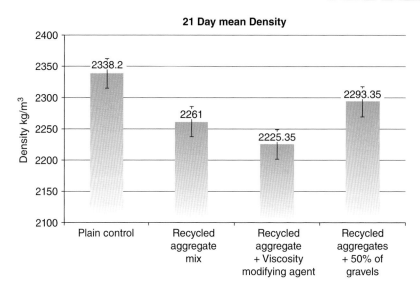

Figure 6.2 Mean density of concrete cubes (kg/m³) per batch type, showing standard error.

deviation, error bars on charts and r^2 values. Error bars are displayed in Figures 6.2, 6.7 and 6.8 and r^2 values are displayed in Figure 6.9. Both error bars and r^2 values display the correlation between the results. A value close to unity makes the statement that the results follow a line of best fit very closely. The results can be plotted using Excel, but a word of warning is due at this point. You may find that you have a straight line and a r^2 value of very close to unity. This does not mean that the results are valid. You must understand the material you are testing and plot the results in accordance with established trends. For example, strength development of concrete may show a linear trend with a good best fit. Plotting what you have recorded is not making a valid statement in this case. Choose the correct plotting profile and equation, such as exponential or logarithmic, and represent the results with a poorer correlation but valid.

The t-test is the most appropriate test for comparing the performance of two dissimilar materials, it is a simple test of significance; however, where more than one variable is used, the ANOVA (analysis of variance – see chapter 9) test is more suitable.

The conclusion should basically be a summary of your findings and what they mean. It would also be appropriate here to suggest how your measurements could be improved upon and how future work in this field might proceed (better experimental design, other studies which might be carried out related to your work).

6.7.1 Conclusions

Conclusions contains a brief summary of the experiment and its objectives, and most importantly a review of all the major values determined (include error bars) compared to literature values. Review any other major conclusions (e.g. identity of unknowns).

6.7.2 Literature references and citations

It is important to adequately reference all the literature sources that you used in preparing the report. There are many ways of providing a reference list of sources used within the research. The assessor/publisher will determine their preferred style. It is always worth checking this out prior to the writing up. Incorrect citations amount to plagiarism and therefore must be inserted with great care into the text. Paraphrasing an author is considered better than using direct quotes for text.

6.7.3 Appendices

Include risk assessments and COSHH data, graphs, tables and figures. Include sample calculations here.

6.8 Health and safety in the laboratory; COSHH and risk assessments

A health and safety questionnaire should be completed to identify any possible issues that may adversely affect the person carrying out the laboratory work. This should be checked by a senior manager and supervisor. Prior to any laboratory work a risk assessment must be carried out by the person who will be carrying out the work. No other person can assess the operative's capability. Skin conditions, asthma, back and joint problems must all be identified, and safe practices adopted prior to starting work. The assessment must be approved by the laboratory manager and the tutor supervising the student. Once signatures have been obtained, the work may progress. The danger posed to operatives by using toxic or caustic materials must be balanced against the likelihood of exposure to them and the effects of the exposure in terms of duration and damage caused. If it is possible to avoid any practices that involve risk, then this is the first line of action. If that is not possible, then appropriate personal protective equipment must be worn and suitable measures put into place to ensure a safe working environment. It is always good practice to have a technician available for first-aid and manual support. It is not recommended that the student works in isolation, as the laboratory can be a relatively dangerous place.

6.8.1 Health and safety induction

Prior to being allowed into the laboratory, a health and safety induction is needed, and the responsibilities of laboratory workers and students are as follows:
   All students are responsible for:

(a)   following all applicable safety rules and practices as outlined by the tutor, supervisor or laboratory manager; lifting with a straight back and clearly identifying the weight of the test specimens
(b)   using and wearing personal protective equipment according to manufacturer's instructions
(c)   reporting all incidents to the laboratory supervisor
(d)   reporting all unsafe conditions to the laboratory supervisor.

6.8.2 Housekeeping

While working in the laboratory, the following housekeeping should be adopted:

(a) Work areas must be kept clean and free of obstructions.
(b) Stairways and halls must not be used for storage. This applies to both equipment and personal property.
(c) Walkways and aisles in laboratories must be kept clear.
(d) Access to emergency equipment or exits must never be blocked.
(e) Equipment and chemicals must be stored properly.
(f) Spilled chemicals must be dealt with immediately and, if safe, cleaned up by the chemical user.
(g) Waste must be placed in appropriate, labelled containers.
(h) Materials no longer used must not be allowed to accumulate and must be disposed of following proper procedures.
(i) Tools and equipment must be cleaned and work areas swept when using cement based materials.

6.8.3 Health questionnaire for students

Appendix B4 includes an exemplar health questionnaire for students. Note this is 'health', not 'health and safety'. Your university should have a similar template. It is really important that you complete the document; that is true whether or not you have any medical issues of which the university should be aware. Accidents do happen – albeit rarely – in laboratories and when undertaking field work. It is clearly very important that people who may need to help you or be aware of your limitations in helping others are aware of important medical history; otherwise unnecessary risks may arise.

6.8.4 Example of risk assessment for laboratory work

When undertaking any laboratory activity, there is always a degree of risk due to the nature of the work being carried out. Identification of risks ensures that they can be managed in a safe manner. For instance, using guns to test concrete for impact/shell resistance is a very high risk test. Expertise should be sought from specialists in this extreme instance, to manage the risk. The management of this situation will probably entail using standard operating procedures from the gun club, army or similar, and building upon these procedures to ensure that the test required can be safely concluded.

Work that is still of high risk but that may not have possible fatal consequences can normally be assessed by the operator and measures put in place to mitigate the risks. Assessing the severity of the process against the likelihood of injury is a key consideration.

Figure 6.3 is colour coded to assist the decision-making process, the traffic light system provides a clear focus as to where the risk resides and whether or not further measures are needed to improve the safe working practices. Once the risk assessment is complete it should signed off by all parties undertaking the work, the laboratory manager should countersign the

	Fatality	Amber	Red	Red	Maroon	Maroon
	Major injury	Amber	Amber	Red	Red	Maroon
Severity	Reportable injury	Green	Amber	Amber	Red	Red
	Lost time injury	Green	Green	Amber	Red	Red
	Minor injury	Green	Green	Green	Amber	Red
		Improbable	Remote	Possible	Probable	Likely
		Probability				

Risk assessment (safety)

Module name: Laboratory work	**Module number/test number**
Activity: Mixing and placing of concrete and handling of materials	**Location: Room C0006 (Lab)**
Method statement + risk assessment made by: include student number	**Date:**

Revision A	**Not yet reviewed**	**Not yet amended**	**First issue**

Figure 6.3 The risk assessment grid and template.

risk assessment and if it is very high risk, then it should be forwarded to the health and safety specialists in your organisation and senior management for approval. If in any doubt, ask for advice and management approval.

Figure 6.4 outlines risks and procedures taken to ameliorate risk. The grid shows the lowering of risk due to safe procedures and the wearing of protective garments. The risk assessment is not a comprehensive overview, and each laboratory experiment will require a bespoke overview and analysis to deal with risks in a satisfactory manner. One size does not fit all, and a new look at an old procedure may show how things change over a period of time. It is time well spent to protect operatives both immediately while the work is in progress and in the longer term with regard to residual health issues, such as breathing in dust.

6.9 Role of the supervisor

A dissertation or project is essentially student-led, and the supervisor is there to provide salient questions for the student to answer and to act as a sounding board. The supervisor treads a fine line between monitoring the progress of the work and advising the student to avoid pitfalls that may be damaging to the overall investigation. This is particularly relevant considering that there may not be time to manufacture and cure twice in the case of concrete. Right

Hazard	Person(s) at risk	Risk level	Control measures	Residual risk
Cement dust/aggregate from the mixer while mixing	Any persons using the electric mixer	MEDIUM (Possible × Lost time injury)	Eye and airway protection to be worn when mixing the concrete Exclusion zone formed around the mixer with restricted access	LOW Improbable and lost time
Skin contact with cement	Any persons that come into contact with cement	MEDIUM (Possible × Lost time injury)	Long-sleeved top and gloves to be worn when handling cement Use a barrier cream	LOW Improbable
Skin contact with mould release agents	Any persons that come into contact with the mould release agents	MEDIUM (Possible × Lost time injury)	Long-sleeved top to be worn when using the mould release agent Waterproof non-absorbent gloves (nitrile) to be worn when using the mould release agents	LOW Improbable

Figure 6.4 Detailed analysis of laboratory procedures.

first time should be the mantra to be adopted. How is this to be achieved? What happens if the student embarks upon a project that is completely out of their taught comfort zone? Firstly should this be permitted? If it is not permitted, then do we investigate what we already know, as this is very stifling with regard to personal development. Most construction management degrees and architectural technology degrees do very little mathematically, when compared to engineering degrees. A technical project can have a broad maths base, and the prior learning very often does not equip students for this leap into the unknown.

From the author's experience, most students are not stretched to their intellectual limit and self and supervisorial belief in their ability to solve complex numerical problems tends to promote student success. Tell a student that they 'can do it' and they will do it. There is nothing like rising to a challenge with good supervisory support to allow a significant degree of personal and career development. Employers must differentiate between job applicants with similar academic degrees, and the dissertation or project allows students to make a statement about themselves. The construction industry needs more technically based employees, and a technical project may be the factor that tips the balance in favour of employment for the student.

The ethical issue is, how much freedom does a supervisor provide to a student to make strategic decisions? This is pertinent if the supervisor can identify a serious problem with the process. Does the supervisor allow the student to fail and have to rethink the work within a very tight timescale? The alternative is to suggest a rethink before the failure occurs. At the end of the day, the document is owned by the student, and the supervisor's role is only to provide good advice which may be accepted or rejected by the student. This situation may require a supervisor with specific knowledge of the subject area. This brings another question of whether or not the student can select a supervisor, or will this be carried out independently with a need to match research to supervisory skills? Supervisors generally have a personal web page. It would be worth students reading these with a view to understanding the strengths and weakness of the person supervising their work.

6.10 Possible research topics for technical dissertations or projects, construction and civil engineering

The range of possible research topics is immense. Favourite themes come and go. New material developments provide a whole new range of comparative tests by comparing the old established material with a new replacement. The question to be addressed is 'Where does a construction civil engineering project become a pure engineering investigation?'

A lightweight steel framed building is composed of various elements bolted together. We can test the effectiveness of an element under load, the performance of the bolt connections or how the frame performs under vibration of earthquake style loading. This is construction with an engineering element.

Recycled timber has huge potential for testing. What effect do nail holes have? Are the timbers notched from previous uses? Is rot present? How far will water penetrate into the timber? What effect will water have upon the timber strength?

Soil investigations offer a student a whole range of possibilities including ground stabilisation, cob wall construction, rammed earth wall technology, microbiological and chemical stabilisation of ground to improve the load bearing capacity.

Structural analysis of building components will depend upon the size of the testing equipment at the lab, but tests can be scaled down. Composite beams made of concrete and carbon fibre, timber and carbon fibre, steel and carbon fibre, or replace the carbon fibre with para-aramid fabric that may or may not be recycled material.

6.11 Examples of research proposals

For research students to obtain an idea of what may be possible for them to investigate, it is strongly recommended that they read conference proceedings in the relevant area of interest. The range of potential tests is enormous and covers all commonly used materials in all situations. It is also recommended that potential researchers examine university websites for ongoing PhD work. Trade magazines will highlight new product development, and this may spark ideas for a new use of existing materials or a new angle of a material's relative performance when compared to other materials that may be used for similar functions. Technical investigations can take many forms and do not have to be material testing. Soils, hydrology and surveying offer many topics that have the potential to produce first class investigations.

There are many opportunities to research hybrid materials such as concrete and carbon fibre, para-aramid and concrete or timber. Other areas worthy of investigation with concrete are cement replacements, aggregate replacement and additive effects with regard to sustainability. A very interesting development is the use of ground-borne bacteria to change the properties of stone and concrete as well as ground conditions. The list goes on: vinylester rebar to replace steel since it is twice the strength of steel. One of the oddest investigations undertaken by the author was to add cigarette filter material to concrete as a plastic shrinkage restraint. It worked! These research examples are discussed in detail in the following section.

6.12 Research objectives and sample findings by the author

This section examines possible research objectives and how the results were displayed for the reader. Many ways of displaying data are provided and these range from tables to scanning electron microscope images. Consistent formatting improves the presentation and this can be difficult to achieve if different computers are used to compile the work. It is also difficult to maintain consistency when research is carried out over an extended period of time. When the work is complete and all of the information is in place, the hard part is presenting it logically and consistently.

6.12.1 Comparative performance of fibre reinforced polymer (FRP) and steel rebar

This research investigated the necessary requirements that concrete reinforcement material must satisfy: the ability to resist tensile forces and have good bond strength, while providing structural qualities of toughness and flexural strength. The required outcome of this test was to determine the relative performance of two beam types – steel reinforced and vinylester (FRP) – identifying a range of performance criteria.

Figures 6.5 and 6.6 display the relative performance of load and extension of steel and vinylester rebar.

Three main standard methods are available that analyse the post crack performance of concrete. These are BS EN 14851, ASTM 1018 and JSCE SF4 (Japan Society of Civil Engineers). The Japanese standard provides Re3 values that equate to 90% of the residual load divided into the maximum load at failure and these are used in TR 34 to calculate floor slab design. TR 34 is under revision to incorporate BS EN 14851 into the design code and new data is currently being sought regarding fibre performance.

ASTM 1018 is a method for calculating toughness of concrete, but it is no longer current. The results were in the form of dimensionless toughness indices to describe the toughness of a sample by analysing the load deflection curve. This, however, cannot easily be applied to industry, as the indices have no dimensional value to use in load calculations for manufacturers or designers. This has now been superseded by ASTM 1399, which operates a two-point loading system that provides residual loads at specific deflections. ASTM was chosen to allow comparison between this work and previous studies. This does not stop the researcher carrying out a different form of analysis to display the data with different parameters.

Load/extension curve steel r.c.

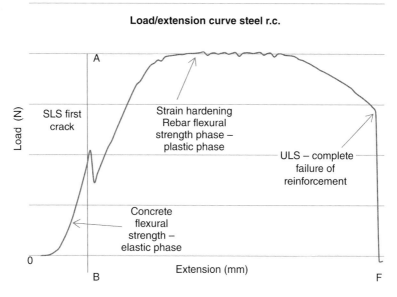

Figure 6.5 Typical steel load/extension curve.

Load/extension curve – specimen FRP

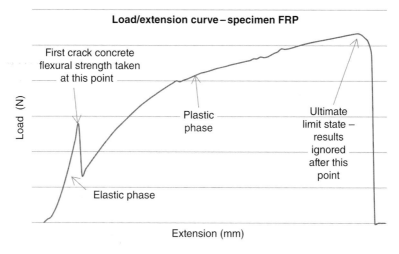

Figure 6.6 Typical FRP load/extension curve.

Table 6.1 shows toughness index values for each test specimen, with their mean values and standard deviation. Toughness can be measured using various standards. These values, as displayed in Table 6.1, follow the procedure outlined in ASTM 1018, where the area of the chart up to failure of the concrete is divided into the total area under the curve.

A parametric unrelated t-test was performed at 95% confidence to compare the variation between the mean values of the two beam types (see chapter 9 for an explanation of t-tests). The null hypothesis for this test is that there will be no difference between the toughness

Table 6.1 Toughness index tables for steel and FRP rebar.

Specimen	Toughness indices	Specimen	Toughness indices
Steel 1	38.34	FRP 1	24.7
Steel 2	37.39	FRP 2	46.06
Steel 3	24.24	FRP 3	49.95
Steel 4	33.88	FRP 4	27.27
Steel 5	26.41	FRP 5	35.86
Steel 6	18.22	FRP 6	74.55
Steel 7	19.48	FRP 7	71.79
Steel 8	23.45	FRP 8	35.22
Steel 9	26.41	FRP 9	37.11
Steel 10	28.2	FRP 10	48.15
Steel 11	24.19	FRP 11	66.56
Steel 12	37.65	FRP 12	52.23
Steel 13	29.28	FRP 13	50.87
Steel 14	22.65	FRP 14	50.79
Mean	27.25	Mean	47.93
Standard deviation	6.81	Standard deviation	15.34

indices of a steel reinforced beam to that of beams with FRP reinforcement bar. The null hypothesis was rejected. We can therefore conclude that the performance between the steel and FRP rebar beams was significant. The FRP samples with a mean toughness index of 47.93 are better than the steel specimens with a mean toughness index of 27.25. The discussion section of the dissertation or project will comment on the importance of this difference of over 20 index points and its potential for industry. There should also be discussion about whether the standard deviations are within acceptable limits. Standard deviation is by and large dependent upon the sample size. As the sample size approaches 30, a more representative standard deviation is obtained. The actual value for a standard deviation for concrete is in the region of 8 if it has not been calculated from test results. Concrete can display large standard deviations if the sample size is small. It would be worth setting a limit for outliers of 2 times the standard deviation and removing them from the test results as they may be due to operator error, especially if the test is a bespoke test such as the one displayed in Table 6.1. Judgement may be required to represent the findings in an unbiased manner depending upon the test design.

There are resource constraints with this work to minimise the likelihood of variation in the test material. To carry out this test with minimal variations to the material as tested, 28 steel beam moulds were needed to be available at one time. Does the lab have that many? Is there a concrete mixer large enough to batch this in one batch, thus reducing the variables associated with multiple batching?

6.12.2 Equating steel and synthetic fibre concrete post crack performance to BS EN 14651:2005+A1:2007

This paper examines the relative pull-out values of two single fibre types, being steel and Type 2 synthetic fibres. The pull-out test results have informed the doses of fibre additions to beams that have been used to equate near equal toughness performance for each fibre type.

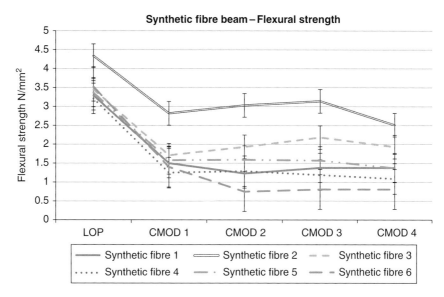

Figure 6.7 Synthetic CMOD – Comparison of individual beam performance, showing error bars.

The results show that synthetic Type 2 fibres, when used at a prescribed dose can provide equal toughness to steel fibre concrete. The results are essentially the same as the steel and vinylester rebar comparison, although a different standard was chosen to represent the findings. The beams as tested in this standard are 150 × 150 mm in cross-section and represent a commonly specified floor slab thickness. The logic for use of this standard was justified, but the health and safety issues associated with handling heavy beams may be a serious constraint with progressing the work, as assistance may be required to carry out the tests.

The BS EN 14851 method, as chosen for this study, uses the load versus crack mouth opening displacement (CMOD) method to give flexural strength values at multiple points of deflection, as well as a value of the limit of proportionality (LOP) at first crack. This allows application of a dimension to results that are usable and easily understood by designers in industry when wishing to control structural and thermal cracking in structural concrete.

Figures 6.7 and 6.8 display the flexural strength values and various crack mouth opening values measured on the base of the beam. These values are very useful to design engineers.

6.12.3 Concrete with crushed, graded and washed recycled construction demolition waste as a coarse aggregate replacement

This research was considered necessary due to current trends in using waste and by-products in concrete which replace binders and aggregates. This trend reduces the impact on the environment and the use of finite natural resources. The research investigated whether or not concrete that included crushed, graded and washed recycled construction demolition waste used as a coarse aggregate can be manufactured to a comparable strength to concrete manufactured from virgin aggregates. Figure 6.9 displays the relative performances of strength

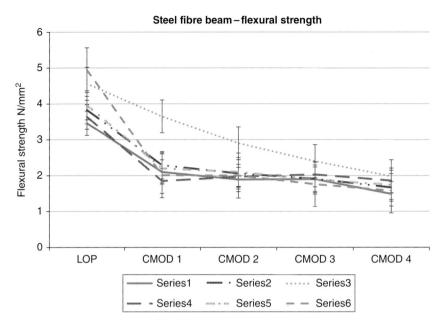

Figure 6.8 Steel CMOD Comparison of individual beam performance.

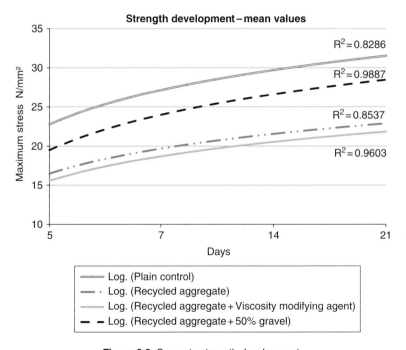

Figure 6.9 Concrete strength development.

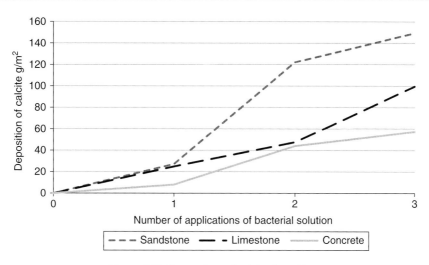

Figure 6.10 Comparison of calcite deposition.

development using different design mixes and comparing the recycled aggregate performance with the virgin aggregate. Each value represented was taken from two cubes, and the average value plotted, the basic requirement used 32 cubes for this test. Using two cubes is a high risk strategy as there are no values to discard if we were to find an outlier. Conclusions can be drawn from this data representation.

The optimum concrete mix design using recycled construction waste was obtained by using a 50–50% mix of virgin gravel and recycled aggregates. Using recycled construction waste as a 100% coarse aggregate replacement produces concrete with a lower compressive strength when compared to concrete made with virgin aggregates. The results are displayed using a log scale for curing – this choice of scale will be informed by the background reading carried out during the literature review. The initial results, when plotted, may well display a linear relationship, but the background reading will inform the researcher of the correct method of displaying the findings where the sample size may produce skewed results.

6.12.4 Surface coating of traditional construction materials using microbially induced calcite precipitation

This research investigates the effect of microbially induced calcite precipitation as a natural treatment for the surface preservation and restoration of historic buildings. The data represented on the deposition chart is supported with scanning electron microscope images to inform the reader. Two stone types and concrete were examined to determine the relative calcite deposition, as displayed in Figure 6.10.

The researcher's considerations are many within this type of research as it may span many different departments. For example, this work spans biology, chemistry, geology, construction and surveying. Background reading could include petrography, history and applied science required to produce the bacterial cultures. A researcher's question that must

Figure 6.11 Calcite formation on the surface of sandstone.

be addressed is 'What budget do I have for this work?' The cooperation between different departments is essential if the research is to be successful. Written approval is needed from the head of department to ensure that all concerned know what is expected of them. A Gantt chart is a very useful tool to show when services are required and to identify key milestone events. This can be easily distributed throughout the research team for effective collaboration.

Figure 6.11 identifies the calcite formation on the surface of sandstone when compared to an untreated sandstone control sample (3000× magnification). The reader can observe the difference in texture and the amount of deposition.

Historic preservation of stone buildings is possible through the application of an external surface treatment to seal micro cracking, coat the surface of the material with calcite and to cement loose surface particles to the firm substrata using microbially induced calcite precipitation (MICP). This treatment is displayed in Figure 6.12.

MICP may be considered to be an environmentally low impact method of surface treatment capable of coating and the re-cementing the surface layer of stonework. Calcite precipitation occurs in microbial metabolic activities in the presence of a nutrient rich alkaline environment. The simplest mechanism for MICP, used by bacillus type bacteria, is through the hydrolysis of urea by the enzyme urease. Calcite can also be induced without the need for an organic producer. Ureolysis can be induced in a laboratory environment by adding urea to the urease enzyme (urea amidohydrolase) which is produced by many organisms.

This research used measurable mass gained from the treatment and photographic evidence to display the findings. The image as shown in Figure 6.12 is marked up to guide the reader to relevant important parts of the evidence that help the findings to be easily understood. The greater the number of links that can be provided to show a trend, the better understanding will be achieved.

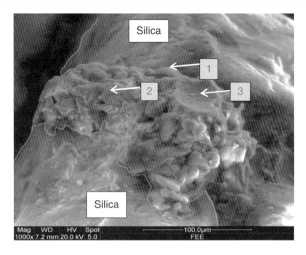

Figure 6.12 Calcite formation bonding sand grains 1000× magnification.

6.12.5 Pull-out performance of chemical anchor bolts in fibre concrete

This work has examined the pull-out force required to cause cone shear failure to structural concrete when resin anchor bolts were subject to an axial load. Anchor bolts were fixed in unreinforced plain and fibre concrete of different fibre dosage and type, for comparisons to be made. The findings show that increased loads can be transferred to the concrete with the use of fibre technology, where fibres are used to reinforce the concrete matrix. When the quantity of steel fibres was increased there was a corresponding increase in the pull-out values of the bolts. Figure 6.13 portrays the nature of the test to the reader. A well drawn clear diagram instantly conveys the nature of the test to the reader, showing what process is being carried out.

Finite element analysis was carried out to assist in determining the slab size for the pull-out test, as shown in Figure 6.14. Using mathematics, the finite element method (FEM) is a numerical technique for finding approximate solutions to boundary value problems for partial differential equations. It uses small areas examined to provide a larger picture of what may occur. Software is available for this purpose, and the time spent is considerable, even using a computer. Once the basic parameters are entered for the component parts, differential and algebraic equations are used to generate a holistic picture.

The stress pattern was analysed to ensure that the fracture plane was contained within the area of the proposed test specimen. This method of pre-manufacturing analysis has the benefit of reducing the likelihood of wasted time, materials and money due to the unsatisfactory failure of a test specimen. This design aspect is of particular importance when an undergraduate dissertation or project is being undertaken.

The results as displayed in Figure 6.14 show an optimum steel fibre dosage of $40\,\mathrm{kg/m^3}$, as the greater cone diameter equates to a greater pull-out force within the same base material. The data is presented in a clear, unambiguous manner that is easily understood by the reader. It is the author's belief that the easier the findings are to read, the more likely that the reader's interest will be maintained, and this is of particular importance for a document or project where the reader/marker may have to read more than one document in one day.

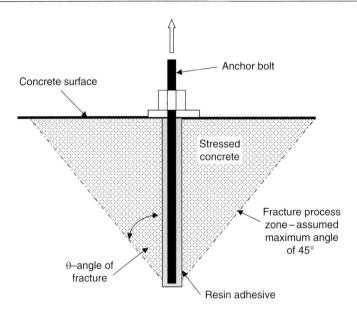

Figure 6.13 Section through anchor bolt under pull-out load.

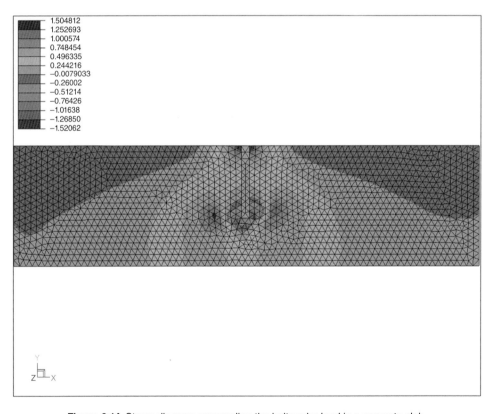

Figure 6.14 Stress diagram surrounding the bolt under load in a concrete slab.

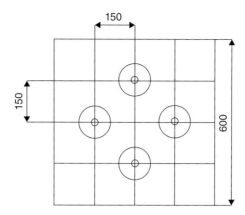

Figure 6.15 Plan view of pull-out test slab (dimensions in mm).

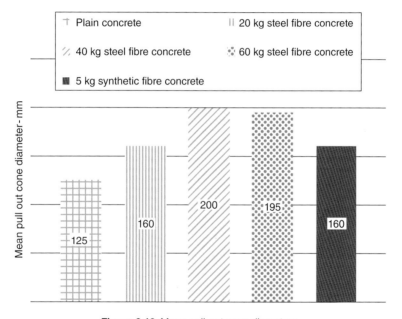

Figure 6.16 Mean pull-out cone diameters.

Four bolts were cast into a single slab as shown in figure 6.15, and this provided the mean pull out values as displayed in figure 6.16 for each concrete type. The finite element analysis of the pull-out shape permitted the concrete pull-out sample to be made to a minimum size, as the fracture zone was estimated with reasonable accuracy.

Bibliography

Coventry, K., Richardson, A.E., McIntyre, C. and Aresh, B. (2011) 'Pull-out Performance of Chemical Anchor Bolts in Fibre Concrete', Fibre Concrete 2011– 6th International Conference, September 8/9, Czech Technical University, Prague, Czech Republic.

Richardson, A.E., Coventry, K.A. and Wilkinson, S. (2012) 'Freeze/thaw durability of concrete with synthetic fibre additions', Cold Regions Science and Technology, Volumes 83–84, December 2012, pp. 49–56 Elsevier. doi: http://dx.doi.org /10.1016/j.coldregions.2012.06.006 Accessed 18.12.2015.

Richardson, A,E., Coventry, K,A., Forster, A, and Jamison, C. (2014) 'Surface consolidation of natural stone materials using microbial induced calcite precipitation', Structural Survey, Vol. 32 Iss: 3, pp. 265–278.

Richardson A.E. (2012) 'Cigarette filter material and polypropylene fibres in concrete to control drying shrinkage', 8th International Conference: Concrete in the Low Carbon Era, CTU, Dundee, UK, 9–11 July 2015.

Richardson, A.E. and Jackson, P. (2011) 'Equating steel and synthetic fibre concrete post crack performance', 2nd International Conference on Current Trends in Technology, NUiCONE, December 8–10, Nirma University, Ahmedabad, India.

Richardson, A.E. and Drew, P. (2011) 'Fibre reinforced polymer and steel rebar comparative performance', Structural Survey, Vol. 29 Iss: 1, Emerald Group Publishing Limited, pp. 63–74.

Richardson, A.E., Allain, P. and Veuille, M. (2010) 'Concrete with crushed, graded and washed recycled construction demolition waste as a coarse aggregate replacement', Structural Survey, Vol. 28 No. 2, Emerald Group Publishing Limited, pp. 142–148.

Richardson A. Editor, (2013) 'Re-use of materials and By-products in Construction, Waste minimisation and re-cycling', Green Energy and Technology, Springer Publishing. (Richardson chapters 1 and 3)

7 Qualitative data analysis

The titles and objectives of the sections of this chapter are the following:

7.1 Introduction; to provide the context for collecting qualitative data
7.2 The process of qualitative data collection
7.3 Steps in the analytical process; to provide a template

7.1 Introduction

Qualitative analysis enables us to better understand the world, and specifically the people within it. It uses words to seek out meaning and explain the *why* to quantitative research's *what*. Qualitative research looks for patterns of relationships between categories, known as themes, and draws on rich data which aims to get insight into how participants view the world. Qualitative research is not an 'easy option' when compared to a quantitative approach – even if the lack of numbers is appealing. Significant amounts of time and effort will be required to carry out the process rigorously: to schedule appointments to gather data, to transcribe it if necessary, to carry out the coding process and then finally carry out the analysis. When using unstructured interviews to collect data the collection process may be relatively quick, but the transcription, coding and analysis will take a considerable amount of time.

As a qualitative researcher, the application of social science to the study and improvement of contemporary life depends upon the intimate understanding of participants (McCracken, 1988). Participants may have difficulty in giving an answer; they may labour to articulate responses, as you may ask questions on subjects they have not previously considered or reflected upon. You may seek to gain access to cultural values held by participants in a certain category, such as craftspeople on construction sites. There are lots of differences in life that need to be teased out, so that diverse societies can have an appreciation of others. In construction, this may be craftspeople seeing the world differently from site managers, or younger workers differently from older workers, or contractors differently from clients, or operational managers differently from strategic managers. There is the danger that you will go into your

Writing Built Environment Dissertations and Projects: Practical Guidance and Examples, Second Edition.
Peter Farrell, Fred Sherratt and Alan Richardson.
© 2017 John Wiley & Sons, Ltd. Published 2017 by John Wiley & Sons, Ltd.
Companion Website: www.wiley.com/go/Farrell/Built_Environment_Dissertations_and_Projects

study with blinkers on; the problem is only as you see it. If you undertake some qualitative work speaking to others, even if it is only modest, it should help you take those blinkers off.

What is discovered in qualitative work cannot be assumed to exist widely. It does not seek to survey the terrain. You will *not* make the inference that what you have found in your very small sample exists in the population. Only quantitative methods with large sample sizes can make inferences. Interviews are used as a substitute for observation: they have limited validity. Interviewees may tell you what they think you want to know, or may speak to show themselves off in the best light. If you want an insight into their practice, observation may be better. However, observation may not be possible; it is potentially time consuming and you may not get permission to do it, other than perhaps if you are a part-time student undertaking action based research.

7.2 The process of qualitative data collection

Interviews are often the data collection technique used in qualitative data analysis. However, qualitative tools can also be applied to data collected in other ways, such as case studies, focus groups, reviewing transcripts of speeches, transcripts of law cases or other literature in various forms. This section assumes the use of interviews, but the analytical process of ordering and coding is applicable to data collected by other approaches.

If you are careful in selecting potential people for interviews, response rates to requests should be in the region of 70–80%. Always follow the ethical code of your institution; provide participants with an information sheet in advance of your interview so that they can have time to digest its contents. Prepare carefully and do not offend participants or breach confidentiality or anonymity. Do not be too ambitious in the level of seniority of people you ask. Interviews do not have to be face to face. It is quite reasonable for telephone or virtual methods such as Skype to be used. A long telephone call is less resource intensive than travelling long distances, though you should pre-book the call with participants. Interviews may be exploratory or may be used as the only method of collecting data. Students often ask 'How many people should I interview for in-depth qualitative work?' There is no definitive answer; it depends on the context of your objective, the precise type of analysis you propose and whether interviews are your only method of data collection. If it is the only method, perhaps eight may be a guide figure for you to start thinking around. You may justify more or less, but since this is an important element of your study, speak to your tutor.

If a qualitative approach forms the main data collection and analytical tool used, with permission of interviewees, ideally they should be digitally recorded. Subsequently you should type up the interviews into verbatim transcripts and follow the detailed analytical process as described in this chapter. However, if you are only using the data to support quantitative work, it is quite reasonable at undergraduate level to just take field notes of your conversations.

Interviews may be considered on the continuum of structured, semi-structured and unstructured. Structured interviews could almost be conducted as though it were an electronic survey. There are no opportunities for follow-up questions or probing, although clarification may be given if the interviewee does not understand something. Structured interviews are moving away from the qualitative domain to the quantitative. Answers to questions may be limited to a number of closed responses that will be converted into quantitative data. Since the proposed analysis will be quantitative, the number of people interviewed may be higher.

The reason for using such interviews is to improve response rates. For example, if your objective were to determine the attractiveness of energy saving technologies to potential home buyers, you may target a sample by asking permission of estate agents to conduct short, structured interviews with members of the public who are browsing their sales stands.

Semi-structured interviews may have a list of definitive questions, but with some pre-prepared possible probe questions and the licence to probe still further and ask for more detail if required. Probing is a technique used to get more information when a response is unclear or incomplete. Probes include gestures such as nodding or neutral questions: 'Can you tell me more?' The answers you get may not answer the question; not that participants are being evasive, as sometimes happens with politicians, but possibly they just misunderstand the questions.

In unstructured interviews you should have some pre-prepared themes that you wish to explore, but the precise nature of questions may be adapted to suit the individual interviewee and the answers you receive. You will derive each question based upon answers you receive from preceding questions; you may go into areas that you could not reasonably have envisaged before the interview, as new lines of enquiry emerge.

As part of your interview preparation, be clear about its 'purpose'. It could be:

(a) at the early stage of your research, to help you explore or define the problem to be investigated
(b) at the early stage, before or after your theory and literature review, to help you set the study objective and define variables
(c) as part of a piloting process
(d) as part of data collection for the main part of your study
(e) at the closing part, to help you interpret findings and develop discussion and conclusions.

Plan the interview structure and questions. Be clear about the generic themes (or objectives or variables) that you wish to explore, and about whether you wish to tease out the breadth or depth of a problem area. Ensure that all your questions are directed towards the themes; there should be no superfluous questions, although inevitably there will be some general introductory questions. You must also ensure that you comply with your university ethical code, and the questions should include confirmation that details in the information sheet are clearly understood. Try to ask questions that will generate long answers, rather than one-line or even single word answers. If you do receive short answers, probe further. Prepare some potential probe questions. Number each question Q1, Q2, Q3 etc. Number any planned probe questions, e.g. two probe questions following Q1 as Q1a and Q1b. In the interview; try to relax, but be alert in your thinking.

To do qualitative work in interviews properly, they need to be recorded digitally; perhaps on your telephone or other mobile device. Be aware that you need to be careful in terms of data management and be sure to delete the original recordings once you have moved them onto your computer. Try to pre-advise that you wish to use digital recording; some interviewees may be reluctant. If they are reluctant, you must respect their wishes. You will no doubt still get valuable data by just taking field notes. The disadvantage of mere notetaking is that in the context that you are trying to collect rich data, you want to be sure that you secure every single word, including intonation. You may be less relaxed and it may take your concentration away from thinking about potential probe questions. Also, you are not able to observe facial expressions that may be telling you more than just the words. If you do only take notes, it is important to write them up in full, ideally on the same day while the mind is fresh. If you have been able to record, you can also make some brief notes during interviews as memory jolts.

Audio data should be transferred to written form verbatim – the creation of interview transcripts. This makes the data easier to manage. There are different ways of transcribing, from simply typing the words spoken, leaving out the 'ers' and 'ums' that you will find pepper everyday talk, to trying to include all pauses and intonations within the speech. A form of this latter type of transcription is known as the Jefferson system (2004). This method of transcription captures not only what was said but also *how* it was said, using standard conventions to represent things like emphasis, overlap of speech, pauses or intonation. This method is very time consuming and only necessary if you are looking to carry out fine-grained content or discourse analysis, where how things are said becomes a vital part of the data. Usually you will need to transcribe word for word to gather the data you need for standard qualitative analysis. However, you can include your own observations at relevant parts; these may be about facial expressions or opinions expressed with passion. Transcripts should include all colloquialisms, slang and anecdotes, but you may wish to dash out any bad language. Be mindful that for every one hour of interview, the writing-up process can be as much as five hours. Imagine the difficulty in playing back the recording for a few seconds, then stopping the tape while you write, then re-winding to replay to ensure your writing is accurate. There is no reason why you should not obtain some support from friends, family or elsewhere to undertake this task, providing you acknowledge the help you have had in the preliminary pages of your document, and that you personally check that the transcripts have been correctly transcribed. Look at the possibility of using free or low-cost voice recognition software or apps that will convert speech to text, such as Dragon Dictation. That may improvee the one to five hour ratio.

The interview questions and interview transcripts will form part of the appendix of your document, and therefore you need to ensure that you comply with your university regulations on ethical procedures and data management. Do not name people interviewed, or give indication of their employer. Person A may be, for example, identified as an architect working for a medium-sized design practice in the south-east of England.

Your interviews may normally be of 30 minutes' duration, though longer interviews may sometimes be justified. You need to respect the time constraints of the interviewees, who are no doubt busy people. Try to be persuasive and informative at the introduction to the interview and then to create smooth conversational flow. At the outset, you will need to set the context for the interview, and explain what you hope to achieve from it. You may wish to ask some personal questions about a interviewee's background, but do not ask for data that may be unnecessarily intrusive.

7.3 Steps in the analytical process

The analytical process is to take the raw data at the beginning and convert it, using a step-by-step process, into a final narrative. The narrative summarises, in a essay form, the literature and the raw data, in preparation for conclusions and recommendations. Busy people such as external examiners, or in an industry context chief executives, may be very interested to read this narrative summary; they may not even have time to browse all the detailed analytical work that goes before it such as verbatim transcripts of interviews. You must therefore write the narrative very carefully. The process described here is qualitative thematic analysis (Seale 2004, p. 314). The data is collected, transcribed, coded and analysed to reveal the dominant themes found within it. It then enables a narrative to be developed that tells the story of why something is like it is, bringing insights to the research subject area.

The following detailed description is aligned to the example in Appendix E. It will help you to refer to that appendix regularly throughout this chapter for examples of the steps in the process as they are discussed. It originates from work by Stott and Farrell (2011). The study aim is to determine whether there is a propensity to put completions before quality within the speculative housing sector. The study is founded on a problem whereby speculative housing builders (or the private housebuilding sector, PHS) can be placed under pressure by their banks to maintain good cashflow. Imagine a situation where a buyer agrees to complete for a house on a Friday and thus the house builder eagerly anticipates receiving £xxx,000s on that day. On the Monday of the same week, it is realised that there is still a lot of work to do, so the house is completed in a rush, leading perhaps to some defects that can only be rectified after the buyer has moved into the property. The study objectives can be found in Appendix E1. The data collected for this study was interviews with ten housing site managers. The questions asked in the interview can be found in Appendix E2. The size of transcripts from such interviews can be large. For brevity, Appendix E does not contain all the data from this study, rather illustrative examples are included as noted.

Data management

The first step in the process is to ensure good data management. Following the data collection, you should now have typed verbatim transcripts of files of your interviews. Appendix E3 contains two verbatim transcripts of interviews with site manager A and site manager B. The decision on how you will manage this data during the analytical process needs to be made now. You could do this manually by photocopying, then cutting and shuffling of pieces of paper. Alternatively, sophisticated computer software could be used, such as QSR International XSight or NVivo, which allow you to upload your transcripts. They can then help you structure the analytical process and produce reports demonstrating what you have done. If you are able to master the use of such software, you should be given credit for it in your mark, but do not let your use of the software overtake the need to immerse yourself in the data and develop a narrative and conclusions with insight. A mid-way approach is to use standard word-processing software, such as Microsoft Word, where you can cut and shuffle electronically. The example given in Appendix E has used Microsoft Word to manage the data. When using an electronic system you must always remember to regularly save your work and back it up carefully.

As you sort your data, you need to ensure that you are able to identify the origins of each transcript and speaker. Within data management software, the computer will do this for you as you upload separate files. If you are using Microsoft Word, you can number your data using a label comprising the interview number, the question number and prompt. Label each answer (Ans) with numbers/letters, e.g. for person A, A-Ans1, A-Ans1a, A-Ans1b, and for person B code each answer B-Ans1, B-Ans1a, B-Ans1b, etc. There is no necessity to label the questions such that they can be traced to the original source, since we will later delete the questions. If you ask probe questions spontaneously in the interview, and thus these questions will not be asked to every interviewee, to distinguish them from pre-prepared probes, start marking them with labels in the second half of the alphabet, e.g. Q1n. Add your own observations from notes you made in the interview; identifying these notes as interviewer's observations, e.g. IntOb1. Numbers for these observations can run numerically across all interviewees. In keeping with this approach, in Appendix E3, Interviewee B's response to question 2c has been labelled B-Ans2c. Alternatively, to simplify the codes, you may use coloured text to distinguish between transcripts for each person; that would mean that instead of a code such as A-Ans1, it could alternatively just be A1 and the transcript is in red text. The transcript for person B to be in blue etc. If you do this, and include transcripts in appendices in your document, be mindful to use colour printing.

Bringing the data together

Once the data within each transcript is in manageable form, you can start to bring it together. Within data management software this will happen as you upload the individual transcripts, making a database for your project. In Microsoft Word, you can now cut and paste all transcripts together into one file.

Content analysis

The simplest form of analysing qualitative data is content analysis; a word count of regularly used key words or phrases. You can do this using the 'find' function in the software you are using; type in a key word or phrase that you observe occurs frequently. Alternatively this must be done by hand. Count and record the number of hits you receive for each; ignore any hits in questions. You may decide that this will be the limit of your analysis. If that is the case, proceed to your narrative write-up, flavouring it with the number of hits you make. For example, transcripts of interviews with site managers may get many hits for 'safety', but few hits for 'money'; the narrative write-up will reflect this balance. Alternatively, this content analysis may be the first part of more detailed analysis, once the data has been coded.

   The frequency counts of some key words for the two site managers within the example data in Appendix E3 is:

Customer/s	33
Subs/subbies/subcontractors	25
Quality	24
Busy	13
Defect/s	12
Snag/s	12
Money	8
Rush	7
Weekend	7
Safety	5
Building inspector	3
Complaints	2

Coding the data

The first stage is to break down or build up the transcripts up into manageable paragraphs of perhaps two to five sentences each. Place a paragraph break where there may be a change in direction by the interviewee. If there are short paragraphs that originally resulted from short answers to main questions, you may merge those answers with the answers given to the probes that you have followed up with. It may be that one answer is itself a manageable paragraph. Hopefully, some answers may be longer answers, and you will break them into more than one manageable paragraph. If using Microsoft Word, ensure that the beginning of each paragraph is marked with a code, e.g. Ans1, Ans2a, Ans3n. If there has been merging of answers, e.g. Ans1 and Ans1a are merged, cut out the Ans1a from the middle of the new paragraph and mark it at the beginning as 'Ans1 and Ans1a'.

You can then start to code your data. Many textbooks point out that coding is not analysis, but whatever approach is made, coding enables you to immerse yourself in your data, and is therefore always part of the analytical process. Coding is the labelling of your data to illustrate the key themes and ideas that you can identify within it. These labels may be single or perhaps two or three words long, and abbreviated to a short code. There are two ways to approach data coding. Either you can start with codes you think may be appropriate from your literature review or content analysis, or you can develop your codes from reading and rereading the data itself, drawing on words and phrases used by the interviewees. Take a careful overview of the transcripts: What are the themes that are arising? Can you spot ideas that are consistent across the transcript data? For example, the interviewees may talk about an idea of 'culture', when asked different questions about time or subcontractor management.

With insight, allocate each paragraph a code; what is the theme that this person is driving at that can be summed up by the code? Type the relevant code at the end of the paragraph. You can use bold text, to distinguish it from the main text. If you feel a paragraph addresses two or more themes, allocate it two or more codes. Put paragraph brackets around the text to which the label is relevant. Some text may be superfluous; that is transcript from interviews that deviated from the study objectives. Do not code this data. Since this superfluous data will not have brackets around it, that will help to distinguish it from labelled data when you take your analysis further – it can be deleted.

Add new codes to your framework if you think it appropriate. Codes can emerge and disappear during the coding process, as you become more familiar with your data. Often what is not included in the answers can be as important as what is. You should aim to immerse yourself in your data; read and reread it many times during the coding process. You will get to know it very well. Keep records of your ideas while coding, and develop a final coding system, the one that you have applied to your data. This allows others to see how you have developed your coding as a process, adding to the transparency and therefore reliability of your research. You can arrange codes that have the same generic theme into groups, and give each code group a generic name, creating a hierarchy in your coding system, allowing more complex ideas to develop and emerge. Some codes may appear and disappear during the process, others may be suggested by the literature and not appear within the data. Keep notes of such things as you code, as these are the beginnings of analysis.

Our example in Appendix E resulted in 17 codes. For ease of manageability and grouping of ideas, these have been collected together into five dominant themes – each under a code label. In our example, three codes 'culture', 'hierarchy' and 'specific periods' have become grouped together into one group and named 'propensity' by the researcher. Give an abbreviation to each code, so that in later searches in Word, it does not generate any hits with the interview transcripts, e.g. PROP1, PROP2 etc. Type up your coding framework, and keep it as a record; you can also include written reflection on the process as to how the final coding scheme developed. For the example in Appendix E, the 17 codes and their abbreviations, sorted into five generic group names, are:

Propensity:
    PROP1 Culture
    PROP2 Hierarchy
    PROP3 Specific periods

Completions:
    COMP1 Workload
    COMP2 Resources
    COMP3 Supply chain
    COMP4 Communication

Quality:
    QUAL1 Defects
    QUAL2 Quality control
    QUAL3 Workforce

Profit:
    PROF1 Budget targets
    PROF2 Repeat business
    PROF3 Bonus payments

Time:
    T1 Build programmes
    T2 Working conditions
    T3 Health & safety
    T4 Sales issues

This coding scheme has been applied to the transcript of site manager A in Appendix E4.

Analysing the data

The next step is to bring like-coded paragraphs together into new files. In one file therefore there will be many paragraphs from different interviewees, but all these paragraphs have the same code, such as culture. You will then reflect with insight; 'what were these interviewees really saying about culture?' Using cut and paste or cut and shuffle, or the functions within your data management software, bring together all the data sources that have been allocated the same code. Within MS Word, save a copy of your coded data file so you have a back-up in place if needed; call this file something like 'coded data for analysis'. Then open up a new blank file, with the intention of cutting data out of your coded data file and rearranging it in the new one. At the end of the process your original coded data file will be a blank file except for superfluous data.

Have both files open at the same time. In the new file, set up a series of separate pages that are blank, and type on each page headings that are each of your codes. In our example we have 17. With the coded transcript file open, take each paragraph in order and cut and paste it to the appropriately headed page in the new file. Do not cut and paste data that is not within brackets, you have decided that this is not relevant. Keep resaving both files as you progress. If your manageable paragraph has two codes (or more; our example in Appendix E4 does not have paragraphs coded more than twice), copy and paste the paragraph twice (or more), into the new file. For example, you may have coded one paragraph with two separate codes 'PROP1' and 'COMP1'. Continue with this process, until your coded data file is blank except for superfluous data, and your new file contains paragraphs arranged under 17 codes. Recognise that this new file is made up of paragraphs bolted together in a clumsy fashion that do not flow in the style of an essay. Delete all the labels at the end of each paragraph.

Now start to review your coded data, read and re-read the content of your codes, which can now be termed themes. What are the aspects of each theme or sub-theme? Refer to your notes made as you coded, and developed the new thematic headings you are now working under. Do all the statements under your themes agree? Or do they challenge each other? Is something striking in its inclusion or absence? How do the themes relate to the literature? Can you spot ideas that are inconsistent? Do people agree or not? Do people focus on one aspect of a phenomenon or take it to mean various different things? Answering these questions can become a complex and thought provoking process. Data management software will help you make notes and bring ideas together as you need to. In Microsoft Word, an easy way to manage it is using a table format. An example of such a table can be seen in Table 7.1. This is one table of five taken from Appendix E6.

The data in the column 'Frequency counts' can be inserted from the earlier initial analysis made of the data, or by revisiting your complete data file or database and carrying out a retrospective process, now that the codes have been developed. For our example, this gives counts of: PROP1 = 11; PROP2 = 15; PROP3 = 6; COMP1 = 16; COMP2 = 4; COMP3 = 1; COMP4 = 2.

Proceed to complete the remaining cells in the table with insight. For the column 'Literature source', go back to your review of the theory and the literature: What authoritative sources or studies address the theme labels that you have identified? Did they agree or disagree with your findings? In our example, there is separate identifiable literature for each of the three labels 'Culture', 'Hierarchy' and 'Specific period'. Therefore, in the 'Literature' column, the three rows are kept separate. However, in Appendix E6, in the group named 'Completions', there is no separate identifiable literature for the four labels 'workload', 'resources', 'supply chain', and 'communication', so therefore the four rows are merged into one cell.

Review your coded data and the relevant parts of the literature. Labour at length over what the literature and your interviewees are really saying, and try to interpret and analyse your data. Make your observations or propositions or explanations – this should be based on the evidence, using your skill, but not giving your personal opinion. Identify where within this complete dataset (your participants and the literature) there are inconsistencies. The inconsistencies may be within the literature (you should have already written about these in the literature review), between the interviewees or between the literature and the interviewees. In the same way also identify similarities. Insert statements in the table.

Table 7.1 Example of using a table for analysis: the private housing sector (PHS).

Data coding number	Main category headings	Sub-category headings	Frequency counts	Literature sources	Observations, implications or interpretations	Data consistencies	Data inconsistencies
1	Propensity	Culture	155	Atkinson (2002;1999), Tam et al. (2000), McCabe et al. (1998), Reason (1998)	If a propensity exists among the site managers then it would seem to stem from higher management targets. Various forms of intimidation are used to pressure the site managers.	Excessive pressure is placed on site managers to achieve company targets. Self-esteem is a prime motivator for site managers.	Job security is not the overriding factor regarding propensity. Some pressure is self-inflicted.
2		Hierarchy	88	Liu (2003), Reason (1995), Morris (1994)	Senior managers and directors operate from a position of self-interest motivated by personal monetary gain via bonus payments. A lack of respect for superiors prevails among site managers.	Most managers experience strained relations with senior managers during busy periods.	A few managers claimed their relationships with hierarchy is unaffected by pressure.
3		Specific periods	105		The PHS as a whole, experiences unusually busy periods in an annual and/or bi-annual cycle. There is a lack of support from higher management during busier times.	Adverse working practices are frequently adopted during such times.	Some become institutionalised by the casual attitudes of hierarchy towards product quality and H&S.

The narrative

The purpose of the narrative is to take the analysed data and open it up into a free-flowing narrative that readers can understand and appraise for themselves. With insight, read your table. Use the table as a framework to write up a final narrative whole, in a essay form. Your narrative should flow, guided by your coding scheme. Your narrative will weave the inconsistencies and similarities together and, as the author, you will state your observations or propositions or explanations. It will bring together in summary form, with the literature, all the important things your interviewees told you. Your narrative may have many caveats and often use 'if' and 'maybe'. Consider whether the narrative should embrace a discussion, or whether a discussion will form a separate chapter. Ensure that the narrative is a platform for the conclusion and recommendations. Recommendations may include a proposal for a quantitative study stemming from your qualitative work. Readers of your document should be very interested to read in detail this narrative, as it will be one of the most interesting and informative parts of your document. The narrative for our example is in Appendix E6.

Incorporating qualitative data within your dissertation or project

You will need to make a judgement about how best to present your data and analysis within your document, or indeed if you need to include some steps at all.

There is the potential to include a separate discussion chapter between the narrative and the conclusions. However, since one role of a discussion is to integrate study findings with literature, and since the literature has already been integrated, there is the possibility not to include a chapter specifically titled 'Discussion'. If you decide not to write a discussion chapter, you should ensure that your narrative has a discussion element to it.

The following suggestions are made:

(a)   The research questions and objectives. These will be in the introduction chapter anyway, and be repeated throughout the document. There is no need to include them again.
(b)   Interview questions and prompts – include as an appendix.
(c)   Verbatim transcripts of interviews – include as an appendix.
(d)   Content analysis frequency counts – include in the main body of the document, in the analysis results and findings chapter.
(e)   Coding scheme – include in the main body or an appendix.
(f)   Analysis tables – include in the main body of the document, in the analysis results and findings chapter.
(g)   The narrative. – include in the main body of the document, in the analysis results and findings chapter, but perhaps as part of the discussion.

If your appendix becomes too bulky, the alternative position is to include a CD or memory stick on the inside back cover, for readers to refer to as they wish.

Summary of this chapter

The collection of qualitative data will attempt to give insights into the world of others, and provide rich data. The method of collecting data for analysis in the middle of your document may be entirely qualitative. Interviews are just one method of collecting qualitative data; they

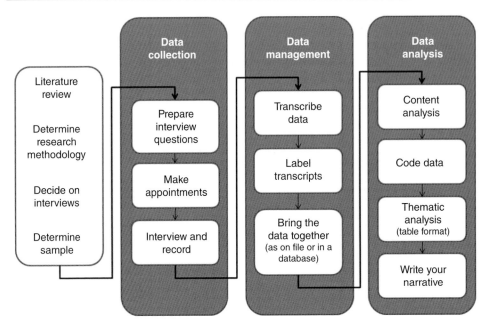

Figure 7.1 Steps in the qualitative data analysis process.

involve few people, and no inferences are made that what is found may be found in the whole population. If you have undertaken a quantitative study, it is better if you support that with some qualitative work to support the development of the problem, or to provide insight for writing the discussion and conclusions. Be sure to follow your university ethical codes. The steps in analysing qualitative data involve having verbatim transcripts of raw data. These are coded and sorted into like themes, and then can be analysed through a tabular format with the literature. Similarities and differences are noted, and observations made. The tables are then used to write a narrative that brings together all the data you have captured, including the literature. It is the narrative that is the final output of the qualitative analytical process. It may also incorporate the discussion. All the steps involved in qualitative data analysis can be seen in figure 7.1. Conclusions must follow; recommendations may include a fully developed proposal for a quantitative study that can be used to make inference across the population.

References

Jefferson, G. (2004) 'Glossary of transcript symbols with an introduction.' In: G.H. Lerner (Ed.) Conversation Analysis: studies from the first generation.: John Benjamins Publishing Company, Amsterdam/Philadelphia.

McCracken, G. (1988) *The Long Interview. Qualitative Research Methods.* Sage, London.

Seale, C. (2004) 'Coding and Analysing Data' In Seale, C. (Ed.) *Researching Society and Culture*, 2nd Ed., Sage Publications Limited, London, pp. 305–321.

Stott, M. and Farrell, P. (2011) New build houses in the private housing sector; cashflow before quality? *10th International Postgraduate Research Conference* (IPGRC). The University of Salford. 14–15 September. pp. 33–44.

8 Quantitative data analysis; descriptive statistics

The titles and objectives of the sections of this chapter are as follows:

8.1 Introduction; to provide context for statistical data collection and analysis
8.2 Examples of the use of descriptive statistical tools; to identify and explain appropriate ways to summarise data, with examples
    8.2.1 Calculation of the descriptive statistics without a spreadsheet and their definitions; missing and treated data.
    8.2.2 Calculation of the descriptive statistics with a spreadsheet
8.3 Ancillary variables; to describe how to record and calculate measurements for them
8.4 Illustration of relevant descriptive statistics in charts; to show examples
8.5 Normal distributions; Z scores; to calculate with an example
    8.5.1 Normal distributions
    8.5.2 The Z score; standard score
8.6 A second variable for descriptive analysis; an IV and a DV; to introduce the concept of measuring two variables in research and doing descriptive analysis for both

8.1 Introduction

Some students may wish to avoid statistical analysis; avoiding some of the complexities in statistical work is perfectly acceptable. It is certainly better to avoid complexities, rather than undertake analysis where there is weak understanding of underpinning principles; misunderstandings are dangerous, so know your limits. In business, manipulation of statistical data is pervasive; it is really difficult to avoid it completely. At a basic level, comparisons of salaries or costs require the analysis of numerical data. At a slightly more advanced level, there is a business need to understand cause and effect issues, and how samples taken from populations are used to predict behaviour or people's buying inclinations.

Analysis dealing with one variable is known as descriptive statistics, whereas inferential statistics are used to find links between two or more variables. Descriptive statistics are often sufficient to allow you to pass your dissertation or project. Their key use is to summarise datasets; a mass of numbers is condensed into easily digested and remembered figures or

Writing Built Environment Dissertations and Projects: Practical Guidance and Examples, Second Edition.
Peter Farrell, Fred Sherratt and Alan Richardson.
© 2017 John Wiley & Sons, Ltd. Published 2017 by John Wiley & Sons, Ltd.
Companion Website: www.wiley.com/go/Farrell/Built_Environment_Dissertations_and_Projects

averages. The three most often cited averages are mean, median and mode. If you undertake statistical analysis, you may use conventional mathematical signs as symbols. Roman letters are often used as abbreviations for statistics related to samples. The English alphabet is based on the Roman, except English has three additional letters; J, U and W. Greek letters are often used to indicate parameters about populations. On a preliminary page of the document, you should provide a glossary of symbols. If you have one or more variables in your research that you have measured quantitatively, you would normally summarise that data. A single score is denoted as X. Ten descriptive tools are given including mean, median and mode. The sequence of this trilogy is interrupted in the list below by percentage, since this measure is often very useful in non-technical work, thus:

Count or sample size	n	lowercase
sum	Σ	Greek capital sigma
mean	$\bar{X}$	of a sample
percentage	%	
median	Md	
mode	Mo	
minimum	min	
maximum	max	
range	–	
standard deviation	SD or s	of a sample

Other statistical symbols used are:

Number of scores in a population	N	uppercase
Mean of a population	μ	Greek mu
Standard deviation of a population	σ	Greek lowercase sigma
Less than or equal to	$\leq$	
Greater than or equal to	$\geq$	
Infinity	∞	

While calculations can be performed longhand, or by use of a calculator, large datasets are best handled in spreadsheets such as Excel. Longhand and calculator work would seem to have more possibilities for error, and make checking calculations more difficult. Numerical data also lends itself to presentation in your document in a spreadsheet. Screenshots of data can be presented using the snipping tool as illustrated in figure 1.1.

8.2 Examples of the use of descriptive statistical tools

Table 8.1 illustrates a raw dataset upon which descriptive analysis is conducted, and includes count, sum and mean. It is presented in Excel, and calculations are illustrated manually. You could present your data in Word; it is just that Excel is more convenient, and rather than calculate manually, once you understand the principles of how to calculate descriptive statistics, you may prefer to use Excel's functions to perform the calculations for you.

The numbers could be any set of numbers, and any variable. It could be based on technical or non-technical work. You will have your variable in your research. In this case, just to give a label to the variable; it is non-technical work and called 'leadership style'. As noted in section 5.4.2, it is based on the work of Douglas McGregor (1906–1964) who used a multiple-item scale to measure leadership style (Business, 2010). It is a multiple-item scale; the original work contained 15 questions; for brevity in table 8.1 only six are shown. Also for brevity in this text, data from just 10 employees/participants is shown. Section 5.3.1 suggests that in a questionnaire survey you may have 25–50 participants or more for undergraduate work.

The raw data is assembled from an electronic questionnaire, as illustrated in figure 5.3, that asks employees to judge their boss. Two leadership styles are examined in this dataset; a task-orientated leadership style which is given the lowest score of zero (0) for each item, and a person-orientated style which is given the highest score of 5. The task-orientated leader speaks to people primarily about the work that is required to be done. The person-orientated leader often speaks to people socially and takes an interest in issues outside work. Which type of leader have you experienced? The research suggests that the person-orientated leader gets best results in the workplace. An objective of the study may be 'to determine the leadership style of construction site managers' or 'partners in design consultancies' etc.

The raw data is presented table 8.1 with the descriptive statistics 'count', 'sum' and 'mean' for each question in columns and each participant in rows. Table 8.2 expands table 8.1 to illustrate the descriptive statistics 'percentage, median, mode, minimum, maximum, range,

Table 8.1 Raw dataset arising from ten participants; the measurement of leadership style of their bosses based on a multiple-item scale of six questions.

Question number >	Q1	Q2	Q3	Q4	Q5	Q6	Count	Sum Q1 to Q6	Mean
	Coding of possible responses: always 5, mostly 4, often 3, occasionally 2, rarely 1, never 0.								
Participant number ∨	My boss asks me politely to do things, gives me reasons why, and invites my suggestions.	I am encouraged to learn skills outside of my immediate area of responsibility.	I am left to work without interference from my boss, but help is available if I want it.	I am given credit and praise when I do good work or put in extra effort.	I am incentivised to work hard and well.	If I want extra responsibility, my boss will find a way to give it to me.			
1	5	4	5	3	2	5	6	24	4.00
2	4	5	4	4	1	3	6	21	3.50
3	0	2	Missing	3	4	2	5	11	2.20
4	3	3	3	4	2	4	6	19	3.17
5	4	5	3	5	1	5	6	23	3.83
6	4	3	2	5	2	2	6	18	3.00
7	2	2	2	0	5	2	6	13	2.17
8	5	5	5	4	1	5	6	25	4.17
9	1	1	0	0	Missing	2	5	4	0.80
10	2	2	2	3	0	3	6	12	2.00
Count	10	10	9	10	9	10			
Sum for each participant	30	32	26	31	18	33			
Mean	3.00	3.20	2.89	3.10	2.00	3.30			

Table 8.2 Expansion of table 8.1 to include missing data (shaded in grey) and percentage (%), median (Md), mode (Mo), minimum, maximum, and standard deviation (SD) for each question in columns.

	Home	Insert	Page Layout	Formulas	Data	Review	View

| | Normal | Page Layout | Page Break Preview | Custom Views | Full Screen | | | |

Workbook Views — Show/Hide — Zoom — 100% Zoom to Selection

☑ Ruler ☑ Gridlines ☐ Message Bar ☑ Formula Bar ☑ Headings

N17 f_x

	A	B	C	D	E	F	G
1	Question number >	Q1	Q2	Q3	Q4	Q5	Q6
2		Coding of possible responses: always 5, mostly 4, often 3, occasionally 2, rarely 1, never 0.						
3	Participant number V	My boss asks me politely to do things, gives me reasons why, and invites my suggestions.	I am encouraged to learn skills outside of my immediate area of responsibility.	I am left to work without interference from my boss, but help is available if I want it.	I am given credit and praise when I do good work or put in extra effort.	I am incentivised to work hard and well.	If I want extra responsibility, my boss will find a way to give it to me.
4	1	5	4	5	3	2	5
5	2	4	5	4	4	1	3
6	3	0	2	2.20	3	4	2
7	4	3	3	3	4	2	4
8	5	4	5	3	5	1	5
9	6	4	3	2	5	2	2
10	7	2	2	2	0	5	2
11	8	5	5	5	4	1	5
12	9	1	1	0	0	0.80	2
13	10	2	2	2	3	0	3
14	Count	10.00	10.00	10.00	10.00	10.00	10.00
15	Sum for each participant	30	32	28.20	31	18.80	33
16	Mean	3.00	3.20	2.82	3.10	1.88	3.30
17	Percentage	60.00	64.00	56.40	62.00	37.60	66.00
18	Median	3.50	3.00	2.60	3.50	1.50	3.00
19	Mode	4.00	5.00	2.00	3.00	2.00	2.00
20	Minimum	0.00	1.00	0.00	0.00	0.00	2.00
21	Maximum	5.00	5.00	5.00	5.00	5.00	5.00
22	Range	5.00	4.00	5.00	5.00	5.00	3.00
23	SD	1.70	1.48	1.54	1.79	1.54	1.34

and standard deviation' for each question. Table 8.3 does similarly for each participant. Very importantly, this last table also shows the overall mean percentage score for leadership style in cell 17K as 57.67%.

This now becomes a very large dataset, and can be too large to fit onto one A4 piece of paper in your document. Therefore, the 'hide' facility in Excel is used in tables 8.2 and 8.3 to cut out data which is not relevant to the context of the results that are being written about in a particular paragraph. Perhaps in your document you can also use the hide facility to show

Table 8.3 Expansion of table 8.1 to include missing data (shaded in grey) and percentage (%), median (Md), mode (Mo), minimum, maximum and standard deviation (SD) for each participant in rows.

Question number >	Q1	Q2	Q3	Q4	Q5	Q6	Count	Sum Q1 to Q6	Mean	Percentage	Median	Mode	Min	Max	Range	SD
	Coding of possible responses: always 5, mostly 4, often 3, occasionally 2, rarely 1, never 0.															
Participant number v	My boss asks me politely to do things, gives me reasons why, and invites my suggestions.	I am encouraged to learn skills outside of my immediate area of responsibility.	I am left to work without interference from my boss, but help is available if I want it.	I am given credit and praise when I do good work or put in extra effort.	I am incentivised to work hard and well.	If I want extra responsibility, my boss will find a way to give it to me.										
1	5	4	5	3	2	5	6	24	4.00	80.00	4.5	5	2	5	3	1.26
2	4	5	4	4	1	3	6	21	3.50	70.00	4	4	1	5	4	1.38
3	0	2	2.20	3	4	2	6	13.2	2.20	44.00	2.1	2	0	4	4	1.33
4	3	3	3	4	2	4	6	19	3.17	63.33	3	3	2	4	2	0.75
5	4	5	3	5	1	5	6	23	3.83	76.67	4.5	5	1	5	4	1.60
6	4	3	2	5	2	2	6	18	3.00	60.00	2.5	2	2	5	3	1.26
7	2	2	2	0	5	2	6	13	2.17	43.33	2	2	0	5	5	1.60
8	5	5	5	4	1	5	6	25	4.17	83.33	5	5	1	5	4	1.60
9	1	1	0	0	0.80	2	6	4.8	0.80	16.00	0.9	1	0	2	2	0.75
10	2	2	2	3	0	3	6	12	2.00	40.00	2	2	0	3	3	1.10
Count	10.00	10.00	10.00	10.00	10.00	10.00				10.00						
Sum for each participant	30	32	28.20	31	18.80	33		–		576.67						
Mean	3.00	3.20	2.82	3.10	1.88	3.30				57.67						
Percentage	60.00	64.00	56.40	62.00	37.60	66.00		–								
SD										**57.67**						

Note: Some rows in this table have been hidden for brevity. The complete table can be accessed online through the companion website of the book (www.wiley.com/go/Farrell/Built_Environment_Dissertations_and_Projects).

key features of your data in the main body of your document. The full dataset can be included in an appendix as an A3 size page or over several A4 pages if necessary.

8.2.1 Calculation of the descriptive statistics without a spreadsheet and their definitions; missing and treated data

Mean, median and mode are measures of central tendency. Standard deviation, range, minimum and maximum are measures of range. Based upon table 8.1, the count, sum and means are shown 'longhand' first.

Count, sum and mean

Count n, is the number of pieces of data. In this case, the number of participants shown in row 14 gives $n = 10$ (except cells 14D and 14F where $n = 9$), and the number of questions in column H is 6 (except cells 6H and 12H where $n = 5$).

Sum Σ, is the addition of the individual pieces of data. For question 1 in column B, $\Sigma = 5 + 4 + 0 + 3 + 4 + 4 + 2 + 5 + 1 + 2 = 30$. For participant 1 in row 4, $\Sigma = 5 + 4 + 5 + 3 + 2 + 5 = 24$.

Mean is the sum divided by the count $= \Sigma X/ n$. For question 1 in cell 15B, the mean response from the ten participants is = cell 15B/cell 14B = 30/10 = 3.00. For participant 1 in cell 4J, the mean score is calculated as cell 4I/cell 4H = 24/6 = 4.00.

The mean figure is most often used as a measure of central tendency, since it does consider the value of each number. The mean may not be appropriate if there is a small dataset, with one or two extreme or rogue values, e.g. a dataset of 2, 2, 3, 4, 5 and 50 is not truly summarised by the mean, which is 66/6 = 11.

Missing data and 'treated' data

Cells 6D and 12F in table 8.1 have missing data; that data is 'treated' in table 8.2. Also, included in table 8.2 is the following descriptive statistics in columns for each question: percentage (%), median (Md), mode (Mo), minimum (min), maximum (max) and standard deviation (SD).

In technical work, there may be a piece of data that you cannot just get at, or you have possibly lost. More likely the case, in surveys involving participants, they may not complete all parts – for whatever reason. In table 8.1, participant 3 has not answered question 3, and participant 9 has not answered question 5. They may not have understood, or may have got distracted while they were completing it. Electronic surveys lessen the risk of missing data over paper-based surveys, since they give you the opportunity not to let participants proceed until all parts are completed. When you have missing data, you need to make a judgement about whether to discard completely all of that participant's responses. If there is too much missing data from a participant, you should do so. Bryman and Cramer (1994) suggest that if more than 10% of data is missing, it should be discarded. Perhaps for undergraduate research more than 20% is acceptable. Since participants 3 and 9 have just one out of six pieces of data missing, that is 16.6%, and is acceptable against the 20% criterion. Hopefully, if your survey has multiple-item scales in your work, you will have more than just six items contributing to the measurement of variables, therefore two or three missing pieces of data will not exceed 20%.

If participants are retained with missing data, the issue then is how to deal with that missing data. We may say that we are to 'treat' the blank cell. The procedure is to assume that the one piece of missing data is considered to be the mean of the other pieces of data that are given. An appropriate calculation needs to be facilitated using Excel or manually. In table 8.1, participant 3 has given five pieces of data thus: 0, 2, 3, 4 and 2 giving a sum of 11. In cell 6D the missing data can be replaced by the mean = 11/5 = 2.20. Participant 9 has one piece of missing data in cell 12F to be replaced by a calculation thus: 1 +1 + 0 +0 + 2 = 4 divided by 5 = 0.80.

To make it clear to readers of your document, you should explain in your methodology chapter how you have dealt with missing data. Also, in a table that summarises your results you should clearly distinguish the missing data that has been 'treated'. In table 8.2 'treated' data is included to two decimal points and is in lightly shaded grey cells. Note that in cells 14D and 14F the count is now 10 (not 9), and in cells 6H and 12H the count is now 6 (not 5).

Percentage, median, mode, minimum, maximum and standard deviation

Table 8.3 is added to show the descriptive statistics Md, Mo, min, max, % and SD in rows for each participant. Both tables 8.2 and 8.3 are used to illustrate the manual calculation of these statistics.

Percentage indicates a score on a 0–100 point scale. The percentage score for questions and individual participants can be calculated. Question 1 has a mean score of 3 (table 8.2, cell 16B) given a range of 0–5. The percentage score is calculated by dividing the actual score by the maximum score and multipling by 100. Thus for question 1 we have 3/5 × 100 = 60% (table 8.2, cell 17B). A percentage scores of 60% is arguably more clearly understood than a score of 3 on a 0–5 scale. Participant 1 has a mean score of 4.00 (table 8.3, cell 4J) given a range of 0–5. A percentage score of 80% (table 8.3, cell 4K) is arguably more clearly understood.

An extremely important summary statistic is the mean of the percentages; in this case with 0 = a task-orientated leader, and 100 = a person-orientated leader, the mean percentage score in cell 17K is 57.67%, say 58%. Since the objective of the study may be 'to determine the leadership style of construction site managers', there is now one key figure that fulfils that objective: 58%. In principle that is the objective fulfilled, but obviously there is the opportunity for a lot more analysis, discussion, conclusions and recommendations. Is 58% better or worse than might be reasonably expected? Chapter 4.12.2 illustrates coding methods for data to ensure that the calculation of percentage scores is possible.

Median is the centrally occurring number when they are all arranged in order of their value. For question 1 in column B table 8.2, the numbers for the 10 participants are rearranged in order thus: 0, 1, 2, 2, 3, 4, 4, 4, 5 and 5. Since there is an even number of participants, the median could be the two numbers 3 and 4. A spreadsheet will take a mean of the two numbers as shown in cell 18B table 8.2, and give the median as 3.5. For participant 1, table 8.3, the dataset rearranged in order is 2, 3, 4, 5, 5, 5. The median response is therefore 4 and 5 or as shown in cell 4L, the mean of these two scores = 4.5.

A disadvantage of the use of the median as an average or a tool to summarise a dataset is that it does not take account of numbers in other parts of the dataset. For example, if the scores for question 1 were 0, 0, 0, 0, 3, 3, 3, 3, 3, 3, a median of 3 does not fairly represent the data. The median is often the preferred measure in reporting statistics such as salaries. For example, the accountancy company KPMG (2013) reported that the median salary of FTSE

100 chief executive officers (CEOs) was £851k, or £3.21M including other benefits. Hay Group (2011) measured pay equality in the UK FTSE 350 companies. It found that the most meaningful comparator was median CEO salaries with median (of the average) of other employees. The range of results was CEO salaries nine times other employees in oil and gas industries where there is a well-educated highly skilled workforce; and 43 times in retail where there are a large number of low-skilled shop-floor workers.

Mode is the number that occurs most often in a dataset. For question 1 the mode = 4 (table 8.2, cell 19B) and participant 1 = 5 (table 8.3, cell 4M). Similarly a disadvantage of the use of the mode is that it too does not take account of numbers in other parts of the dataset. Again if the scores for question 1 were 0, 0, 0, 0, 3, 3, 3, 3, 3 and 3, a mode of 3 does not fairly represent the data. In this non-technical work, if the mode were at the extremities of the scale 0 to 5 scale, that may indicate some problems with the questions or the response scale. A data-set may have two different modes at different ends of a range of figures e.g. 0, 0, 1, 2, 3, 4 5 and 5; mode = 0 and 5.

Minimum is the lowest number in a dataset, for question 1 = 0 (table 8.2, cell 20B) and participant 1 = 2 (table 8.3, cell 4N). In a survey of this type, there is a reasonable expectation that some participants would score 0; nothing unusual. If the minimum score was 3 or 4 given a range of 0–5, there may be something wrong with the questions or the response scale. If this was a technical dataset, a minimum score that is higher or lower than expected might be interesting or set some alarm bells ringing.

Maximum is the highest number in a dataset, for question 1 = 5 (table 8.2, cell 21B) and participant 1 = 5 (table 8.3, cell 4O). If the maximum score was only 2 or 3 given a range of 0–5, again there may be something wrong with the questions or the response scale. If this was a technical dataset, a maximum score that is higher or lower than expected might be interesting or again set some alarm bells ringing.

Range is the difference between the maximum and the minimum. For question 1, range = 5–0 = 5 (table 8.2, cell 22B) and for participant 1 = 5–2 = 3 (table 8.3, cell 4P). For a single question, it would be normal that different participants answer at the extremities, and therefore a range of 5 might be expected; a narrow range may be of concern. For technical work, a wide range in a set of results might be of great concern; for example, if one set of six cubes with the same mix design failed in compression with a range of $10\,N/mm^2$.

Standard deviation SD is a measure of 'spread'. The mean, median or mode chosen as measures of central tendency on their own cannot accurately summarise a dataset. It may be possible that many of the numbers are clustered closely around the mean, with few at the extremities. Alternatively, the frequency count of numbers close to the mean may be low, and the count for numbers at the extremities is high. To support the understanding of the SD, a knowledge of normal distributions is required; an explanation follows in section 8.5. In tech-nical work, if one test of the same type of material was undertaken six times, there would be an expectation that the SD score would be low; that is also to say most of the scores would be close to the mean. In university assessment results, often quoted expected scores for large groups of students are means of circa 58% with an SD of 10%; that is also to say that most of the scores would be 58% ± 10% or in the range of 48 to 68%. Problems may be evident if the SD score is substantially higher or lower than 10%. These figures, by experience, may be benchmark standards. A mean score substantially more than 58% may indicate an assessment that has been too easy, or less than 58% too difficult. A standard deviation of more than 10% may indicate a class where students who are good in a subject have been allowed to flourish, but less capable students have been left behind. A standard deviation of less than 10% may indicate a class where the more capable students have not been allowed to flourish.

SD is the most often used measure of spread, particularly if the dataset is from a sample representing a population. The unit of measure for the standard deviation is the unit being used to measure the variable on the horizontal scale when the data is presented in a frequency histogram, e.g. percentage score, or £s, or millimetres. The formula for calculating the standard deviation for a sample is:

$$SD = \sqrt{\frac{\Sigma(X - \bar{X})^2}{n - 1}}$$

The formula uses all scores, and compares how far they all deviate from the mean. This is performed by this part of the formula: $\Sigma(X - \bar{X})^2$. It is the longest part of the calculation, and is illustrated in table 8.4 using six steps. The data used is the percentage scores from table 8.3 column K. The mean of 57.67 is deducted from each individual score, the difference is squared, and then the sum (total) for all differences is calculated as 4148.67; that completes steps 1, 2, 3 and 4. In step 5 the sum from step 4 is divided by $n - 1$. The calculation is completed in step 6, by taking the square root of the figure in step 5. The SD for the set of data = 21.47%.

Table 8.4 Calculation of the standard deviation of a sample in six steps from a column of ten raw scores.

	A	B	C
	Step 1: list the raw scores 'X' in a column ∨	Step 2: deduct the mean of 57.67 from the raw score 'X' e.g. 80.00 - 57.67 = 22.33	Step 3: square the result of step 2 e.g. 22.33 squared = 22.33 x 22.33 = 498.63. Note a negative number squared = a positive (minus x minus = positive)
1			
2	80.00	22.33	498.63
3	70.00	12.33	152.03
4	44.00	-13.67	186.87
5	63.33	5.66	32.07
6	76.67	19.00	360.87
7	60.00	2.33	5.43
8	43.33	-14.34	205.54
9	83.33	25.66	658.61
10	16.00	-41.67	1736.39
11	40.00	-17.67	312.23
12	Step 4: take the sum (total) of the ten scores from step 3 Σ		4148.67
13	Step 5: divide the sum from step 4 by n-1 = 10 -1 = 9. 4148.67 / 9 =		460.96
14	Step 6: take the square root of step 5 √ 460.96. The result is the standard deviation (SD)		**21.47**

To mark the 21.47 as a percentage is important, since that is the 'unit of measure'. In the dataset given, the majority of marks are in the range of the mean plus or minus one standard deviation. Therefore is this example 57.67% plus or minus 21.47% or 57.67± 21.47 = 36.20 to 79.14%. In section 8.5 it will be suggested that 'majority' is equal to circa 68% of the scores.

Tables 8.2 and 8.3 now give a wealth of information, arguably too much. There is certainly too much to comment on it all, so the process in the discussion chapter is to write about the important statistics such as the mean of the percentage scores of 57.67%, but also outliers and unexpected scores: why does participant 7 have a range of 5? Why is the mean percentage score for question 5 only 37.6% when other scores are around 60%?

8.2.2 Calculation of the descriptive statistics with a spreadsheet

The lead dropdown menus in Excel are in the sequence of 'Formulas', 'More Functions' and then 'Statistical' as shown in figure 8.1. The Excel terms matching the descriptive statistics are:

Descriptive statistic	Excel term
Count	COUNT
Sum	SUM
Mean	AVERAGE
Median	MEDIAN
Mode	MODE
Minimum	MIN
Maximum	MAX
Standard deviation	STDEV.S

Excel will not calculate the sum, range or percentage figure using a function argument. You will need to use the 'ΣAutoSum' function that is shown below 'Home' at the upper left hand corner in figure 8.1.

An explanation is given here to determine the mean, using Excel; other descriptive statistics are determined similarly. Before starting calculations, place the cursor in the cell where the answer is required e.g. in cell 16B to determine the mean for question 1. Scroll through 'Formulas', 'More Functions', 'Statistical' and then for the mean select 'AVERAGE'. The dialogue box as shown in figure 8.2 will appear. In the prompt labelled 'Number1' either type in manually 'B4:B13' or perhaps better click and drag over those cells. The answer '3' will appear in the centre of the dialogue box; click 'OK' and the answer will appear in cell 16B. Use the copy and paste facility extensively; place the cursor in cell 16B – copy and then paste into cells 16C to 16G.

8.3 Ancillary variables

Section 4.10 refers to ancillary variables, variables not primarily related to the description of the problem, but which are easily collected. Analysis can be undertaken to determine if ancillary variables have an influence on main study variables. Tables 4.4 and 4.5 gave examples of

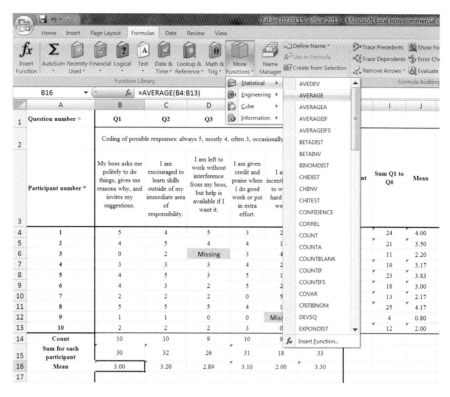

Figure 8.1 Using Excel to calculate descriptive statistics: 'Formulas', 'More Functions' and then 'Statistical'.

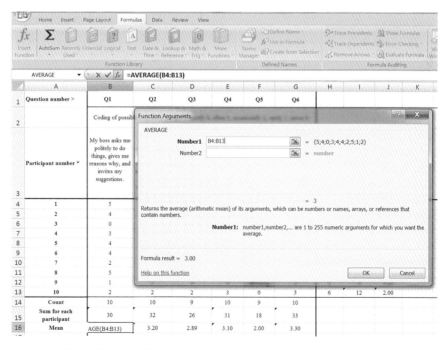

Figure 8.2 Using Excel to calculate the mean of a column of numbers.

Table 8.5 Ancillary variables, gender, age and qualifications of participants.

Home | Insert | Page Layout | Formulas | Data | Review | View

Insert Function | AutoSum | Recently Used | Financial | Logical | Text | Date & Time | Lookup & Reference | Math & Trig | More Functions — Function Library

Name Manager | Define Name | Use in Formula | Create from Selection — Defined Names

Trace Precedents | Trace Dependents | Remove Arrows | Show Formulas | Error Checking | Evaluate Formula — Formula Auditing

Y24 | fx

	A	B	I	J	K	L	M	N	O	P	Q	R	S	T	U	V
	Question number	Q1	Coding of possible responses: always 5, mostly 4, often 3, occasionally 2, rarely 1, never 0.	Male = M, female = F	18 to 25 years = A, 26 to 37 = B, 38 to 55 = C, over 55 = D	Age reclassified: 18 to 55 years = E, over 55 = F	Do you have the following qualifications; no = N, yes = Y — A levels or equivalent	First degree or equivalent	Post graduate degree or equivalent	Member of a professional body	Percentage score for each of two gender groups — Male	Female	Percentage score for each of two age groups — Age group E: 18 to 55 years	Age group F: over 55 years	Percentage score for whether a member of professional body — No, not a member	Yes; a member
	Participant number	My boss asks me politely to do things, gives me reasons why, and invites my suggestions.	Percentage to two decimal places. Sum divided by max score of 30 *100 e.g 24/30 * 100 = 80.00%	What is your gender?	What is your age?											
1			80.00	M	A	E	Y	N	N	N	80.00		80.00		80.00	
2			70.00	M	B	E	Y	N	N	Y	70.00		70.00			70.00
3			44.00	F	B	E	N	N	N	N		44.00	44.00		44.00	
4			63.33	M	D	F	N	N	N	Y	63.33			63.33		63.33
5			76.67	M	D	F	N	N	N	N	76.67			76.67	76.67	
6			60.00	M	D	E	N	N	N	N	60.00		60.00		60.00	
7			43.33	F	C	E	Y	Y	N	Y		43.33	43.33			43.33
8			83.33	F	C	E	Y	N	Y	Y		83.33	83.33			83.33
9			16.00	M	D	F	Y	N	N	N	16.00			16.00	16.00	
10			40.00	M	D	F	Y	N	N	Y	40.00			40.00		40.00
Sum			576.67								406.00	170.66	380.66	196.00	276.67	299.99
Percentage to two			57.67								58.00	56.89	63.44	49.00	55.33	60.00

Countif M = 7 | Countif A = 1 | Countif E = 6 | Countif N = 4 | 9 | 8 | 5
Countif F = 3 | Countif B = 2 | Countif F = 4 | Countif Y = 6 | 1 | 2 | 5
Countif C = 2
Countif D = 5 | Countif 3 = 0

The Excel function 'hide' used to remove columns C to H. Columns A, B and I retained to illustrate the origin of the data. This text inserted by using 'text box'.

Note: Some columns in this table have been hidden for brevity. The complete table can be accessed online through the companion website of the book (www.wiley.com/go/Farrell/Built_Environment_Dissertations_and_Projects).

how to arrange data to consider the ancillary variable 'age'. In table 8.1 only one main study variable is shown, leadership style of construction managers. Table 8.5 is an expansion of table 8.1, which shows data for the ancillary variables 'gender', 'age' and 'qualifications'. Columns C to H are 'hidden' to allow the data to fit onto one page. Columns A, B and I are retained from table 8.1. Column I is important since that gives the percentage score for leadership style. The 'COUNTIF' function is used in Excel to determine the number of participants in each group; for example, in cell 17J there are shown 7 males (Formulas > statistics > COUNTIF).

There is now the opportunity to do more analysis that may help obtain more marks. Three more RQs/OBs/Hypotheses arise, expressed as objectives thus:

(1)   To determine whether the gender of leaders influences their leadership style.

The data in column I is reallocated to two columns Q and R, based upon gender detailed in column J. Cells 15Q and 15R show that the mean leadership style of male leaders is 58.00%, and of female leaders 56.89% respectively. Just eyeballing these two figures suggests that it does not appear that gender does influence leadership style.

(2)   To determine whether the age of leaders influences their leadership style.

In column K, the age of leaders is given in four groups: 18–25 years = A, 26–37 = B, 38–55 = C, over 55 = D. There is potential to calculate a mean leadership score for each group. However, the sample size of 10 is small in this exemplar; it would lead to group sizes that are meaningless, for example only one person in the age group 18–25 years. In your research if you have perhaps 50 responses you may make the judgement that three or four groups are possible. In figure 8.4, the four groups in column K are reclassified into two groups in column L. To ensure as near as possible a similar number in each group, given that n in group A = 1, group B = 2, group C = 2 and group D = 5, it is decided to reclassify groups A to C into one group relabelled 'E' and classify group D as the other, now relabelled group 'F'. Cells 15S and 15T show that the mean leadership style of young leaders is 63.44%, and of older leaders is 49.00%. Just eyeballing these two figures, it does appear that age influences leadership style. If you judge that it is rather surprising that young people are better, that is for your discussion chapter. Chapter 9 illustrates how to do inferential statistical tests (in this case it would be a Mann–Whitney test) to allow you to assert whether or not the differences between the two groups has occurred due to chance. The statistical tests involve a little more complexity; if you can do them you should do so, since that gives the opportunity for more marks still. If you cannot do inferential tests that is fine; just state your findings based on the two mean scores.

(3)   To determine whether being a member of a professional body influences leadership style.

While data is collated about four qualifications, a simplified approach is to consider only whether leaders are qualified as practitioners. The data in column I is reallocated to two columns U and V, based upon whether a member of a professional body in column P. Columns U and V show that the mean leadership style of leaders who are not members is 55.33%, and of leaders who are members 60.00%. Since the difference between these two mean scores is 'close but not very close' it is difficult to come to a judgement by just eyeballing whether being a member of a professional body influences leadership style. In your discussion chapter say that. The best position, if you can, is to do the Mann–Whitney test as demonstrated in chapter 9.

8.4 Illustration of relevant descriptive statistics in charts

Figure 8.3 illustrates the range of charts in Excel. Using the dropdown menu called 'Insert', seven charts are illustrated: column, line, pie, bar, area, scatter and other. There are five options under 'other': stock, surface, doughnut, bubble and radar.

A judgement needs to be made about whether a table conveys data adequately, or perhaps a chart can support communicating the message to readers with greater clarity. In figure 8.4 the scores for leadership style are rearranged in Excel to aid the production of a frequency histogram. Scores are allocated to 10 percentage point class intervals 10 to 20, 20 to 30 etc. Does this histogram more clearly illustrate the data in column I of table 8.1?

It is arguable whether this small dataset is adequately communicated in column K of table 8.3, but chapter 9 illustrates examples of where frequency histograms can be extremely useful if doing inferential statistics. A frequency histogram or bar chart is also possible for the data in row 17 of table 8.2; the percentage scores for each of questions 1 to 5. You need to decide if this is necessary in the context of your work. Pie charts are often used to illustrate data such as age of participants as shown in column K of table 8.5.

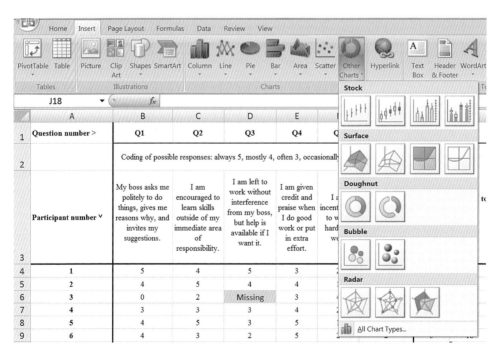

Figure 8.3 Using charts in Excel to communicate your data.

Class interval for leadership style	Frequency counts
10 to <20	1
20 to <30	0
30 to <40	0
40 to <50	3
50 to <60	0
60 to <70	2
70 to <80	2
80 to <90	2

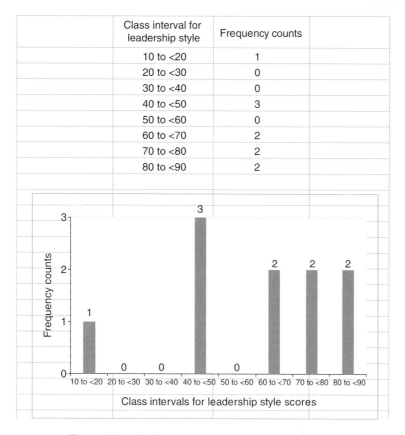

Figure 8.4 Using Excel to illustrate a frequency histogram.

Figure 8.5 illustrates how a chart can be used to communicate a dataset more clearly. It is a line diagram of construction tender price indices and general building cost indices over a 14 year time frame 2006 to 2019, sourced from the Building Cost Information Service (BCIS, 2015). Indices on the vertical scale were anchored at a score of 100 in 1985. Indices for years 2016 to 2019 are forecast figures as at 2015. The line diagram clearly indicates a trend where tender indices changed in a period of world economic problems starting in 2008, and that while tenders levels were decreasing, costs to contractors were increasing. Using the line diagrams, a discussion can perhaps be more easily articulated. Chapter 9 gives examples of scatter diagrams that are essential in the interpretation of data.

8.5 Normal distributions; Z scores

8.5.1 Normal distributions

The bell-shaped normal distribution is represented in figure 8.6. The horizontal scale is the unit of measurement for the variable being considered, and the vertical scale is the frequency or number of counts for each value of the variable. Normal distributions might be expected in

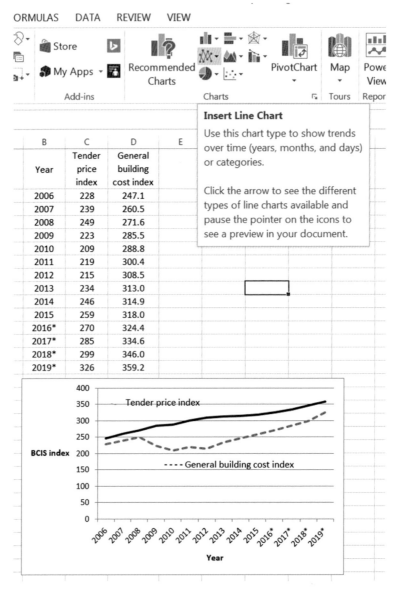

Figure 8.5 Data taken from a table in Excel to produce a line diagram.

the measurement of many variables to do with living organisms. The normal distribution represents frequency counts of measures across whole populations; thus frequency counts are very high, often many millions. Since table 8.1 has an n value of just 10, it cannot be expected that its frequency distribution in figure 8.4 will resemble a normal distribution. However, with a sample size of perhaps 25, or better 50 or even better 100, eyeballing of a frequency distribution may suggest the points at the head of each column, may be joined by a curve that is approximately bell-shaped.

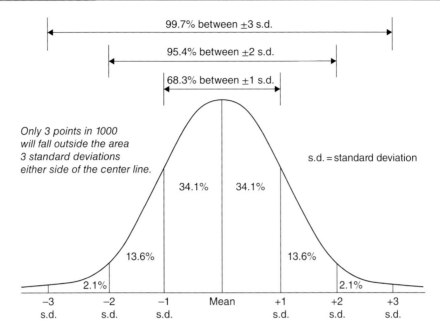

Figure 8.6 The bellshape and numerical features of the normal distribution.

The features of a normal distribution are that:

- the mean, median and mode are all the same value
- 68.26% of scores should lie within ±1 standard deviation of the mean
- 95.44% of scores should lie between ±2 standard deviations of the mean.

You will not be required to estimate or plot a normal distribution based upon results from your sample. To support understanding, consider three hypothetical examples of variables that could be measured on the horizontal axis:

Example 1: The ONS (2014) reports that in 2013 there were 1,151,800 (1.15 million; the population) people employed by private construction contractors in Great Britain. A research objective is to measure job satisfaction levels of this population of construction workers on a percentage scale. The horizontal axis is 0–100.

Example 2: The cost per m² of new build houses is plotted on the horizontal axis in £s, perhaps with a range of £0 to £8000/m². The Building Cost Information Service database (BCIS, 2015) reports that at 2015 prices for a sample of 628 projects, the mean price is £1223/m², with the minimum price being £47/m² and the maximum price being £7258/m². The Department for Communities and Local Government (DCLG, 2015) reports that 'Annual housing completions in England totalled 125,110 in the 12 months to March 2015'. The population n would therefore be 125,110.

Example 3: A brick manufacturer produces Class B engineering bricks from the same quarry over many years. It accumulates several thousand test results for water absorption measured by percentage increase in mass and compressive strength in N/mm². Minimum criteria are: water absorption ≤7% and compressive strength ≥50N/mm² (Ibstock, 2015). Distributions are possible for each of the two measures. The horizontal axis for water absorption will be in % mass increase, with a range of perhaps 2–12%. The horizontal axis for compressive strength will be N/mm² with a range of perhaps 40–90 N/mm². In each case the vertical axis will be frequency.

Reverting to example 1, job satisfaction of construction workers, on the horizontal axis at the polar ends, 0 = extremely unsatisfied and 100 = extremely satisfied. There are 1,151,000 frequency accounts across 101 possible values of the variable 'job satisfaction' (0 to 100, including zero). It will be shown that on the vertical axis, the number that occurs most often (the mode) is 58%, and it has a frequency count of 45,842; this will be the highest number on the vertical axis.

Given a population size of 1.15 million, assume that a survey is completed with perhaps a sample size of just 50 people. The mean satisfaction score is 58% and the frequency histogram resembles a normal distribution. The standard deviation is calculated at 10%.

It is inferred that the results from the sample of 50, reflect the results that would have been obtained if the whole population of 1.15 million people had been surveyed. For the population, it can be inferred that the three measures of central tendency are calculated thus:

- Mean: the sum of all scores added together (ΣX) would be expected to be 67,210,400. This number divided by 1,151,800 gives the expected mean score of 58%.
- Median: the score that would be in the middle position, if the 1,151,800 individual scores were arranged in ascending or descending order, the score that is expected to appear in the 575,900th (half of 1,151,800) position is 58%.
- Mode: as the score that occurs most frequently is 58% which has an expected frequency count of 45,842.
- The standard deviation is assumed to be the same as that calculated for the sample: 10%. It can also be inferred that:
- 68.26% of scores should lie between ±1 standard deviation of the mean = 1,151,800 × 0.6826 = 786,219 people
- 95.44% of scores should lie between ±2 standard deviations of the mean = 1,151,800 × 0.9544 = 1,105,958 people

8.5.2 The Z score; standard score

Z scores are used to predict the frequency count at each value of the variable, each value 0 to 100. The area under a normal distribution curve can be considered equal to 1.0, and is linked to probabilities and percentages. The areas to the left and right of the central mean, median and mode position each have an area of 0.50. Z scores allow you to take an individual score on a horizontal scale and calculate what percentage of scores lie to the left and right of that score on the distribution. This can be done, irrespective of what the measure is on the horizontal scale; perhaps job satisfaction of construction workers or the cost per m² of new build houses in England or the water absorption of bricks. Z scores below the mean are negative. Consider the job satisfaction scores presented in table 8.6.

We have already inferred from our sample that the mean of the population is 58% and the standard deviation 10%. We have stated that a feature of a normal distribution is that 68.26%

of scores should lie between ±1 standard deviation of the mean. Therefore a job satisfaction score of 48% should have 15.87% of scores less than this (100 − 68.26 = 31.74 then divided by 2 = 15.87), and a job satisfaction score of 68% should have 15.87% of scores more than this. To prove this, the procedure is first to calculate the Z score. Then use the standard normal distribution tables in appendix F1 to determine the probabilities of finding scores higher or lower than calculated Z scores. The formula to calculate Z is:

$$Z = \frac{x - \mu}{\sigma}$$

where
  Z = the standardised score
  x = the single score
  μ = mean
  σ = standard deviation
  For the single score of 48%, mean of 58% and standard deviation of 10%, the Z score computes as:

$$Z = \frac{48 - 58}{10} = -1.00.$$

Ignore the minus sign. Looking at the values in the table in appendix F1, the probability z score of 1.0 is 0.1587, thus indicating that 15.87% of scores in a normal distribution would be expected to be lower than 48%. If the calculation were repeated for the score of 68%, that would similarly give a Z score of 1.0, indicating 15.87% of scores in a normal distribution would be expected to be higher than 68%. The percentage of scores expected to be between 48% and 68% would thus be 100 − (15.87 +15.87) = 68.26%.

Take the arbitrary figure of 40%; the computation is:

$$Z = \frac{40 - 58}{10} = -1.80.$$

Looking at the values in the table in appendix F1, the probability Z score of 1.80 is 0.0359; thus indicating that 3.59% of scores in a normal distribution would be expected to be lower than 40%. If a qualitative judgement were made that a score of 40% indicates low job satisfaction, that would indicate that 3.59% of workers were in this category or worse. Table 8.6 illustrates in cell 43D 41,350 workers were expected to score 40% or less, and that in cell 42D, 33,057 were expected to score 39% or less; cell 43E shows that 41,350 − 33,057 = 8293 people expected to score exactly 40%.

Columns F and G illustrate how Z scores calculate that 68.26% of scores lie between ±1 standard deviation of the mean (786,219 people; cell 105F) and 95.44% of scores lie between ±2 standard deviations of the mean (1,105,958 people; cell 105G). Figure 8.7 shows the predicted distribution of the data, based upon data in table 8.6 columns A and E.

Not all datasets resemble a normal distribution, therefore mean median and modes do not always have the same score. They can all individually summarise a dataset but, depending on the distribution of the numbers, one of them may be better than the other two. Eyeball observation

Table 8.6 Z scores and predicted or inferred n values for job satisfaction in UK construction workers; n = 1.15 million.

	A	B	C	D	E	F	G
1			N = 1,151,800 (1.151 million) private contractors total employment in Great Britain				
	Each possible value of the variable on 1 0 - 100 point scale	Z score	From appendix F1; probability of Z score	Column C multiplied by N = 1,151,800. From 0 to 58; number of people score this mark or less. From 58 to 100; number of people scoring this mark or more	Number of people scoring this mark; the number in column D minus adjacent number e.g. for score at 40%, 41,350 minus 33,057 = 8,293	Scores between 48 and 68. 1SD either side of the mean	Scores between 38 and 78. 2SDs either side of the mean
2							
3	0	-5.8	0	0	0		
4	1	-5.7	0	0	0		
20	17	-4.1					
21	18	-4					
22	19	-3.9	0	0	0		
23	20	-3.8	0.0001	115	115		
24	21	-3.7	0.0001	115	0		
37	34	-2.4	0.0082	9445	2304		
38	35	-2.3					
39	36	-2.2					
40	37	-2.1	0.0179	20617	4607		
41	38	-2	0.0228	26261	5644		
42	39	-1.9	0.0287	33057	6796		6796
43	40	-1.8	0.0359	41350	8293		8293
44	41	-1.7	0.0446	51370	10021		10021
45	42	-1.6	0.0548	63119	11748		11748
46	43	-1.5	0.0668	76940	13822		13822
47	44	-1.4	0.0808	93065	16125		16125
48	45	-1.3	0.0968	111494	18429		18429
49	46	-1.2	0.1151	132572	21078		21078
50	47	-1.1	0.1357	156299	23727		23727
51	48	-1	0.1587	182791	26491		26491
52	49	-0.9	0.1841	212046	29256	29256	29256
53	50	-0.8	0.2119	244066	32020	32020	32020
54	51	-0.7	0.242	278736	34669	34669	34669
55	52	-0.6	0.2743	315939	37203	37203	37203
56	53	-0.5	0.3085	355330	39392	39392	39392
57	54	-0.4	0.3446	396910	41580	41580	41580
58	55	-0.3	0.3821	440103	43192	43192	43192
59	56	-0.2	0.4207	484562	44459	44459	44459
60	57	-0.1	0.4602	530058	45496	45496	45496
61	58	0	0.5	575900	45842	45842	45842
62	58	0	0.5	575900	45842	45842	45842
63	59	0.1	0.4602	530058	45496	45496	45496
64	60	0.2	0.4207	484562	44459	44459	44459
65	61	0.3	0.3821	440103	43192	43192	43192
66	62	0.4	0.3446	396910	41580	41580	41580
67	63	0.5	0.3085	355330	39392	39392	39392
68	64	0.6	0.2743	315939	37203	37203	37203
69	65	0.7	0.242	278736	34669	34669	34669
70	66	0.8	0.2119	244066	32020	32020	32020
71	67	0.9	0.1841	212046	29256	29256	29256
72	68	1	0.1587	182791	26491		26491
73	69	1.1	0.1357	156299	23727		23727
74	70	1.2	0.1151	132572	21078		21078
75	71	1.3	0.0968	111494	18429		18429
76	72	1.4	0.0808	93065	16125		16125
77	73	1.5	0.0668	76940	13822		13822
78	74	1.6	0.0548	63119	11748		11748
79	75	1.7	0.0446	51370	10021		10021
80	76	1.8	0.0359	41350	8293		8293
81	77	1.9	0.0287	33057	6796		6796
82	78	2	0.0228	26261	5644		
83	79	2.1					
84	80	2.2					
85	81	2.3	0.0107	12324	2880		
103	99	4.1	0	0	0		
104	100	4.2	0	0	0		
105	Sum				1151800	786219	1099278

Note: Some rows in this table have been hidden for brevity. The complete table can be accessed online through the companion website of the book (www.wiley.com/go/Farrell/Built_Environment_Dissertations_and_Projects).

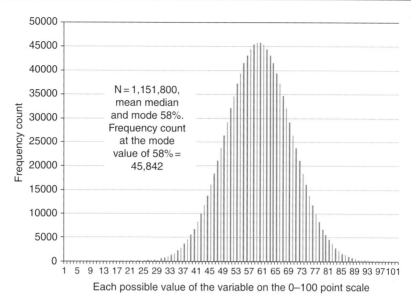

Figure 8.7 Predicted normal distribution for job satisfaction of workers in the UK construction industry.

of the frequency histogram is necessary to decide which is most appropriate. Figure 8.8 indicates six hypothetical datasets plotted in frequency histograms with n of 50; an n of 50 is perfectly acceptable in undergraduate research. There are six potential shapes: normal distribution, skewed distribution, flat distribution, stepped, U-shaped and irregular with no pattern. Whatever the distribution in your dataset, you should comment upon it in your discussion chapter.

Chapter 9 describes that in some inferential statistical tests, it is necessary to make judgements about whether a dataset resembles a normal distribution. While there are statistical tests that can do that, at undergraduate level it is sufficient that you produce a frequency histogram of your data and merely observe how close it mirrors a bell-shaped curve.

8.6 A second variable for descriptive analysis; an IV and a DV

Table 8.1 illustrates one variable only: leadership style. The research may also include a second variable 'motivation of workers'. The objective may be thus: to determine if leadership style (IV) influences motivation of workers (DV). Table 8.7 includes some data for this DV. It is possible to calculate motivation descriptive statistics for each participant and each question: count, mean, median, mode, minimum, maximum, range and standard deviation. However, for brevity in this text, that is not done in table 8.7. Sum and percentage scores are calculated, and the motivation level of workers is shown to be 66.5%. It is possible to descriptively examine whether there may be a relationship between these two variables using a scatter diagram, as shown in figure 8.9. Chapter 9 will deal with calculating a correlation coefficient to illustrate the strength of any relationship and a probability (p) value to indicate whether any

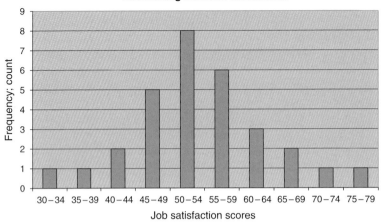

Resembling a normal distribution

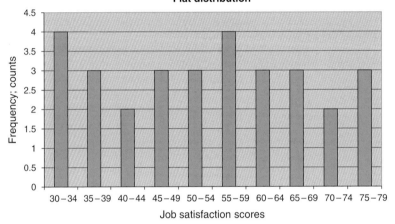

Flat distribution

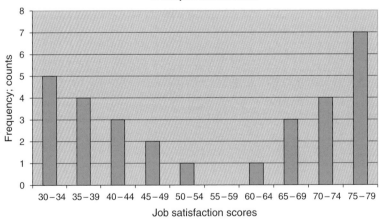

U shaped distribution

Figure 8.8 Six hypothetical datasets plotted in a frequency histogram with *n* of 50: normal distribution, skewed distribution, flat distribution, stepped, U-shaped and irregular with no pattern.

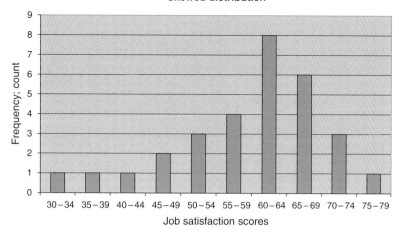

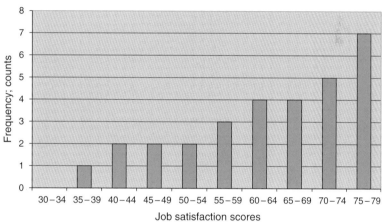

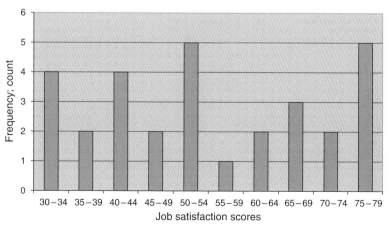

Figure 8.8 (Continued)

Table 8.7 Data added to table 8.3 to show a dependent variable (DV) 'motivation of workers.'

	A	B	I	J	K	L	M	N	O	P
		dependent variable: leadership styl					Dependant variable: Motivation			
1	Question number >	Q1		Q1	Q2	Q3	Q4	Q5	Q11 to Q15 tot	Q11 to Q15 %
2		Coding of possible responses: never 0, rarely 1,	Percentage to two decimal places. Sum divided by max score of 30*100. e.g. 24/30* 100 = 80.00%	I tend to do the minimum amount of work necessary to keep my boss and my team satisfied.	When working on my goals, I put in maximum effort and work even harder if I've suffered a setback	When an unexpected event threatens or jeopardizes my work, I can tend to walk away, and leave someone else to finish the task	I believe that if I work hard and apply my abilities and talents, I will be successful.	My biggest reward after completing something is the satisfaction of knowing I've done a good job		
3						Coding of possible responses: Strongly agree 0, agree 1, unsure 2, disagree 3, and strongly disagree 4				
Participant number ˅									Q11 to Q15 tot	Q11 to Q15 %
4	1		80.00	3	3	4	4	4	18	90.00
5	2		70.00	4	4	4	3	2	17	85.00
6	3		44.00	1	1	1	1	2	6	30.00
7	4		63.33	4	4	3	4	4	19	95.00
8	5		76.67	4	3	3	3	3	16	80.00
9	6	4	60.00	2	2	3	2	3	12	60.00
10	7	2	43.33	2	2	2	1	2	9	45.00
11	8	5	83.33	4	4	4	4	4	20	100.00
12	9	1	16.00	1	1	2	0	0	4	20.00
13	10	2	40.00	3	2	3	2	2	12	60.00
14	Sum	30	576.67	28	26	29	24	26	133	665.00
15	Percentage to two decimal places e.g. 30/50*100 = 60.00%	60.00	**57.67**	2.8	2.6	2.9	2.4	2.6	13	**66.50**

Note: Some columns in this table have been hidden for brevity. The complete table can be accessed online through the companion website of the book (www.wiley.com/go/Farrell/Built_Environment_Dissertations_and_Projects).

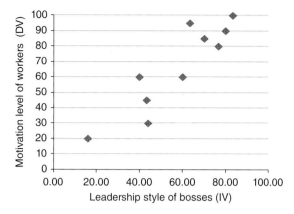

Figure 8.9 Use of a scatter diagram to illustrate a relationship between an IV (leadership style) and a DV (motivation).

relationship may have occurred due to chance. Put these two issues to one side for the moment and just look at the scatter diagram in figure 8.9. It does appear that those leaders with a low score for leadership style have poorly motivated workers and those with a high score for leadership style have more highly motivated workers. If you can design a data collection tool to collect information about two variables (an IV and a DV), as well as ancillary variables, you should do it. If you have got as far as being able to plot a scatter diagram, ideally you should go on to make inferences using the correlation coefficient and p value. If you find yourself short of time, and cannot do the latter, that is fine; just write descriptively about the scatter diagram. When you have read section 9.9.4, you should be able to use Excel to determine that the correlation coefficient for this dataset, based on column K of table 8.3 and column P of table 8.7, is 0.92

Summary of this chapter

It is likely that, even if your study is predominantly qualitatively based, you will collect some statistical data, if only in the literature review. Descriptive statistics are frequently used in industry and academia to summarise large datasets with much fewer numbers and diagrams. Greek letters and italics are used as symbols to abbreviate tests. Computer spreadsheets will help you to handle large sets of data and perform calculations. You should select the descriptive tool appropriate to the data that you have. It may be helpful to arrange data into class intervals and produce frequency histograms; these histograms can be compared to normal distributions. Standard deviation is the most often used measure of spread. The mean is the single figure most often used to summarise a dataset; median and mode are also useful. Z scores allow you to compare your bespoke dataset to the normal distribution curve. If the dataset is such that a mean percentage score can be calculated, that single figure is a powerful statistic on which to base findings and discussion.

References

BCIS (2015) Building Cost Information Service. Available only by subscription at: www.service.bcis. co.uk/BCISOnline/Analyses/AnalysesView?originatingPage=AnalysesResults Accessed 20.07.15.

Bryman, A. and Cramer, D. (1994) Quantitative Data Analysis for Social Scientists. Routledge, London.

Business (2010) Business Balls. Available at: http://www.businessballs.com/freepdfmaterials/ X-Y_Theory_Questionnaire_2pages.pdf Accessed 21.12.15.

DCLG (2015) House Building: March Quarter 2015, England Housing Statistical Release 21 May 2015. Department for Communities and Local Government. https://www.gov.uk/government/uploads/ system/uploads/attachment_data/file/428601/House_Building_Release_-_Mar_Qtr_2015.pdf Accessed 14.08.15.

Hay Group (2011) Getting the balance right: the ratio of CEO to average employee pay and what it means for company performance. https://www.haygroup.com/downloads/uk/Getting-the-balance-right.pdf Accessed 10.08.15.

Ibstock (2015) BS EN 771-1 and PAS 70 – Guide to the Standards. www.ibstock.com/pdfs/technical-support/BSEN771-1-PAS-70-A-guide-to-the-Standards.pdf Accessed 14.08.15.

KPMG (2013) Guide to directors' remuneration. https://www.kpmg.com/UK/en/IssuesAndInsights/ ArticlesPublications/Documents/PDF/Tax/guide-directors-remuneration-2013.pdf p. 15 Accessed 10.08.15.

ONS (2014) Construction Statistics No. 15, 2014 edition, www.ons.gov.uk/ons/rel/construction/ construction-statistics/index.html Accessed 04.09.15.

9 Quantitative data analysis; inferential statistics

The titles and objectives of the sections of this chapter are the following:

9.1   Introduction; to provide context for inferential statistical data collection and analysis
9.2   Probability values and three key tests: chi-square, difference in means and correlation; to explain the p values concept and introduce the tests
    9.2.1 The p value of ≤ 0.05
    9.2.2 Setting the significance level of p; alternatives to 0.05
9.3   The chi-square test; to explain and illustrate manual and computer calculations
    9.3.1 Chi-square by Excel
    9.3.2 The consequence of a different spread of numbers and sample size
    9.3.3 More complex chi-square
    9.3.4 More simple one row chi-square
9.4   Determining whether the dataset is parametric or non-parametric; to explain and provide an example
    9.4.1 How calculation procedures in parametric and non-parametric tests are different
9.5   Difference in mean tests; the t-test; to introduce this test
    9.5.1 Unrelated or related data
9.6   Difference in means; the unrelated Mann–Whitney test; to explain and illustrate manual and computer calculations
    9.6.1 Mann–Whitney by Excel
    9.6.2 Frequency histogram of mean scores
    9.6.3 The consequence of larger sample size
9.7   Difference in means; the related Wilcoxon t-test; to explain and illustrate manual and computer calculations
    9.7.1 Wilcoxon test by Excel
    9.7.2 Frequency histogram of mean scores
9.8   Difference in means; the parametric related t-test; to explain and illustrate manual and computer calculations
9.9   Correlations; to explain and illustrate manual and computer calculations
    9.9.1 The scatter diagram
    9.9.2 The correlation coefficient and its p value
    9.9.3 Longhand calculation for Spearman's rho
    9.9.4 Correlation test by Excel
9.10 Using correlation coefficients to measure internal reliability and validity in questionnaires; to explain and illustrate manual calculations

Writing Built Environment Dissertations and Projects: Practical Guidance and Examples, Second Edition.
Peter Farrell, Fred Sherratt and Alan Richardson.
© 2017 John Wiley & Sons, Ltd. Published 2017 by John Wiley & Sons, Ltd.
Companion Website: www.wiley.com/go/Farrell/Built_Environment_Dissertations_and_Projects

9.11 Which test? to explain how to choose between the different types of test
9.12 Confidence intervals; to describe and illustrate with examples
9.13 Summarising results; to illustrate how to summarise

9.1 Introduction

This chapter builds on the descriptive statistical analysis in chapter 8. Inferential statistical tools allow inferences to be made, that is that results and characteristics from samples reflect those in populations. They also contribute to theory building, and establishing cause and effect relationships between variables.

Better marks may be given if you demonstrate an understanding of inferential statistics. This is justified because there is more complexity in inferential statistics than descriptive statistics, and one facet of dissertations and projects is to test your ability in complex data analysis. Better marks for inferential statistics are also justified in a business sense; since after all, life is about answering the question, 'If I change A, what will happen to B?' An understanding of descriptive statistics is required before you move on to inferential tests, and an understanding of the principles of normal distributions is also required. If you do not feel able to grasp the concepts of inferential statistical analysis, that is fine; concentrate on presenting your descriptive work as best as you can.

The key tests that will be referred to, with their symbol from the Greek or Roman alphabet are the following:

Pearson's chi-square	χ^2	Greek chi
Mann–Whitney statistic (difference in means t-test)	U	
Wilcoxon's statistic (difference in means t-test)	T	
Related t-test	t	
Spearman's rho (correlation)	ρ	Greek rho
Pearson's product moment coefficient (correlation)	r	

Other symbols are:

Significance level	α	Greek alpha
Probability value	p	
The null hypotheses	H_0	
The alternative hypothesis	H_1	
Degrees of freedom	df	

Many tests are named after the statisticians who developed them, e.g. Karl Pearson, (1857–1936); Charles Spearman, (1863–1945); William Gosset, (the t-test) (1876–1937); Frank Wilcoxon, (1882–1965); Ronald Fischer, (1890–1962); Frank Yates, (1902–1994); Andrei Kolmogorov, (1903–1987); Henry Mann, (1905–2000); Maurice Kendall, (1907–1983); Milton Friedman, (1912–2006); Lee Cronbach, (1916–2001); William Kruskal, (1919–2005); Allen Wallis, (1912–1998); John Tukey, (1915–2000).

In this chapter, tests are initially illustrated longhand to show the principles underlying them. The same tests are then performed in Excel. In your document, you need to make a decision about which way to present your results. If you wish to demonstrate that you have understanding of underlying principles, you may illustrate at least one of each test longhand in the main body. If you then need to perform the same type of test again, and that may often be the case even with small datasets, you can put calculations in an appendix. Alternatively, there is no need to put longhand calculations in an appendix; just do one of each type of test longhand and the rest in Excel. It would not be appropriate that tests are completed in Excel, when you are not able to demonstrate understanding of their meaning.

Excel is demonstrated in preference to SPSS or Minitab since Excel will complete most tests that are required at undergraduate level and even above. SPSS (Statistical Package for the Social Sciences; also known as PASW – Predictive Analysis Software) and Minitab are only spreadsheets like Excel, but they are more sophisticated and will do more tests. However, it is another piece of software that you would need to take time to learn. It is unlikely that you will need to use SPSS or Minitab in your career or other study, but very likely that you will need to use Excel or similar. If you want to learn either, that is fine. There are specialist texts such as that by Gray and Kinnear (2014) that will support your learning for SPSS.

There are three key examples used in this chapter to illustrate tests. They are all around the way clients procure work from main contractors: What methods do clients use when they place orders or commission work from the construction industry? Are projects in the construction industry successful? The tests illustrated are also used for comparing differences or looking for relationships in technical data arising from laboratory experiments and fieldwork.

Following on from the work of Latham (1994) and Egan (1998, 2002) many captains of industry argue that best outcomes are achieved if clients partner with their supply chains. This may involve some negotiation about project objectives and the price to be paid. Other literature recognises that industry often needs to demonstrate to its stakeholders that it has attained lowest market price, and this is best proven by competitive bidding processes. Method of procurement is the label for the independent variable (IV) measured at the categorical level with two values: competition or negotiation. The dependent variable (DV) is project success measured by cost predictability. In the first test (the chi-square) the DV is measured at the categorical level with two values: on or under budget, and over budget. In the second test (the Mann–Whitney t-test) the DV is measured at the interval level with many possible percentage values (negative if under budget, zero on budget and positive if over budget).

The project success theme is maintained in a Wilcoxon t-test. At inception and design in traditional procurement methods, construction consultants provide clients with indicative cost estimates. Those estimates are updated as design work proceeds, and design teams will have budgets to work to before prices are sought from contractors. The accuracy of those budgets is important. At the stage where design teams go out to tender they will estimate how much projects will cost, termed 'pre-tender estimates' (stage 1 estimates). Bids are then received from contractors (stage 2 price); hopefully those bids will be at or below the stage 1 'pre-tender estimates'. It is a problem if stage 2 prices are higher than stage 1 estimates. As work proceeds on site, changes are often made. The amount paid to contractors will reflect those changes, and 'final accounts' agreed (stage 3 prices). The difference between contractor bid prices (stage 2) and final account prices (stage 3) is termed 'cost predictability'. Again it is a problem if final accounts are higher than bid prices. The Wilcoxon test will seek to determine if variances between pre-tender estimates (stage 1) and contractor's bids (stage 2) are different from those between contractor's bids (stage 2) and final accounts (stage 3). Figure 9.1 illustrates how estimates and prices may vary; hopefully in practice the figures shown will not increase; the construction process should be one where savings are made if possible.

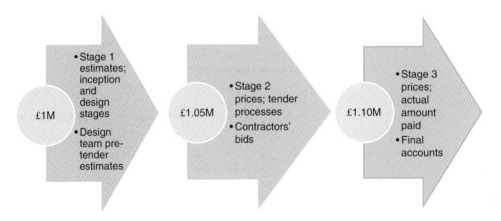

Figure 9.1 Potential for significant differences in prices; cost predictability.

A correlation test is based on two variables: cost predictability and time predictability. Is it the case that projects that have good cost predictability also finish on time, and those that go over cost predictions finish late? Data from completed projects is used to determine if there may be a relationship. The objective is to 'Determine whether there is a relationship between time and cost predictability of projects'. Time predictability is named as the IV and cost predictability as the DV; it could make for an interesting debate about whether it might be the other way round. The objective emerged from the articulation of a problem, which may have raised issues about whether time and cost go hand in hand, and therefore whether energy would be better expended controlling time rather than, as seems at present on many projects, controlling cost.

9.2 Probability values and three key tests: chi-square, difference in means and correlation

Chapter 4 explains that the way we measure things can be put on a continuum of (a) categorical or nominal, (b) ordinal, (c) interval, (d) ratio. These are levels of measurement. The sequence is important: the richness of the data increases as the scale moves from (a) to (d). Similarly, some of the statistical tests can be put on the continuum of (a) chi-square, (b) difference in means, (c) correlations. The richness and power of the tests increases as they move from (a) to (c). These tests are inferential tests, and for a given set of data it is only necessary to perform one of the tests. For all tests, there needs to be two variables, an IV and a DV (or an ancillary variable). One issue that drives which type of test can be used is the level of the data. For example, if the IV and DV were at the categorical level, the appropriate test would be chi-square; if the IV and DV were at the interval level, the appropriate test would be correlation. Indicatively, the richer the level of the data, the greater the possibility to use the richer tests. In the same way that interval data gives greater insight than categorical data, correlation tests give greater insight than chi-square tests. However, in some studies, it is only appropriate to measure variables at the categorical level, and therefore also appropriate that the chi-square test or difference-in-means test are undertaken.

Before you proceed to inferential analysis you need to be clear about the variables you have measured, the possible values for each variable and their measurement level: categorical, ordinal or interval. If you are not clear about these, stop and seek advice.

For each of the three types of test (chi-square, difference in means, correlation) there are two stages in the analysis and two outputs at the end which comprise the results. The chi-square test has a contingency table and then a p value. The difference-in-means test produces mean scores and then a p value. A correlation test has a correlation coefficient and then a p value. So that you do not under-emphasise the first output of each test, it may be useful to think of it as the primary output (that is, depending on the type of test, the contingency table or the mean score or the correlation coefficient) and the p value as the secondary output.

Statistics textbooks seem to create a lot of fuss around the concept of probability or p values, also known as significance levels. As a concept, it is arguably made overly complex, and as a secondary output, inappropriately often takes precedence over the primary output. Students sometimes only report their p values in results sections, and do not report or comment on the primary contingency tables, mean scores or correlation coefficients; nor do they report on the direction of any relationships. It is as though studies are completely overtaken by p values, and Utopia is achieved if p is found to be ≤ 0.05 or some other figure. You need to steer away from this possibility. Whichever test you may use, you must present your results with the primary contingency table, or mean scores, or correlation coefficient, and then, in all cases, the p value.

The p value or probability value is about 'chance'. To emphasise, chance, chance and chance. As a number, it always lies between zero and +1.00; it does not have negative values. When people take risks, there may be a 5 in 100 chance of a particular event occurring, and 5 in 100 is the same as $p \leq 0.05$. In the chi-square, difference in means and correlations tests, we are testing for links between the IV and the DV; that is, does the IV have an association with or causal effect on the DV? In the three tests, the first stage of looking for relationships or associations is to eyeball the data. In the chi-square test, eyeball the contingency table; you will be able to observe whether the dataset is unbalanced, and therefore whether it appears that the IV is influencing the DV or not. In the difference-in-means test, eyeball the mean scores; if there is a large difference in mean scores, that may suggest that the IV influences the DV. In the correlation test, eyeball the scatter diagram; if the data appears to be tightly clustered around a straight line, and though it could be surrounded by a narrow elliptical shape, that may suggest that the IV influences the DV. However, in all cases the spread of the data may make it difficult to assert with confidence that the IV influences the DV. It may be a close call. Therefore, the second stage of the analysis is to perform the statistical test and calculate the p value, to determine whether, to the extent that there may be a relationship, did it occur due to chance?

9.2.1 The p value of ≤ 0.05

The significance level or alpha for tests is often set at ≤ 0.05. At the end of your calculation you will get a p value. If your calculation is undertaken by computer you will get a precise p value to several decimal places. Alternatively, if your calculation is performed manually, you will compare the penultimate figure in the calculation to figures in tables in appendices F1 to F8 to determine whether your p value is ≤ 0.05. If the p value is ≤ 0.05, you will be able to assert that you 'reject the null hypothesis (see section 4.8) and the IV *does* influence the DV, with p set at 0.05'.

It may help to think of this as that you reject the null or the notion that there is no association or no difference or no relationship between the IV and the DV; there is a relationship (potential cause and effect) between the IV and the DV, and you are 95% sure of that. Alternatively, the p value may be >0.05. Therefore, you will assert that you 'cannot reject the null hypothesis, the IV *does not* influence the DV, with p set at 0.05'. It may help that you think of this as that you cannot reject the null or the notion that there is no association or no difference or no relationship between the IV and the DV; it may be that there is no cause and effect between the IV and the DV. If the p value >0.05, some students write words such as 'accept the null hypothesis'. By convention those types of words should not be used. Statistical tests are undertaken against the null and the finding is either, 'reject the null' or 'cannot reject the null'. To emphasise this:

> If the outcome is ≤ 0.05; reject the null hypothesis – there is a relationship or association between the variables.
>
> If the outcome is >0.05; cannot reject the null hypothesis – there is no relationship or association between the variables.

If you find the p value is ≤ 0.05, and you assert 'reject the null hypothesis, with p set at 0.05', you do need to recognise that there is a 5 in 100 *chance* that you could be wrong. The numbers in your calculations could be such that, to the extent that a relationship between the IV and DV is found, it could be a fluke occurrence. If a subsequent study found that your claim for a relationship or association between the variables is incorrect, you will have made a 'type 1 error'. You need to recognise in your conclusions that you could be making a 'type 1 error'. Alternatively, if the p value >0.05 and you assert you 'cannot reject the null hypothesis', there is a possibility that a subsequent study may find that there is a link between the variables. In that case you have made a 'type 2 error'.

The statistical calculations can only manipulate the raw numbers, so whether the p value that is 'spat out' at the end of the calculations is 0.05 or some other number is derived solely from the original raw numbers. The factors that drive the p value result are (a) sample size, (b) differences between numbers, and (c) the spread or distribution of the numbers.

If you find $p \leq 0.05$ in the context of one study, that is not a definitive link; you should recommend more work to test and re-test, with perhaps slightly different methodologies. If the result of your study is that the p value is >0.05 and you 'cannot reject the null hypothesis', it is reasonable that you do not suggest that there is definitely no link between the variables. It is correct that you have not found a link in your data. However, you must have been pointed in some way towards the possibility of a link by your own experience or the literature. You may therefore make a recommendation to re-test, using a larger sample size or slightly different methodology. Only if testing, re-testing and re-re-testing by others does not identify a link, should 'we' abandon that exploration and move on to other potential links to the DV.

9.2.2 Setting the significance level of p; alternatives to 0.05

The terms 'p value' and 'significance level' are sometimes used interchangeably and together. In the media, it may be reported that a particular study has achieved significant results. What lies behind the words 'significant results' is that statistical testing has been undertaken, and

p has been found to be ≤ 0.05 or perhaps another figure. 'Significant results' is a phrase that the general public may loosely associate itself with; it simplifies the academic concepts and language and avoids needing to report to the public that p has been found to be ≤ 0.05.

You should distinguish between significance levels and p values. At the start of your research you should set your significance or alpha level; that is, set it at the outset of your study, and state it in the introduction. You will benchmark your study against the significance level that you set. This is sometimes also called 'setting alpha'. Lots of students blindly get hung up around the 0.05 level. It does not have to be ≤ 0.05. This 0.05 is a level followed by convention in much social science work. It is supported by Siegel and Castellan (1988), but they also state that levels of significance should be set to reflect the 'perceived consequences of the application of the results'. Alternative cut-off points used by convention are:

$p \leq 0.001$; equivalent to a 1 in 1000 chance
$p \leq 0.01$; equivalent to a 1 in 100 chance
$p \leq 0.10$; equivalent to a 10 in 100 or 1 in 10 chance.

The first two are very widely used in many professions. In medical work, for example, with life or death decisions a recurring issue, $p \leq 0.001$ may be appropriate. In non-technical construction research, the issues are often not of such importance and $p \leq 0.05$ may be acceptable. It may be only about a business decision to invest money, and companies may be willing to risk relatively small sums with only a 5 in 100 chance of getting it wrong. Companies may, however, be prepared to invest a larger sum only with a 1 in 100 chance of getting it wrong, and therefore in business p may be set at ≤ 0.01. In undergraduate research, it is unlikely that the 'perceived consequences of the application of the results' is high. It is unlikely that construction companies will invest substantial sums based solely upon your work, or take decisions that may influence issues such as health and safety. Therefore it may seem reasonable at the start of your study, that you set p at ≤ 0.10. If you do this, you need to substantiate that the 'perceived consequences of the application of the results' do not pose business or other risks. Most textbooks are silent on the possibility of setting p at ≤ 0.10; it is often $p \leq 0.05$ or nothing.

There are issues about why 0.05 is the often used cut-off point. Why not 0.03 or 0.08 or 0.15 or 0.21362? Matthews (1998) states that 0.05 was selected by Fischer in the 1920s because it was 'convenient', and that is that. There was a proposal by the British Psychological Society in 1995 to re-examine the use of p values, but it was disbanded because it was thought it 'would cause too much upheaval' for the academic community. Values of p are used in all countries, in all disciplines; they are universal.

At some point there has to be a decision, such as (a) to invest or not to invest the money, (b) to change or not to change the system of work, or (c) to jump or not to jump into the river. As Kerlinger (quoted in Miles and Huberman, 1994) stated, 'everything is either 1 or 0', therefore we often choose to jump at $p \leq 0.05$. However, even after you have set your significance level, be it at ≤ 0.01 or ≤ 0.05 or ≤ 0.10, that still leaves the possibility of some 'greyness'. Imagine a study that was set up to test against a significance level of 0.05, and the outcome would influence a business decision to invest money. Suppose the p value results were just over 0.05, say 0.06 or even up to 0.09. A narrow view would be to find 'cannot reject the null hypothesis', the IV does not influence the DV, do not invest the money. However, a business may want to look more closely at the result. There may be lots of potential advantages if the investment is successful. When you write up your findings, discussion and conclusions, you cannot retrospectively change the significance level from 0.05 to 0.10; that would be cheating. However, a p value that is close to the established significance level will have substantial impact on the way that you write up the latter part of your study, including your recommendations. If an experimental study were set up with ≤ 0.10, and the result was 0.11, that p value of 0.11 gives you greater incentive to continue study in this area, than if a much higher p value, say 0.40, were found.

At the end of your calculations you should give the primary result (contingency table, means scores or scatter diagram) and then state secondly the p value, stating whether it is equal to or lower than the set significance or alpha level. For example, if the result of a statistical test is $p=0.04$, you should assert that: 'p was calculated at 0.04 and with the significance level set at $p \leq 0.05$, the null hypothesis is rejected'. While you may report the actual figure (to two decimal places), the headline issue is whether or not p is ≤ 0.05.

If you set the significance at $p \leq 0.05$, and your result also is within the $p \leq 0.01$ cut off point, you may state that p was calculated at 0.01 and with the significance level set at $p \leq 0.05$, the null hypothesis is rejected. The result is also significant at the conventionally set cut-off point of $p \leq 0.01$. The headline issue remains that you reject the null hypothesis with p set at ≤ 0.05. A significant difference is not an indicator of the size effect, or on the importance of the difference. Judgements on the importance of statistical differences must be made subjectively. A difference between two means of 10% with a small sample size, may not be so important as a difference in means of 5% with a larger sample size, even though both datasets may have p calculated at ≤ 0.05.

9.3 The chi-square test

The Pearson chi-square test, is often just called the chi-square test, without Pearson's name. It is 'square' not 'squared'. Chi is pronounced as 'Ki', rhyming with 'fly'. Sometimes it is represented by the Greek letter χ, and for 'squared' expressed as χ^2, but spelt out it is chi-square. It compares the actual distribution of a set of frequency counts with theoretical or expected frequency counts. If the actual frequencies are 'significantly' different from the expected frequencies, it is inferred that the difference is due to an association between the variables; the null hypothesis will be rejected. If the differences between actual frequencies and expected frequencies are small, it is inferred there is no association between the variables; the null hypothesis cannot be rejected. In this case, to the extent that there are differences between actual frequencies and expected frequencies, these differences only occur due to chance.

Table 9.1 illustrates a 2×2 chi-square contingency table. From the perspective of clients, the objective is to 'determine whether method of procuring work influences cost predictability'. Section 9.1 explains the context of this objective. As a null hypothesis for testing, this is written 'method of procuring work (IV) *does not* influence cost predictability (DV)'. The significance level is set at $p \leq 0.05$. On the one hand the study is intended to be inferential, so that results could be inferred to replicate the whole population. However, a sample of convenience is taken, comprising data from completed projects in your own organisation. This sample of convenience effectively classifies the research as a case study; that is fine, providing its limitations are recognised.

The data is based on the assumption that you are a part-time student who works for a client that procures buildings from the construction industry. Data is obtained from 30 completed projects, selected randomly (30 is the sample size, often referred to as n). If it were possible to get data for more, you should do so. It is not necessary to get equal numbers of 15/15 for the IV in each group, but the more closely the numbers are to each other, the better. To have 16 in one group and 14 in the other is fine, and 17/13, 18/12 is also fine. If the balance starts to tip to 19/11 or 20/10, the sample size of 10 for one group is arguably becoming too low. Group sizes of 25/5 would be completely inappropriate.

The IV 'method of procurement' is in rows, and has two values: competition coded 'A' and negotiation coded 'B'. The DV 'cost predictability' is in columns, and it too has two values: completion on or below budget coded 'C' and completion over budget coded 'D'. If, as a client, you were looking to procure a new project and would like it to complete on or under budget, would you place yourself in the competition row or the negotiation row? Just by eyeballing the data, since 10 out of 13 negotiated projects are successful and only 6 out of 17 competition projects are successful, it would appear that negotiation is best. It is important that you write about the direction of the relationship, not just to say that method of procurement influences cost predictability. You should write that the data suggests that negotiation is best.

The next stage is to determine whether – to the extent that there would appear to be a relationship between the IV and the DV – this relationship could be due to chance. Alternatively, could the data spread be a fluke occurrence? Could it be that by chance we have stumbled on 30 projects that show this relationship? Could it be that if another 30 projects were selected, a relationship might be shown the other way round? The chi-square test will provide the answer, and provide a probability level (p value) indicating whether the relationship can be considered genuine, or not.

The chi-square test is used if both the IV and DV are measured at the categorical/nominal level. Another often used label by statisticians, which can be helpful in thinking through the principles involved, is 'groups': both the IV and DV have 'values', which can be placed into groups. While table 9.1 is a simple 2×2 contingency table, it may be larger, e.g. 2×3, 3×3 etc., or if measures are available in more than two categories. Perhaps a third method of procurement could be introduced for the IV: design and build. As contingency tables get too large, it makes interpretation of the results more difficult. A 2×2 is often adequate for undergraduate work, but perhaps a 2×3 is also appropriate. Single row chi-square tests are also possible (section **9.3.4**). Six stages are presented for the manual calculation of the chi-square test, as follows.

Stage 1: Assemble the raw data

To derive the contingency table in table 9.1, the first step is to assemble the raw data in a spreadsheet, as table 9.2. Table 9.1 has been shown first in this text, just to give an initial overview of the test.

The projects are numbered 1 to 30 in column A. For columns B and C, label each project based on its method of procurement and whether it was completed on budget. Then in column D, in preparation for transferring the data into a 2×2 contingency table, allocate each project to one of four cells. Each project will be one of the following:

- competitive bid, completed on or below budget, coded 1,1
- competitive bid, completed over budget, coded 1,2
- procured by negotiation, completed on or below budget, coded 2,1
- procured by negotiation, ad completed over budget, coded 2,2

Table 9.1 Frequency counts in a 2×2 contingency table; $n = 30$.

	A	B	C	D	E
1			DV: cost predictability		Totals
2			On or below budget: 'C'	Over budget: 'D'	
3	IV: method of procurement	Competition: 'A'	6	11	17
4		Negotiation: 'B'	10	3	13
5	Totals		16	14	30

Table 9.2 Raw data for 30 completed projects. Type of procurement method and whether completed to budget; allocated to one of for cells.

	A	B	C	D
1	Project number	The IV: method of procurement. Competitive bidding = A; negotiation = B	The DV: cost predictability. On or below budget = C, over budget = D.	Contingency table cell; lower case a, b, c, or d.
2	1	B	D	2,2
3	2	B	C	2,1
4	3	A	C	1,1
5	4	A	C	1,1
6	5	A	D	1,2
7	6	B	D	2,2
8	7	A	D	1,2
9	8	A	D	1,2
10	9	A	D	1,2
11	10	B	D	2,2
12	11	B	C	2,1
13	12	A	D	1,2
14	13	B	C	2,1
15	14	A	C	1,1
16	15	B	C	2,1
17	16	A	C	1,1
18	17	B	C	2,1
19	18	A	C	1,1
20	19	A	D	1,2
21	20	A	C	1,1
22	21	B	C	2,1
23	22	B	C	2,1
24	23	A	D	1,2
25	24	A	D	1,2
26	25	A	D	1,2
27	26	A	D	1,2
28	27	A	D	1,2
29	28	B	C	2,1
30	29	B	C	2,1
31	30	B	C	2,1
32		Countif A =	Countif C =	Countif 1,1 =
33		17	16	6
34	COUNTIF frequency counts	Countif B =	Countif D =	Countif 1,2 =
35		13	14	11
36				Countif 2,1 =
37				10
38				Countif 2,2 =
39				3

The frequency counts in column D (rows 33, 35, 37 and 39) figures 6, 11, 10 and 3 are calculated using the Excel function: Formulas > statistics, COUNTIF.

Stage 2: Transfer the raw data to the contingency table

The codes used in column D of table 9.2 are transferred in the contingency table, as illustrated in table 9.3:

- codes 1,1 allocated to the top left cell
- codes 1,2 allocated to the top right cell
- codes 2,1 allocated to the bottom left cell
- codes 2,2 allocated to the bottom right cell

The contingency table is illustrated in table 9.1, based upon tables 9.2 and 9.3.

If you feel unable to go to stage 3 of the analysis, just do stages 1 and 2. If you can write in your document to illustrate your understanding of the contingency table in stage 2 and table 9.1, that may be sufficient to achieve a pass grade, but talk to your supervisor about it. More marks are likely to be awarded if you are able to complete stage 3 of the analysis.

Stage 3: Calculate the expected frequencies

The principle underpinning the chi-square test, is to calculate the frequency counts in each cell that would have been expected, if the IV does not influence the DV; and then compare the actual frequencies (your data) with the expected frequencies. If there are small differences between the actual and expected frequencies, the last stage of the calculation will show the p value higher than your set significance level; your finding will be 'cannot reject the null hypothesis'. Alternatively, if there are large differences between the actual and expected frequencies, the last stage will show the p value lower than your set significance level; your finding will be 'reject the null hypothesis'.

The formula for calculating the expected frequencies in each cell is:

row total, multiplied by column total, divided by overall total.

Thus from table 9.1, for cell '1,1', the row total is 17, column total 16, and for all cells the overall total is 30. Therefore, the expected frequency for cell 'A' is $17 \times 16/30 = 9.07$. As a theoretical number, it is usual to express it to two decimal places. Expected values for cells, 1,1; 1,2; 2,1 and 2,2 are shown in table 9.4.

Table 9.3 Codes for allocating raw data to the contingency table.

	A	B	C	D	E
1			DV: cost predictability		Totals
2			On or below budget	Over budget	
3	IV: method of procurement	Competition	Cell 1,1	Cell 1,2	Total row 1
4		Negotiation	Cell 2,1	Cell 2,2	Total row 2
5	Totals		Total column 1	Total column 2	Overall total

Table 9.4 Expected frequency counts if the IV did not influence the DV.

	A	B	C	D	E
1			DV: cost predictability		Totals
2			On or below budget	Over budget	
3	IV: method of procurement	Competition	9.07	7.93	17
4		Negotiation	6.93	6.07	13
5	Totals		16	14	30

Table 9.5 Four steps to calculate the chi-square value χ^2.

	A	B	C	D	E	F
	Cell	**Step 1**: calculate the expected frequencies for each cell, by 'row total, multiplied by column total, divided by overall total'	**Step 2**: deduct expected and actual frequencies from each other. No negative signs.	**Step 3**: deduct 0.50 from step 2 answers	**Step 4**: square step 3 answers then divide by expected frequency	
1						
2	1,1	17 x 16 / 30 = 9.07	9.07 - 6 = 3.07	3.07 - 0.50 = 2.57	2.57^2 / 9.06 =	0.73
3	1,2	17 x 14 / 30 = 7.93	7.93 - 11 = 3.07	3.07 - 0.50 = 2.57	2.57^2 / 7.93 =	0.83
4	2,1	13 x 16 / 30 = 6.93	6.39 - 10 = 3.07	2.07 - 0.50 = 1.57	1.57^2 / 6.93 =	0.95
5	2,2	13 x 14 / 30 = 6.06	6.06 - 3 = 3.06	3.06 - 0.50 = 2.56	2.56^2 / 6.06 =	1.08
6	Step 5: sum the answers to step 4 to obtain the chi-square value					3.59

Stage 4: Calculate chi-square χ^2

Chi-square is calculated from the mathematical formula, thus:

$$\text{Chi-square} = \frac{\sum\left(\text{observed frequencies} - \text{expected frequencies} - 0.5\right)^2}{\text{expected frequencies}}$$

where Σ is the Greek letter sigma, representing 'the sum of'.

For simplicity, 'stage 4' is separated into five steps in a manual calculation, as presented in table 9.5, and included again in table 9.5 as step 1 is the calculation of the expected frequencies. It will be noted that in step 3, there is a deduction of 0.50. Deducting 0.50 is known as Yates' correction for continuity. Yates' correction is also used if the expected frequency (not actual frequency) in one cell or more is less than 5. Not all authors agree that Yates' correction should be applied; they think it over-corrects. Its effect is to reduce the chi-square value and thus increase the p value. Note, the chi-square value in table 9.5, without Yates' correction, would be:

$$3.07^2 / 9.06 + 3.07^2 / 7.93 + 2.02^2 / 6.39 + 3.06^2 / 6.06 =$$
$$1.04 + 1.18 + 0.61 + 1.54 =$$
$$4.37.$$

The chi-square statistic in table 9.5 is 3.59.

Stage 5: Evaluate the calculated chi-square statistic; degrees of freedom

Compare the calculated value with the critical values of chi-square in the table in appendix F3; state whether the finding is to reject or not to reject the null hypothesis. The operation of the table is based on the chi-square value in your calculation being equal to or exceeding the pre-determined critical values in appendix F3. These values were determined as part of the original design of the test by Karl Pearson. To use the table requires an explanation about the concept of 'degrees of freedom'. The range of chi-square values in the table is given in rows for varying numbers of 'degrees of freedom'. A degree of freedom can be defined as:

> the total number from a sample which has to be known, when the overall total is known, to be able to determine the missing data.

To illustrate this, table 9.6 shows a version of table 9.1 but with data from cells '1,2', '2,1' and '2,2' missing. Given the data in cell 1,1 and the overall totals, we must 'determine the missing data'. Cell '1,2' must be 11 $(17-6)$. If cell '1,2' is 11, cell '2,2' must be 3 $(14-11)$. If cell '1,1' is 6, cell 2,1 must be 10 $(16-6)$. Therefore, with only one 'number from the sample ... known', that is cell '1,1' at 6, and the overall totals are known, that is 17, 13, 16 and 14; it is possible to determine the missing data. The degrees of freedom in this case is 1. The degrees of freedom for a 2×2 contingency table is always 1.

Table 9.7 illustrates a 2×3 table, where the DV, cost predictability, has three values or is in three groups: (a) on or below budget, (b) up to 10% over budget, (c) more than 10% over budget. In a 2×3 table, following through the explanation for table 9.6, the degrees of freedom is always two. The total number from the sample which has to be known is 2; in this case, cells '1,1' and '1,2' are known, to be able to determine the missing data in cells '1,3', '2,1', '2,2' and '2,3'. Rather than draw up a table and think the logic through if you have contingency tables larger than 2×3, the formula for chi-square is:

degrees of freedom = (number of rows – 1) × (number of columns – 1)
for a 2×2 table: $(2-1) \times (2-1) = 1$
for a 2×3 table: $(2-1) \times (3-1) = 2$

Now to the evaluation of the chi-square statistic. For the stated significance level of $p \le 0.05$, the chi-square value in the table in appendix F3 for one degree of freedom is 3.84. Our value of 3.59 does not exceed 3.84, so we cannot reject the null hypothesis with the significance level set at ≤ 0.05. Note that if the significance level had been set at ≤ 0.10, it would have been significant.

Table 9.6 Degrees of freedom; only one piece of data is needed (plus overall totals), to be able to determine the values of the missing data.

	A	B	C	D	E
1			DV: cost predictability		Totals
2			On or below budget: 'C'	Over budget: 'D'	
3	IV: method of procurement	Competition: 'A'	(Cell 1,1) n = 6	(Cell 1,2) missing	17
4		Negotiation: 'B'	(Cell 2,1) missing	(Cell 2,2) missing	13
5	Totals		16	14	30

Table 9.7 A 2×3 contingency table; the DV in three groups.

	A	B	C	D	E	F
1				DV: cost predictability		Totals
2			On or below budget: 'C'	Upto 10% over budget: 'D'	More than 10% over budget: 'E'	
3	IV: method of procurement	Competition: 'A'	(Cell 1,1) n = 6	(Cell 1,2) n = 11	(Cell 1,3) missing	20
4		Negotiation: 'B'	(Cell 2,1) missing	(Cell 2,2) missing	(Cell 2,3) missing	14
5	Totals		16	14	4	34

Stage 6: formally state the finding and direction

'The null hypothesis cannot be rejected with p set at ≤ 0.05; the method of procurement *does not* influence cost predictability; differences noted by eyeball observation in table 9.1 suggest that negotiated projects may give better cost predictability than competitively procured projects; those differences are not significant.'

9.3.1 Chi-square by Excel

Excel will complete stages 4 and 5 above. Stages 1 to 3 still need to be performed manually. Table 9.8 illustrates the actual frequencies and expected frequencies. The Excel procedure is then:

Formulas > More Functions > Statistical > CHITEST. A 'Function arguments' dialogue box will appear. Place the cursor into the 'Actual range' box and drag over cells C3:D4. Then, place the cursor into the 'Expected range' and drag over cells C9:D10.

The chi-square statistic will appear as the 'formula result'. In table 9.7 that is 0.023375841; rounded to two d.p. 0.02. It will be noted that this is a lower p value than that calculated manually in stage 5 above; the p value in that case was between 0.05 and 0.10. The reason for this is that Excel does not use Yates' correction. If the non-use of Yates' correction is accepted, the finding and direction in stage 6 becomes:

'The null hypothesis is rejected with p set at ≤ 0.05; method of procurement *does* influence cost predictability. Negotiated projects give better cost predictability than competitively procured projects.'

9.3.2 The consequence of a different spread of numbers and sample size

Table 9.9 illustrates data where the differences in rows is not so great as in table 9.1. In table 9.1 the frequency counts for cells 1,1; 1,2; 2,1 and 2,2 are 6, 11, 10 and 3 respectively. In table 9.8 the frequency counts are 7, 10, 9 and 4. The differences between the numbers is smaller. The row and column total are the same. The consequences of these smaller differences is that the actual numbers are closer to the expected frequencies and thus the p value moves closer to 1.00, albeit only a marginal move. Calculated by Excel it is 0.127122 or 0.13 to two d.p.; the finding is that the null hypothesis cannot be rejected. With frequency counts of 6, 11, 10 and 3 the p value by Excel was 0.02. As expected and actual frequencies get close to each other, the p value will always move closer to 1.00; in the unlikely event that expected and actual frequency counts were identical, the p value would be 1.00.

Table 9.8 The chi-square calculation in Excel.

	A	B	C	D	E	F
1	Actual frequencies		DV: cost predictability		Totals	
2			On or below budget	Over budget		
3	IV: method of procurement	Competition	6	11	17	
4		Negotiation	10	3	13	
5	Totals		16	14	30	
7	Expected frequencies		DV: cost predictability		Totals	
8			On or below budget	Over budget		
9	IV: method of procurement	Competition	9.07	7.93	17	
10		Negotiation	6.93	6.07	13	
11	Totals		16	14	30	

Function Arguments   ? ☒

CHITEST

Actual_range   C3:D4   🔢 = {6,11;10,3}

Expected_range   C9:D10   🔢 = {9.07,7.93;6.93,6.07}

= 0.023375841

Returns the test for independence: the value from the chi-squared distribution for the statistic and the appropriate degrees of freedom.

Expected_range   is the range of data that contains the ratio of the product of row totals and column totals to the grand total.

Formula result =   0.023375841

Help on this function   OK   Cancel

Table 9.9 A contingency table with smaller differences between frequency counts than table 9.1; $p = 0.13$.

	A	B	C	D	E
1	Actual frequencies		DV: cost predictability		Totals
2			On or below budget	Over budget	
3	IV: method of procurement	Competition	7	10	17
4		Negotiation	9	4	13
5	Totals		16	14	30

Table 9.10 A contingency table with frequency counts increased tenfold over table 9.8; $p = 0.00$.

	A	B	C	D	E
1	Actual frequencies		DV: cost predictability		Totals
2			On or below budget	Over budget	
3	IV: method of procurement	Competition	70	100	170
4		Negotiation	90	40	130
5	Totals		160	140	300

To illustrate the consequence of larger sample size, table 9.10 illustrates frequency counts from table 9.9 increased tenfold such that they are 70, 100, 90 and 40. The p value, calculated by Excel, is $1.405566\text{E-}06 = 1.405566 \times 10^{-6} = 0.000001405566$. To two d.p that is 0.00; the finding is that the null hypothesis is rejected. Since there is a larger sample size, the p value is lower, but be mindful that if using negotiated methods of procurement, based upon the 5, 10, 9 and 4, or 50, 100, 90 and 40 datasets, only 9 out of 13 projects are successful.

9.3.3 More complex chi-square

The 2×2 contingency table is called the simple chi-square. A contingency table larger than 2×2 is a complex chi-square. Table 9.7 uses a 2×3 contingency table to illustrate the principle of degrees of freedom. As a reminder, the DV, cost predictability, has three values or is in three groups: (a) on or below budget, (b) up to 10% over budget, (c) more than 10% over budget. It could be a 2×4 table if data were available to be able to classify the DV as (a) on or below budget, (b) up to 5% over budget, (c) more than 5%, up to 10% over budget, (c) more than 10% over budget. The IV, method of procurement, could possibly be split into three or more groups for a 3×3 or 3×4 contingency table, since some projects may be awarded using design and build systems. The formula and calculation procedure for a complex chi-square is the same as for the simple chi-square, but remember that the degrees of freedom will be higher.

9.3.4 More simple one row chi-square

A 'more simple' chi-square is the one row chi-square. The concept that is different in the one row chi-square is that it compares actual frequencies with theoretical frequencies, the latter ideally from the literature. Two examples are provided.

Firstly, suppose that data is available for completion of projects to time; as a DV this is again expressed as 'time predictability'. It can be in two groups, or have two values, e.g. (a) on time or better, (b) late, or in three or more groups as in table 9.7. The objective is to compare projects completed in a sample, to industry norms; the sample may be projects in a single company (a case study) or in a sector of industry. To the extent that there may be differences between the sample and the industry norm, are those differences so large that they are 'significantly' different; they have not occurred due to chance? The null hypothesis to be tested is 'the data in the sample for completion of projects on time is not different from industry norms'. Table 9.11 illustrates the data.

In the sample, data on 200 projects is available: 98 are completed on time or better, and 102 are completed late. The UK Industry Performance Report related to construction KPIs (Glenigan, 2015) states that in 2015 only 40% of projects came in on time or better. The respective cells are:

- 1,2 = 98 projects actually completed on time or better
- 1,2 = 102 projects actually completed late
- 1,3 = 80 projects (200×0.40) that theoretically should have been completed on time or better
- 1,4 = 120 projects ($200 - 80$) that theoretically would have been completed late.

The formula and calculation procedure for the one row chi-square is the same as the simple chi-square; the degrees of freedom is the number of categories minus 1; for table 9.11 that is

$2-1=1$. At the start of the research it is decided to set alpha or the significance level at $p \leq 0.10$ not 0.05 as described in section 9.2.2. The p value, calculated by Excel is 0.01. The finding is:

The second example of the one-row chi-square is based on a traffic engineer who is interested in road design. Secondary data is available about the number of motor bicycle accidents

> reject the null hypothesis with p set at ≤ 0.10; the data from the sample are significantly different to industry norms published by Glenigan (2015). The performance in the sample is better.

in a locality. The data is shown in table 9.12. The actual numbers of accidents that have taken place are categorised by age of driver in seven age band categories; the degrees of freedom therefore is 6. Also available is an estimate of the age of the UK population of riders in the same age bands, for example 31% of riders are aged 30–39. Expected frequencies are calculated by percentage of riders in each category multiplied by the total number of accidents, e.g. age $30-39 = 31\% \times 395 = 122.45$. The chi-square calculation compares the actual frequencies in row 2 with the expected frequencies in row 4. The p value, calculated by Excel is 0.00. The null hypothesis to be tested is 'the actual number of accidents per age group of riders is not different the theoretical spread'. The finding is:

> reject the null hypothesis with p set at ≤ 0.05; the actual ages of riders involved in accidents are significantly different from expected ages. The data shows that, proportionally, more young riders are involved in accidents.

The important element in the one row chi-square is that the theoretical frequencies should come from the literature, and ideally not an informal assumption or guess (e.g. 50%/50% split for a variable). If you do make an 'informal assumption', you should go as far as possible to substantiate it.

Table 9.11 The one row chi-square. Completion of projects to time.

	A	B	C	D
1	Actual project completions		Theoretical project completions based on the UK Industry Performance Report for 2015 (Glenigan, 2015)	
2	Completed on time or better	Completed late	Completed on time or better	Completed late
3	98	102	80	120

Table 9.12 The one row chi-square. Motorcyclist accidents.

	A	B	C	D	E	F	G	H	I
1	Age of cyclists in years →	<16	16-19	20-29	30-39	40-49	50-59	≥60	Totals
2	Actual number of accidents	8	116	75	86	67	30	13	395
3	Percentage of motor cycling population	1	9	10	31	28	13	8	100
4	Expected number of accidents in each age category given a total of 395	3.95	35.55	39.5	122.45	110.6	51.35	31.6	395

9.4 Determining whether the dataset is parametric or non-parametric

Before progressing to difference in means and correlation tests, an understanding of the difference between parametric and non-parametric data is required. This is not an issue for chi-square tests, since the data is always non-parametric.

Many undergraduate students do not attempt to classify their data as parametric or non-parametric; since there is some modest complexity in it, as an issue it is ignored. In such cases, they may select a non-parametric test or parametric test without justifying their choice and without demonstrating insight into what they are doing. If you can address it as an issue, and illustrate that you have appropriate understanding, it gives you the opportunity to gain extra marks. But if you ignore it or get it wrong it is arguably not catastrophic.

The word parametric is defined in dictionaries as 'boundaries'. In a statistical sense, does the data fall within certain boundaries or parameters? If they do, the data is parametric; if they do not, they are non-parametric. The three boundaries adapted from Bryman and Cramer (2005), are:

(1) the data must be at the level of interval or ratio scales
(2) the distribution of the population scores must be nearly normal
(3) the standard deviation of the variables being investigated should be similar.

Criterion (1) can be assessed by reference to chapter 4.

Criterion (2) judgement can be made in the following two ways.

The first is to plot the data in a frequency histogram, and make a visual or eyeball judgement. Consider the scenario in example 1 of section 8.5.1. A survey is proposed of 50 construction workers to determine their job satisfaction levels on a scale of 0–100. Some hypothetical results are inserted into Excel shown in table 9.13. The data is arranged into class intervals of 5 marks, though starting with a six-wide range of 0 to 5, and then 6–10, 11–15 etc. up to 95–100. Excel is then used to plot the frequency histogram as illustrated in figure 9.2. It could be argued that the frequency histogram in figure 9.2 approximates to a normal distribution.

The second and more accurate way to make a judgement about criterion (2) is the Kolmogorov–Smirnov test. If your dataset does not resemble a normal distribution, the result of the test will be a p value less than your stated significance level. If your dataset does resemble a normal distribution, the result of the test will be greater than the stated significance level. However, given that the Kolmogorov–Smirnov test is best executed by specialist software such as SPSS, it is probably sufficient at undergraduate level to merely use the eyeball test, that is, to plot a frequency histogram, and compare it visually to a normal distribution.

Criterion (3) is often cited as saying that the variance of the variables being investigated should be similar, since the standard deviation is derived from variance. They are both measures of spread; standard deviation is the square root of the variance, e.g. an SD of 10 will have a variance of 100. If the distributions are similar, the ratio of the two standard deviations will be close to 1. The ratio is the larger number standard deviation over the smaller. The F test is used to compare the ratio of the actual variances (or standard deviations) with critical values of F in the table at appendix F2. If the calculated ratio is smaller than the value in the table, it is deemed that any differences between the standard deviations of the (two) variables is not significant. If the calculated ratio is higher than the value in the table, differences are significant, and the dataset is deemed to be non-parametric.

Table 9.13 Frequency counts for job
satisfaction of construction industry
workers in five-point class intervals.

	A	B
	Class intervals	Frequency counts in each class interval
1		
2	0-5	0
3	6_10	0
4	11_15	0
5	16-20	1
6	21-25	2
7	26-30	0
8	31-35	0
9	36-40	4
10	41-45	7
11	46-50	9
12	51-55	8
13	56-60	6
14	61-65	5
15	66-70	6
16	71-75	2
17	76-80	0
18	81-85	0
19	86-90	0
20	91-95	0
21	96-100	0
22	Total	50

To use appendix F2, a brief explanation is required of degrees of freedom in the context of interval data. Stage 5 in section 9.3 defines degrees of freedom as:

> the total number from a sample which has to be known, when the overall total is known, to be able to determine the missing data.

The dataset best used to illustrate degrees of freedom is table 8.3 column K, which measures the leadership style of bosses. There are 10 pieces of data and an overall percentage total of 576.67 in cell 15K. If the data in cell 4K were missing, you would need to know 9 data values in cells 5K to 13K to determine that the missing data in cell 4K is 80: $df = n - 1$. In table 8.3, the value of n is 10, therefore the df is 9. The standard deviation for the data in column K is calculated at 21.47% in table 8.4.

To illustrate how Criterion (3) is judged, in addition to table 8.3 also consider table 8.7, which measures motivation of workers. The data in both tables brought together seeks to fulfil the objective 'to determine if leadership style of bosses influences motivation of workers'. The same workers who scored their bosses leadership style also scored their own motivation.

By Excel, the standard deviation of motivation of workers from the data in cells 5P to 14P is 27.99%. For table 8.7, n is also 10 and $df = 9$. The ratio of the standard deviations of the two

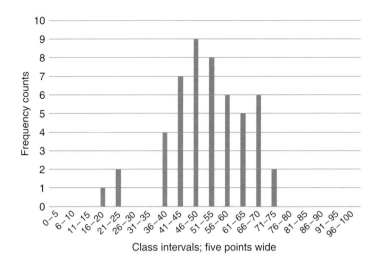

Figure 9.2 Frequency counts: job satisfaction of people working in the UK construction industry where 0 = extremely unsatisfied and 100 = extremely satisfied; n = 50.

variables is 27.99/ 21.47 = 1.30. The critical value of F from appendix F2 is 3.18, so the standard deviations can be deemed similar, and against criterion (3), the dataset is deemed parametric. It will be noted that such a high ratio of 3.18 would not seem to indicate similarity; something much closer to 1.00 would be expected. This is because, in the example used, n = 9 is very small. It will be noted from table F2 that as the value of n increases for each variable, the critical value of F becomes closer to 1.00.

9.4.1 *How calculation procedures in parametric and non-parametric tests are different.*

Parametric tests are more powerful and robust than non-parametric tests, and therefore more desirable. Parametric tests find differences in the data that non-parametric tests may not find. This is because the calculations underpinning the parametric tests are based on the absolute differences between sets of scores. This is not the case for the non-parametric tests, because differences in scores are merely ranked. To illustrate the difference between how parametric and non-parametric tests manipulate data, take for example three scores:

3, 5 and 19.

The parametric test will calculate the difference between 3 and 5 as 2, and the difference between 5 and 19 as 14. These real absolute differences of 2 and 14 will be carried forward to the next step in the calculations.

However, the non-parametric test executed on the same numbers will rank in sequence 3, 5 and 19, with the lowest number ranked number 1:

3 becomes 1
5 becomes 2
19 becomes 3

The non-parametric test will calculate the difference between ranks 1 and 2 as 1, and the difference between ranks 2 and 3 as 1. These diluted rank score differences are carried forward to the next step in calculations. If there were another dataset with the numbers 3, 5 and 6 (instead of 3, 5 and 19), the non-parametric test would still rank the number 6 as 3. The number that is substantially diluted is number 19, but in a parametric test it stands out and will make a real impact on the p value. In the non-parametric test, 19 carries the same weight as the number 6.

There is conflict in the literature about the need to distinguish between parametric and non-parametric data. Some authors argue that since parametric tests are robust, it is not absolutely essential that the data being tested meets parametric requirements. Therefore, even if you judge your data to be non-parametric, arguably you can still go ahead and do parametric tests. However, Bryman and Cramer (2005) argue that the non-parametric Mann–Whitney test is 95% as powerful as the equivalent parametric t-test, which suggests that there is no need to do this.

If you can justify that your dataset is parametric, do the parametric tests. If there is doubt about whether your dataset is parametric, taking the safe position as a measure of prudence, you may choose to do both types of test; that is especially the case if you are using software to perform your calculations. You may find the p value results (and correlation coefficients for correlation tests) from both tests are similar, but if they are not similar, write about that openly in your discussion and suggest how this impacts on your findings and conclusions. If you are clear that your data is non-parametric, do the non-parametric tests. However, to demonstrate your understanding, you may also do the parametric tests. If the p values are not similar for both non-parametric and parametric tests, write about that in your findings and conclusions.

9.5 Difference in mean tests; the t-test

Tests that are sometimes called difference-of-mean tests are richer than the chi-square test. They actually do more than compare differences in means; they compare the distribution of numbers in a dataset. Mean scores, which are a key output from the tests, are more informative than contingency tables. One type of difference-in-means test, often referred to, is the t-test. To use difference-in-means tests, one of the variables (the IV or the DV, it does not matter which) needs to be at the categorical/nominal level. The label 'groups', similar to the concepts of things being arranged into categories, is useful in thinking through principles about when to use difference-in-means tests. For the other variable (the IV or the DV), the level of measurement needs to be at the ordinal or interval level.

The principle of groups is often used in experiments, sometimes but not always involving people. In such cases two (or more) groups are considered and some 'treatment' is administered to one group. The two groups are called, 'A' a control group that does not have treatment, and 'B' an experimental group that is subjected to a treatment. Though strict ethical procedures have to be followed, in medical research the treatment may be some kind of drug, where the IV is treatment, and it has two values: no treatment to group 'A' and treatment to group 'B', and the variable measured for each group could be recovery from illness (the DV). A treatment may not have to be administered since some groups are naturally occurring in society, e.g. group 'A', people who do not smoke, and group 'B', people who do smoke, or group 'A' white collar workers, and group 'B' blue collar workers or the most naturally occurring, group 'A' males and group 'B' females. Lots of work in health-related sectors is based on data being analysed in groups. In construction, methods of work, sector of work, type of materials and lots of ancillary variables can be classified in groups.

9.5.1 Unrelated or related data

In addition to determining whether the data is parametric or non-parametric, there is one more small issue of complexity to consider in the difference-in-means tests that do not appear in the chi-square or correlations: Is the raw data unrelated or related? You will need to classify your data around both issues, such that your data may be:

- unrelated and non-parametric
- related and non-parametric
- unrelated and parametric
- related and parametric.

The above is important since, alongside knowing the number of 'groups', it determines which statistical test is appropriate for use.

The language used by statisticians can be really unhelpful; different words or phrases are often used to describe the same concepts. We just have to accept that, and be able to use these words and phrases interchangeably. The term unrelated is also called by some authors: unmatched, or different, or between subjects, or repeated measures. The term related is also called by some authors: matched, or same, or within subjects. For emphasis, this can be illustrated as follows:

Most construction data is likely to be unrelated. The Mann–Whitney t-test is the appropriate test if there are just two groups for one of the variables, but only if the data is non-parametric. If the data is parametric, the appropriate test is the unrelated t-test, which does not have a person's name assigned to it. Some statisticians call the Mann–Whitney test the 'non-parametric unrelated t-test'.

> Meaning the same thing: related, matched, same, between subjects, repeated measures
> Also meaning the same thing: unrelated, unmatched, different, within subjects

The example in table 9.14 collects data about the cost predictability of 30 projects. An explanation about cost predictability is given in section 9.1. It is a collection of data or numbers that cannot be paired up with any other number in the collection; the data is 'unrelated'. The projects are split into two groups: those procured by competition ($n = 17$) and those procured by negotiation ($n = 13$). Project 1 in column A is procured by negotiation, and it is not related to project 3 (or to any other project) which is procured by competition. There are four extra projects in the competition group.

Related data is where two or more measures are taken for each participant or project. The example in table 9.17 columns E and F collects two pieces of cost predictability data (not one piece of data) for 30 projects: for each project, percentage changes between 'pre-tender estimates' (stage 1 estimates) and contractors' bids (stage 2); and between contractors' bids (stage 2) and 'final accounts'. Statistical textbooks report that measures of the same variable can be of two kinds: (a) matched subjects: scores for a subject can be paired up with scores for another subject, e.g. often used in medical studies involving twins, and (b) repeated measures: two scores are obtained from every subject. Repeated measures may involve collecting data before and after in some kind of experiment; perhaps measures of change. Related data is highly valued since to obtain significant results, smaller sample sizes are required.

If you misinterpret unrelated data for related data, or vice versa, you will have a fundamental (and catastrophic) error in your work. It is therefore important that you are clear on this issue; ideally you should be clear about this before you collect your data. If in doubt, ask for help.

9.6 Difference in means; the unrelated Mann–Whitney test

To illustrate a Mann–Whitney difference-in-means test, the same scenario and study objective is used from the chi-square test in table 9.1. However, in this hypothetical example, assume the data for the DV cost predictability is available at the interval level.

For the chi-square test, cost predictability data was only available at the categorical/nominal level, in two groups: (a) on or below budget, (b) over budget. For the difference-in-means test, cost predictability data is available at the interval level: the percentage figure above or below budget. Since this is a hypothetical example, for the chi-square test, assume that data for the DV was not available at the interval level; if it had been available, the data would have been used at the interval level, and the difference-of-mean test would be undertaken in preference to the chi-square test. The raw dataset is now in table 9.14 in three columns: in column A the project number that has been assigned randomly, in column B the method of procuring the project (which still has two values) and column C cost predictability in percentages. Whether the calculation is to be completed longhand or by Excel is irrelevant; it is best to use Excel to at least set out the data since it facilitates manipulation of numbers and letters in the longhand calculation.

Repeating important information used in the chi-square test, the objective remains to 'determine whether method of procuring work influences cost predictability'. As a null hypothesis for testing, this is written as 'the method of procuring work does not influence cost predictability'. The significance level is set at $p \leq 0.05$. The IV is method of procuring work, which has two potential values or is in two groups A competition, and B negotiation. The DV is cost predictability measured on completion of projects. It is measured at the interval level in terms of percentages. Data is obtained from 30 completed projects, selected randomly.

The data from table 9.14 is copied into a new table 9.15. The longhand procedure is laid out in table 9.15. Two Mann–Whitney U values are calculated. The larger number, which is known notionally as U', pronounced 'U prime', is discarded. The smaller number, which is just called U, is used for interpretation of the p value in the table in appendix F5. The p value will be used, alongside the mean scores and other descriptive statistics, in interpretation of the result. The formula to calculate the first Mann–Whitney U is:

$$\text{Mann-Whitney}U = NaNb + \frac{Na(Na+1)}{2} - \sum Ra$$

where:

Na is the number of scores in group A
nb is the number of scores in group B
Ra is the sum of ranks in list A, smallest score to be ranked 1
$\sum$ is the Greek letter sigma, representing 'the sum of'
Mann–Whitney U is the critical value to be assessed for a p value in the appendix H.

Table 9.14 Raw data for the Mann–Whitney test. Method of procurement with two values or in two groups.

	A	B	C
	Project number	The IV; method of procurement. Competitition = A, negotiation = B	The DV; cost predictability. Percentage change on budget
1			
2	1	B	5
3	2	B	0
4	3	A	0
5	4	A	-1.1
6	5	A	15.3
7	6	B	2.5
8	7	A	1
9	8	A	8.3
10	9	A	6.2
11	10	B	4.8
12	11	B	-3
13	12	A	4.5
14	13	B	-2.8
15	14	A	-7.2
16	15	B	-1
17	16	A	0
18	17	B	-0.1
19	18	A	0
20	19	A	6.6
21	20	A	-0.2
22	21	B	-2.2
23	22	B	-5.2
24	23	A	8.2
25	24	A	2
26	25	A	1.2
27	26	A	3.2
28	27	A	10
29	28	B	-3
30	29	B	-2.2
31	30	B	-2.9
32	Sum		47.9
33	n	30	30
34		Countif = A	Mean
35		17	1.60
36		Countif = B	
37		13	

The second U is calculated by:

NaNb – the result of the calculation from the first U.

The formula for the first Mann–Whitney U can be split into three steps:

(1) NaNb: from table 9.14, Na = 17 and Nb = 13; therefore $17 \times 13 = 221$

(2) $\dfrac{Na(Na+1)}{2}$: from table 9.14, Na = 17; therefore $\dfrac{17(17+1)}{2} = \dfrac{17 \times 18}{2} = 153$

(3) $\sum Ra$: This is the 'tricky' part, the sum of the ranks of list A. The sum of the ranks for group B is not required. Using the Excel functions 'Sort and filter', 'Sort smallest to largest' and 'Expand selection', columns A, B and C are now rearranged such that first listed is project number 14 with percentage change at –7.2%, and last listed is project 5 with percentage change of 15.3%. Column D lists percentage change on budget for only group A projects, and column F for only group B projects. Columns E and G rank the scores.

When ranking, the lowest number is ranked 1. The one issue of complication is where scores are the same, and thus have joint ranks. It is dealt with in the same sort of way that golfers with equal scores are ranked in a tournament. Joint ranks are averaged: if there is an even number of scores that are the same, a half point ranking rises. In table 9.15, projects 11 and 28 are joint 3rd with –3%. There will be no 3rd and 4th rank; they are jointly ranked 3.5. After 3.5 the next rank is 5. Projects 2, 3, 16 and 18 are joint 13th. There is no 13th, 14th, 15th or 16th rank; they are all joint 14.5. If there are an odd number of projects with the same score, the same principle applies, and the mean rank will be a whole number.

It is important to note that when ranking numbers, the 30 projects are treated as one group, which is why project 14 is ranked 1 in column E, but then a jump across to column G to rank project 22 as 2.

The outcome from table 9.15 is that $\sum Ra = 326.5$. The table is also used to calculate the mean scores for group A as 3.41%, group B as – 0.78% and the overall mean as 1.60%. Eyeballing these mean scores suggests that the IV may be influencing the DV; method of procurement may influence cost predictability; the differences in the mean scores seems quite large.

1), 2) and 3) Bringing together the three parts of the first Mann–Whitney U:
$221 + 153 – 326.5 = 47.5$

The second Mann–Whitney U = NaNb – the result of the calculation from the first U
$= 221 – 47.5 = 173.5$. Since this is larger than 47.5, it becomes U' and is discarded.

Taking the value of U at 47.5 to appendix F5, the value in the table for $n = 17$ and $13 = 63$. The value of U must be equal to or smaller than the number in the table for significance: 47.5 is smaller than 63. The finding is:

reject the null hypothesis with p set at ≤ 0.05, method of procurement influences cost predictability. The mean scores are competitively bid projects +3.52%, negotiated projects = –0.78%. Negotiated working projects are more likely than competitively bid projects to complete projects on or below budget. Looking at Appendix F4, Mann–Whitney U statistics with $p = 0.01$, the critical value is 49. An incidental finding is that it is also possible to reject the null hypothesis with p set at ≤ 0.01.

Table 9.15 Raw data rearranged to facilitate the ranking process for the Mann–Whitney test.

	A	B	C	D	E	F	G
	Project number	The IV; method of procurement. Competitition = A, negotiation = B	Sort and filter smallest to largest (expand the selection). The DV; cost predictability. Percentage change on budget	Percentage change on budget for competition projects. Group A	Group A rank	Percentage change on budget for negotiated projects. Group B	Group B rank
1							
2	14	A	-7.2	-7.2	1		
3	22	B	-5.2			-5.2	2
4	11	B	-3			-3	3.5
5	28	B	-3			-3	3.5
6	30	B	-2.9			-2.9	5
7	13	B	-2.8			-2.8	6
8	21	B	-2.2			-2.2	7.5
9	29	B	-2.2			-2.2	7.5
10	4	A	-1.1	-1.1	9		
11	15	B	-1			-1	10
12	20	A	-0.2	-0.2	11		
13	17	B	-0.1			-0.1	12
14	2	B	0			0	14.5
15	3	A	0	0	14.5		
16	16	A	0	0	14.5		
17	18	A	0	0	14.5		
18	7	A	1	1	17		
19	25	A	1.2	1.2	18		
20	24	A	2	2	19		
21	6	B	2.5			2.5	20
22	26	A	3.2	3.2	21		
23	12	A	4.5	4.5	22		
24	10	B	4.8			4.8	23
25	1	B	5			5	24
26	9	A	6.2	6.2	25		
27	19	A	6.6	6.6	26		
28	23	A	8.2	8.2	27		
29	8	A	8.3	8.3	28		
30	27	A	10	10	29		
31	5	A	15.3	15.3	30		
32			Overall mean	Mean group A	Sum of ranks group A (ΣRa)	Mean group B	Sum of ranks for group B not required
33			1.60	3.41	326.5	-0.78	

9.6.1 Mann–Whitney by Excel

The data assembled in columns D and F in table 9.15 are used to calculate the *p* value. The data is represented in table 9.16. The Excel procedure is: Formulas > More Functions > Statistical > TTEST. A 'Function arguments' dialogue box will appear with four boxes to complete:

array 1; block in cells D2:D31
array 2; block in cells F2:F31
tails; choice of 1 or 2; see section 4.8. Normally 2 tail. Type in '2'.

Table 9.16 Using Excel to calculate the *p* value in the Mann–Whitney t-test.

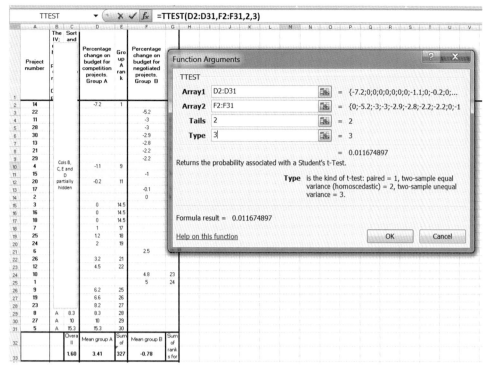

Note: Some columns in this table have been hidden for brevity. The complete table can be accessed online through the companion website of the book (www.wiley.com/go/Farrell/Built_Environment_Dissertations_and_Projects).

Type: paired =1 (this is the Wilcoxon test), two-sample equal variance = 2 (this is the unrelated t-test), two sample unequal variance = 3 (this is the Mann–Whitney test). Type in '3'.

The formula result (the *p* value) is 0.011674897; to two d.p. = 0.01. Reject the null hypothesis.

9.6.2 Frequency histogram of mean scores

A frequency histogram of the data can be drawn to support the eyeball observation and discussion. However, to produce the histogram in a meaningful format, the data needs to be arranged into appropriate class intervals. In this case, 4% class intervals are used, as illustrated in figure 9.3. The frequency histograms are shown. It is visually clear that, for competitively procured projects, some finish on or under budget, but most finish over budget. Also, for projects procured by negotiation, some finish over budget, but the vast majority are on or under budget.

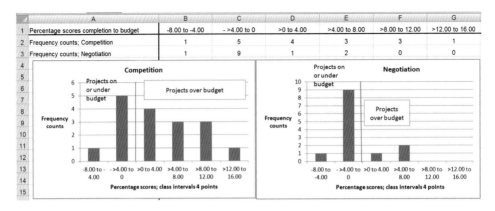

	A	B	C	D	E	F	G
1	Percentage scores completion to budget	-8.00 to -4.00	- >4.00 to 0	>0 to 4.00	>4.00 to 8.00	>8.00 to 12.00	>12.00 to 16.00
2	Frequency counts; Competition	1	5	4	3	3	1
3	Frequency counts; Negotiation	1	9	1	2	0	0

Figure 9.3 Comparison of two groups of data using frequency histograms.

9.6.3 The consequence of larger sample size

Consider again the data in table 9.15. Suppose data is available for a further 30 projects, such that $n = 60$. Suppose instead of 17 competitively procured projects and 13 negotiated projects, there is now data for 34 and 26 projects respectively. Suppose hypothetically, by some fluke, the dataset for the further projects is identical to the original dataset. That would mean that mean scores, standard deviation and other descriptive statistics are the same. Since the sample size is higher, the p value for the Mann–Whitney test, calculated by Excel, for $n = 34$ group A and $n = 26$ group B $= 0.0002$; to two decimal places $= 0.00$. The p value for $n = 16$ group A and $n = 13$ group B $= 0.01$. Therefore with a similar distribution of scores, but with higher sample size, we can be more confident that differences that may occur are not due to chance.

9.7 Difference in means; the related Wilcoxon t-test

As noted in section 9.1 and briefly again in section 9.5.1, the Wilcoxon t-test is used to determine if variances between pre-tender estimates (stage 1 estimates) and contractors' bids (stage 2) are significant; also between contractors' bids (stage 2) and 'final accounts' (stage 3). Excel is used to facilitate the longhand calculation. The data is collated in table 9.17. The Wilcoxon test is used because, importantly, the data is paired or related; there are two pieces of data for each project. As a part-time student, you may be able to access this data from company files, ensuring that you follow ethical protocols. The null hypothesis to be tested is 'the variance between stage 1 and 2 bids is not different from the variance between stage 2 and 3 bids'.

The Wilcoxon test is spilt into ten steps:

(1) Assemble the raw data; columns A, B, C and D.
(2) Calculate percentage changes in columns E and F. The formulas for those calculations are given in cells 1E and 1F.
(3) Determine the differences between each piece of paired data; this is done in column G, and minus signs are maintained. In this column, minus figures have shaded cells.
(4) Remove the minus signs; this is done in column H, though those figures that originated as minus figures remain in shaded cells.

Table 9.17 Raw data for the Wilcoxon test and the calculation procedure to determine t.

Project number	Estimate at pre-tender £ Stage 1	Successful contractor's bid £ Stage 2	Final account £	Percentage change on budget between stages 1 and 2 %. (B-C) / B x 100 x -1	Percentage change on budget between stages 2 and 3 %. (C-D) / C x 100	Difference. Shaded boxes indicates negative figures	Minus signs removed. Shaded boxes indicates figures that were originally negative	Sort and filter figures in column H; smallest to largest. Shaded boxes indicates figures that were originally negative	Ranks for negative differences	Ranks for positive differences
1	121000	119526	156231	-1.22	30.71	-31.93	31.93	0.28		1
2	80000	100080	100080	25.10	0.00	25.10	25.10	0.63		2
3	852000	822632	950220	-3.45	15.51	-18.96	18.96	1.22		3
4	650000	526232	603315	-19.04	14.65	-33.69	33.69	1.44	4	
5	623000	621000	864892	-0.32	39.27	-39.60	39.60	3.06	5	
6	85000	85000	87934	0.00	3.45	-3.45	3.45	3.45	6	
7	899526	901326	858961	0.20	-4.70	4.90	4.90	3.56	7	
8	750000	742056	974830	-1.06	31.37	-32.43	32.43	4.54	8	
9	450000	525100	557658	16.69	6.20	10.49	10.49	4.56	9	
10	424000	418562	472080	-1.28	12.79	-14.07	14.07	4.90		10.5
11	363000	358235	369798	-1.31	3.23	-4.54	4.54	4.90	10.5	
12	190000	192562	204595	1.35	6.25	-4.90	4.90	6.50	12	
13	220000	195800	224029	-11.00	14.42	-25.42	25.42	7.58	13	
14	565000	575585	534124	1.87	-7.20	9.08	9.08	8.10	14	
15	565000	558232	559568	-1.20	0.24	-1.44	1.44	8.64		15
16	575000	574000	590569	-0.17	2.89	-3.06	3.06	9.08		16
17	100000	98222	99971	-1.78	1.78	-3.56	3.56	10.49		17
18	90000	95232	80222	5.81	-15.76	21.57	21.57	12.26	18	
19	650000	650666	693610	0.10	6.60	-6.50	6.50	14.07	19	
20	120000	130123	129863	8.44	-0.20	8.64	8.64	18.96	20	
21	195000	250555	245043	28.49	-2.20	30.69	30.69	21.51	21	
22	200000	190855	180931	-4.57	-5.20	0.63	0.63	21.57		22
23	800000	800780	866444	0.10	8.20	-8.10	8.10	24.35	23	
24	825000	815222	867342	-1.19	6.39	-7.58	7.58	25.10		24
25	440000	450665	456073	2.42	1.20	1.22	1.22	25.42	25	
26	725000	699856	826116	-3.47	18.04	-21.51	21.51	30.69		26
27	890000	862232	1045244	-3.12	21.23	-24.35	24.35	31.93	27	
28	125000	120562	131057	-3.55	8.71	-12.26	12.26	32.43	28	
29	510000	500231	513450	-1.92	2.64	-4.56	4.56	33.69	29	
30	555000	550123	543770	-0.88	-1.15	0.28	0.28	39.60	30	
Sum				30.05	219.34				328.5	136.5
Mean				1.00	7.31					t statistic

(5)   Arrange the differences in column H into order (with minus sign removed). Excel will do this using the functions 'Sort and filter' and Sort smallest to largest'. The results are shown in column I. Again those figures that originated as minus figures remain in shaded cells.

(6)   Rank the differences in two separate columns J and K. Column J is for those figures that originated as a minus figure, and column K for positive. The lowest number is ranked at 1. Note that there are two values the same at 4.90%; in lieu of the ranks 10 and 11, these have the shared ranking of 10.5 as explained for the Mann–Whitney test in section 9.6.

(7)   Sum the ranks in columns J and K; in cells 32J and 32K shown as 328.5 and 136.5 respectively.

(8)   Take the smaller of the two totals as the value of t, i.e.136.5.

(9)   Assess the value of t in appendix F6. For the stated significance level of $p \le 0.05$, the critical t value in the table in appendix F6 for $n = 30$ is 137. The calculated value is required to be equal to or less than the stated value to be significant. Our value of 136.5 is less than the critical value, and we can reject the null hypothesis with the significance level set at ≤ 0.05.

(10)   State the result and finding, thus:

> reject the null hypothesis with p set at ≤ 0.05; 'stage of estimate influences cost predict-ability'. The variance between stage 1 and 2 bids is significantly different from the variance between stage 2 and 3 bids. Estimates at pre-tender stage are 1.0% under-contractors' bids; contractors' bids are 7.31% under final accounts.

9.7.1 Wilcoxon test by Excel

The data assembled in columns E and F of table 9.17 can be used to calculate the p value. The Excel procedure is similar to the Mann–Whitney except for the last step 'type'. It is illustrated in table 9.18. Formulas > More Functions > Statistical > TTEST. A 'Function arguments' dialogue box will appear with four boxes to complete, thus:

Array 1; block in cells E2:E31
Array 2; block in cells F2:F31
Tails; choice of 1 or 2; see section 4.8. Normally 2 tail. Type in '2'.
Type: paired $= 1$ (this is the Wilcoxon test), two-sample equal variance $= 2$ (this is the unrelated t-test), two sample unequal variance $= 3$ (this is the Mann–Whitney test). Type in '1'.

The formula result (the p value) is 0.025040969; to two d.p. $= 0.03$. Reject the null hypothesis.

9.7.2 Frequency histogram of mean scores

As noted in section 9.6.2. for the Mann–Whitney test, a frequency histogram of the data can be drawn to support discussion.

9.8 Difference in means; the parametric related t-test

This test is shown longhand to illustrate the difference between non-parametric and parametric procedures. A key issue to note in the parametric tests is that the computation requires the calculation of real differences between numbers, whereas the non-parametric tests require

Table 9.18 Using Excel to calculate the *p* value in the Wilcoxon t-test.

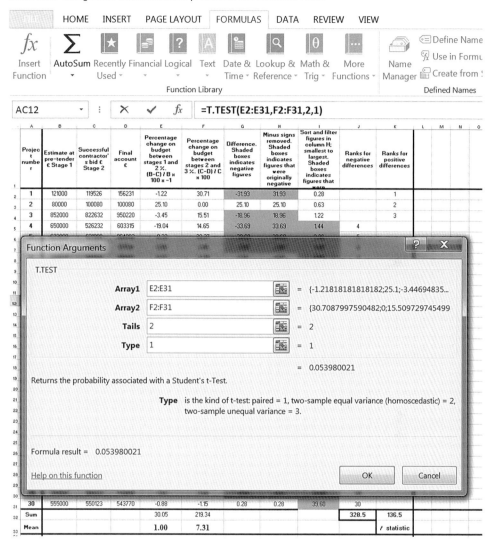

only ranking. Since real differences are used, this parametric related t-test is considered more powerful than the non-parametric Wilcoxon test, where differences are diluted by ranking as illustrated in table 9.17.

The dataset in table 9.17 columns A to F is represented in table 9.19; two new columns G and H are added. Column G shows the real difference between columns E and F.

Do not be put off by the formula for the related *t* test, which is thus:

$$t = \frac{\left(\bar{X} - \bar{Y}\right)}{\sqrt{\dfrac{\sum D^2 - \dfrac{\left(\sum D\right)^2}{N}}{N\left(N-1\right)}}}$$

where:

$\bar{X}$ = the mean of the first list
$\bar{Y}$ = the mean of the second list
d = the difference between paired $\bar{X}$ and $\bar{Y}$ scores
n = the number of score
ΣD^2 = the differences squared, then totalled
$(\Sigma D)^2$ = the differences totalled, then squared.

To simplify the calculation, it is undertaken in eight steps; for ease the first three parts are presented in table 9.19.

Step 1 = $\bar{X} - \bar{Y}$ = cell 32F − 32E = 7.31 − 1.00

$\qquad = 6.31$

Step 2 = ΣD^2 = from cell 32H = 9779.51

Step 3 = $(\Sigma D)^{2/N}$. The sum of D from cell 32G = −189.28

$\qquad = -189.28^{2/30}$

$\qquad = 35{,}826.91 / 30$

$\qquad = 1194.23$

Step 4 = $N(N-1)$

$\qquad = 30(30-1)$

$\qquad = 30 \times 29 = 870$

Step 5 = step 2 − step 3 / step 4

$\qquad = 9779.51 - 1194.23 / 870$

$\qquad = 8585.28 / 870$

$\qquad = 9.86$

Step 6 = square root of part 5 = $\sqrt{9.86}$

$\qquad = 3.14$

Step 7 : the critical value of t = part 1 / part 6

$\qquad = 6.31 / 3.14$

$\qquad = 2.01$

Step 8: observe the stated critical value of t, from the table in appendix F7. Calculated value of t must be equal to or more than the stated value to be significant. For the significance level of $p \le 0.05$, the critical t value for degrees of freedom of 29 $(n-1)$ = 2.045.

Our value of 2.01 does not exceed the critical value. The finding is:

Cannot reject the null hypothesis with p set at ≤ 0.05; 'stage of estimate does not influence cost predictability'. The variance between stage 1 and 2 bids are not significantly different to the variance between stage 2 and 3 bids. Estimates at pre-tender stage are 1.0% under contractors' bids; contractors' bids are 7.31% under final accounts.

Table 9.19 Raw data for the parametric related t-test and the calculation procedure to determine t.

	A	B	C	D	E	F	G	H
	Project number	Estimate at pre-tender £ Stage 1	Successful contractor's bid £ Stage 2	Final account £ Stage 3	Percentage change on budget between stages 1 and 2 %. (B-C) / B x 100 x -1	Percentage change on budget between stages 2 and 3 %. (C-D) / C x 100	Difference: Column E - column F	Difference² or Difference squared
2	1	121000	119526	156231	-1.22	30.71	-31.93	1019.33
3	2	80000	100080	100080	25.10	0.00	25.10	630.01
4	3	852000	822632	950220	-3.45	15.51	-18.96	359.36
5	4	650000	526232	603315	-19.04	14.65	-33.69	1134.97
6	5	623000	621000	864892	-0.32	39.27	-39.60	1567.77
7	6	85000	85000	87934	0.00	3.45	-3.45	11.91
8	7	899526	901326	858961	0.20	-4.70	4.90	24.01
9	8	750000	742056	974830	-1.06	31.37	-32.43	1051.57
10	9	450000	525100	557658	16.69	6.20	10.49	110.01
11	10	424000	418562	472080	-1.28	12.79	-14.07	197.93
12	11	363000	358235	369798	-1.31	3.23	-4.54	20.62
13	12	190000	192562	204595	1.35	6.25	-4.90	24.01
14	13	220000	195800	224029	-11.00	14.42	-25.42	646.04
15	14	565000	575585	534124	1.87	-7.20	9.08	82.39
16	15	565000	558232	559568	-1.20	0.24	-1.44	2.07
17	16	575000	574000	590569	-0.17	2.89	-3.06	9.37
18	17	100000	98222	99971	-1.78	1.78	-3.56	12.66
19	18	90000	95232	80222	5.81	-15.76	21.57	465.47
20	19	650000	650666	693610	0.10	6.60	-6.50	42.22
21	20	120000	130123	129863	8.44	-0.20	8.64	74.57
22	21	195000	250555	245043	28.49	-2.20	30.69	941.86
23	22	200000	190855	180931	-4.57	-5.20	0.63	0.39
24	23	800000	800780	866444	0.10	8.20	-8.10	65.65
25	24	825000	815222	867342	-1.19	6.39	-7.58	57.43
26	25	440000	450665	456073	2.42	1.20	1.22	1.50
27	26	725000	699856	826116	-3.47	18.04	-21.51	462.64
28	27	890000	862232	1045244	-3.12	21.23	-24.35	592.70
29	28	125000	120562	131057	-3.55	8.71	-12.26	150.20
30	29	510000	500231	513450	-1.92	2.64	-4.56	20.78
31	30	555000	550123	543770	-0.88	-1.15	0.28	0.08
32	Sum				1.00	7.31	-189.28	9779.51
33	Mean				Mean list $\overline{X}$	Mean list $\overline{Y}$	$\sum D$	$\sum D^2$

It is noted that this outcome is different from the non-parametric Wilcoxon test; this is fine. On the understanding that the data is parametric, this related t-test result is more powerful than the Wilcoxon test. Excel will not execute the parametric related t-test; you would need to calculate longhand (using Excel to help, as table 9.20) or use SPSS/Minitab.

Table 9.20 Raw data for Spearman's rho and calculating $\sum D^2$ in a table. Raw data in columns B and E; $\sum D^2$ in cell 32I = 2643.

	A	B	C	D	E	F	G	H	I
	Project number	Pre-start percentage change in cost predictability	Sort and filter figures in column B; smallest to largest	Rank cost predictability from column B	Pre-start percentage change in time predictability	Sort and filter figures in column E; smallest to largest	Rank time predictability from column E	Difference: column D - column G	D² or difference squared
2	1	30.71	-15.76	28	1	-3.3	13	15	225
3	2	0.00	-7.20	8	5	-2	21.5	-13.5	182.25
4	3	15.51	-5.20	25	0	-1	8	17	289
5	4	14.65	-4.70	24	1.1	0	14	10	100
6	5	39.27	-2.20	30	10.5	0	26	4	16
7	6	3.45	-1.15	15	0	0	8	7	49
8	7	-4.70	-0.20	4	0	0	8	-4	16
9	8	31.37	0.00	29	12.5	0	28	1	1
10	9	6.20	0.24	16	3	0	18.5	-2.5	6.25
11	10	12.79	1.20	22	12	0	27	-5	25
12	11	3.23	1.78	14	0	0	8	6	36
13	12	6.25	2.64	17	8.5	0	25	-8	64
14	13	14.42	2.89	23	0	1	8	15	225
15	14	-7.20	3.23	2	-2	1.1	2	0	0
16	15	0.24	3.45	9	0	2	8	1	1
17	16	2.89	6.20	13	-3.3	2	1	12	144
18	17	1.78	6.25	11	2.1	2.1	17	-6	36
19	18	-15.76	6.39	1	5	3	21.5	-20.5	420.25
20	19	6.60	6.60	19	-1	3	3	16	256
21	20	-0.20	8.20	7	0	5	8	-1	1
22	21	-2.20	8.71	5	5	5	21.5	-16.5	272.25
23	22	-5.20	12.79	3	0	5	8	-5	25
24	23	8.20	14.42	20	15	5	29	-9	81
25	24	6.39	14.65	18	8	8	24	-6	36
26	25	1.20	15.51	10	2	8.5	15.5	-5.5	30.25
27	26	18.04	18.04	26	5	10.5	21.5	4.5	20.25
28	27	21.23	21.23	27	20	12	30	-3	9
29	28	8.71	30.71	21	2	12.5	15.5	5.5	30.25
30	29	2.64	31.37	12	3	15	18.5	-6.5	42.25
31	30	-1.15	39.27	6	0	20	8	-2	4
32		7.31			3.81				2643
33		Mean Score			Mean Score				Σ D2

9.9 Correlations

Correlations are richer than chi-square and difference-of-mean tests, and they are more inform-ative than contingency tables or mean scores. You should firstly assemble the raw data and plot a scatter diagram with the independent variable on the x or horizontal scale, and the dependent variable on the y or vertical scale. Then calculate the correlation coefficient and its p value.

To use correlation tests, both of the variables need to be at the ordinal or interval/ratio level. Scores are paired; a score on one variable is automatically paired with a score on another variable. The issue of whether the datasets are non-parametric or parametric needs to be considered for both variables. Both variables need to be parametric to use the Pearson correla-tion. If one or both are non-parametric, the Spearman rho test is appropriate.

The IV and the DV may be measured using the same units, e.g. interest rates and inflation rates both expressed as percentages. Alternatively, they may have completely different measurement units, e.g. cement content in concrete in kg/m^3, and compressive strength of concrete in N/mm^2. Measures such as interest rates and inflation again (there are lots of others too) may be paired by time periods, e.g. each calendar month over several years, or average annual rates paired over many years. The two variables often measure different concepts; they may, however, measure exactly the same concept, but perhaps in two different countries, e.g. construction unemployment figures over a given period.

There is the possibility of taking linear correlations to the next stage of analysis, linear regression. This involves calculating a line of best fit and plotting it on the scatter graph. This will allow, given the score of one variable, a prediction of a score of the other variable. If you wish to do this, refer to a specialist statistical text such as Hinton (2004).

As a precursor to starting the correlation tests, as always, you should undertake the descriptive analysis around both variables; that is, count, sum, mean, percentage, median, mode, minimum, maximum, range and standard deviation. There is a lot of potential for discussion around the mean scores in their own right: are they higher or lower than might reasonably be expected?

9.9.1 The scatter diagram

Figure 9.4 illustrates six potential scatters as part of the first stage of analysis.

Scatter 1: The data may tightly cluster around a theoretical straight line, as though close to a thin elliptical shape. This indicates a strong relationship. A given manipulation of the IV may result in a strong movement of the DV.

Scatter 2: The data may have a modest cluster around a theoretical straight line, though a fatter elliptical shape. This indicates a relationship, though not so strong. A given manipulation of the IV may result in a modest movement of the DV.

Both scatters 1 and 2 go up left to right indicating a positive relationship. Manipulating the IV up similarly moves the DV up (or manipulating the IV down similarly moves the DV down). Correlation coefficients in such cases are positive between 0 and 1; for scatter 1 it is calculated by Excel to be +0.95, and scatter 2 = +0.63.

Scatter 3: This has a good cluster around a theoretical straight line, but it is a negative relationship. Manipulating the IV up moves the DV down (or manipulating the IV down moves the DV up). For example, convention in economics is that as interest rates move up, inflation rates will move down. Correlation coefficients in such cases are negative between −1 and 0; for scatter 3 it is calculated by Excel to be −0.83.

Scatter 4: There may be outliers, and although the general spread may indicate a relationship, there may be one or two measures that are not consistent with the majority. Consideration may be given to whether such measures are spurious; they may be just bad or faulty data. Analysis and findings can be undertaken with and without the outliers present in the dataset.

Scatter 5: The shape and pattern of the dataset may not cluster; it may be completely random or square or close to circular. This indicates no relationship.

Scatter 6: The data may be parabolic or curve linear; this would require careful interpretation.

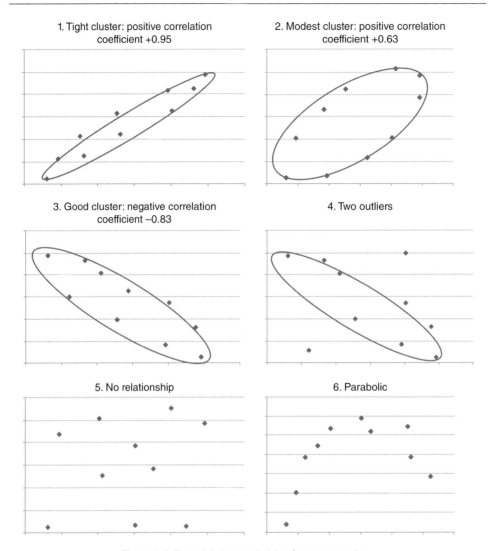

Figure 9.4 Potential shapes deriving from scatter plots.

9.9.2 The correlation coefficient and its p value

Correlation coefficients always lie in the range −1.00 to +1.00 (note the negative figure). Values of p always lie in the range 0.00 to +1.00 (note there is no negative). The two figures must be reported, and must not be confused; their meanings are fundamentally different. Strong correlations are close to −1 or +1, and they resemble a straight line relationship. A correlation coefficient of 1.0 or −1.0 is perfect; correlation coefficients of 1.0 or close to 1.0 rarely occur. Studies may report success, with correlation coefficients of circa 0.30 or −0.30. Though a correlation coefficient is a numerical value on the scale of −1.0 to +1.0, qualitative labels are often loosely ascribed to the quantitative correlation coefficients, e.g.:

0.00 to 0.25 (or −0.25) = little or no relationship
0.26 to 0.50 = fair degree of relationship
0.51 to 0.75 = moderate to good relationship
0.75 to 1.00 = very good to excellent relationship

Alternatively, Cohen and Holliday (1996, pp. 82–83) offer their 'rough and ready guide':

0.00 to 0.19 = a very low correlation
0.20 to 0.39 = a low correlation
0.40 to 0.69 = a modest correlation
0.70 to 0.89 = a high correlation
0.90 to 1.00 a very high correlation

Alongside the correlation coefficient you should also calculate the correlation coefficient squared (or r^2), sometimes called the coefficient of determination; it always lies between 0 and +1.00. The r^2 value indicates the 'percentage of the variation in one variable that is related to the variation in the other'. A correlation coefficient of 0.90 or −0.90 thus becomes +0.81, or 0.50 becomes 0.25; in interpretation, they are then expressed as a percentage, e.g. 25%, meaning that 25% of the movement in one variable (the DV) is due to movement in the other variable (the IV). This concept is useful in understanding that it is not only the IV that influences the DV; it sits very well alongside the phrase in chapter 4, that there are 'lots of variables at large'. If the DV is, for example, profit there are clearly lots of variables that influence that; if your study can just identify one variable that has a very modest influence on profit, that would be worthwhile. You should note that some statisticians argue that it is not appropriate to compute r^2 from the non-parametric Spearman correlation coefficient.

When interpreting the test, if a p value is found equal to or less than the stated significance level, the null hypothesis will be rejected; there is a significant relationship between the variables. If a p value is found to be more than the stated significance level, the null hypothesis cannot be rejected; there is no relationship found between the variables.

9.9.3 Longhand calculation for Spearman's rho

As noted in 9.1, the example used to complete a longhand correlation test is based on two variables, cost predictability and time predictability. As a part-time student you may take data from completed projects to meet the objective to 'determine whether there is a relationship between time and cost predictability of projects'; be careful to follow ethical guidelines.

Spearman's rho (sometimes written ρ or r_s) is calculated from the mathematical formula, thus:

$$Spearman's\ rho = 1 - \frac{6\sum D^2}{\left(N^3 - N\right)}$$

where:

Spearman's rho is the critical value to be assessed for a *p* value in the table
Σ is 'the sum of'
D is the differences between ranked scores
N is the number of paired scores

Table 9.20 illustrates the raw data and calculation of ΣD^2. Cost predictability percentages are copied from column F in table 9.19. For the sake of brevity in table 9.20, the originating data for time predictability percentages is not provided; the method of calculating percentages is simply: actual duration – contract duration/contract duration × 100. Thus a project with actual duration of 55 weeks and contract duration of 50 weeks, the calculation would be: $(55 - 50)/50 \times 100 = 10\%$. The scatter diagram is illustrated in figure 9.5.

Eyeball observation suggests a positive relationship; the majority of the points on the graph could be surrounded by a 'thinnish' elliptical shape. The relationship does appear to be close to linear, not curved or parabolic. There are two potential outliers: project 1 was 30.71% over budget, but only 1.00% over time, and project 18 has a 15.76% cost saving, but went 5.00% over time.

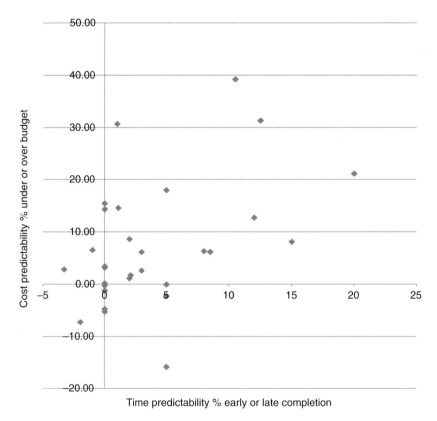

Figure 9.5 Scatter diagram to illustrate the relationship between the IV time predictability, and the DV cost predictability.

The stage 2 process is to calculate the correlation coefficient and its p value. There is a ranking procedure, as explained in section 9.6 for the Mann–Whitney test. Raw data for cost predictability is in table 9.20 column B; it is rearranged into order in column C and ranked in column D. Similarly for time predictability in columns E, F and G. Differences between ranks are calculated in column H and those differences are squared in column I.

Supported by Excel, the calculation can be completed in five steps, thus:

Step 1: ΣD^2, is calculated in table 9.20 cell 32I as 2643

$$\text{Step 2}: 6\sum D^2 = 6 \times 2643$$
$$= 15,858$$

$$\text{Step 3}: \left(N^3 - N\right) = \left(30 \times 30 \times 30\right) - 30$$
$$= 27,000 - 30$$
$$= 26,970$$

$$\text{Step 4}: \text{Step 2 divided by step 3} = 15,858 / 26,970$$
$$= 0.59$$

$$\text{Step 5}: 1 \text{ minus step } 4 = 1 - 0.59$$
$$= 0.41$$

Step 6: Determine the p value and state the finding. From appendix F8, the critical value of Spearman's rho with the significance level set at 0.05 $= 0.364$. The calculated value of 0.41 must be equal to or more than the critical value. As 0.41 is more than 0.364, so the finding is:

> Reject the null hypothesis with p set at ≤ 0.05; there is a relationship between time predictability and cost predictability. The correlation coefficient is 0.41; this can be classified as modest. Projects that beat time targets may beat budgets. Projects that complete on time may complete to budget. Projects that overrun in terms of time may also overrun budget. The mean percentage cost overrun is 7.31%, and the mean percentage time overrun is 3.81%.

As in all statistical calculations, in correlation calculations larger sample sizes are likely to give more significant results.

9.9.4 Correlation test by Excel

Excel will perform the Pearson's correlation; it does not perform Spearman's. If you want to use Spearman's you will need to do it long hand using Excel as in table 9.20 (it is quick to do) or use SPSS. The data assembled in columns B and E of table 9.20 are used. The Excel procedure is similar to the Mann–Whitney except for the last step 'type'. Formulas > More Functions > Statistical > CORREL. It is illustrated in table 9.21. A 'Function arguments' dialogue box will appear with four boxes to complete, thus:

Array 1; block in cells B2:B31
Array 2; block in cells E2:E31

Table 9.21 Using Excel to calculate Pearson's correlation coefficient.

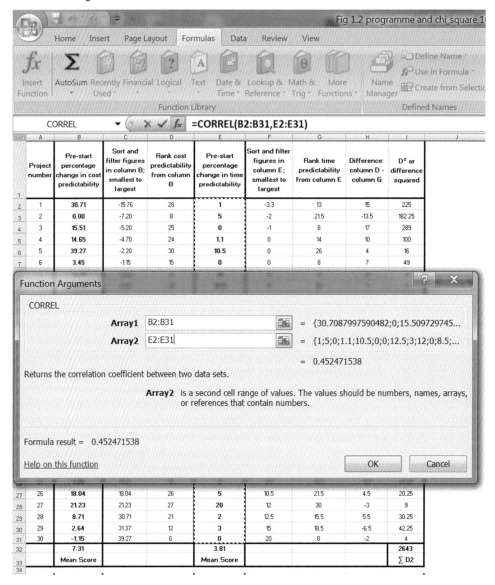

The formula result is 0.452471538; to two d.p. = 0.45. This is the correlation coefficient, which can be classed as modest. To determine the *p* value refer to appendix F9. The critical value of Pearson's r with the significance level set at 0.05 and $n - 2$ (say 25) = 0.381. The calculated value of 0.45 must be equal to or more than the critical value. As 0.45 is more than 0.381, the finding is, as above, reject the null hypothesis with *p* set at ≤ 0.05.

9.10 Using correlation coefficients to measure internal reliability and validity in questionnaires

Using multiple-item scales inherently improves reliability. Single item questions about one issue are unreliable. It is better to use multiple-items that combine to measure one variable. Figure 5.3 gives an example of a multiple-item scale to measure leadership style of bosses. There are six questions used to make up the scale. All six questions should be sub-measures of the overall measure of leadership style. It follows that:

participants who score highly for one question should score highly for others
participants who score mid-range for one question should score mid-range for others
participants who score low for one question should score low for others.

It is possible to check the internal reliability of the multiple-item scale, that is to check each question individually to ascertain if its answers are correlating with the others. Table 9.22 sets out the data based on table 8.2. Columns I to N are added. Column I is for question 1, and is the total score in column H minus its own score in column B. A scatter diagram showing the relationship between columns B and I is shown in figure 9.6. By observation it seems that question 1 seems to be internally 'reliable'; participants who scored high for Q1 scored high in the total for the other five questions, and participants who scored low for Q1 scored low in the total for the other five questions. There is just one outlier; that is participant 6. The correlation coefficient between columns B and I calculated by Excel is 0.78, which is deemed as a high correlation.

The correlation coefficients for Q2 to Q6 are shown in row 15. All appear to be high positive correlations, with the exception of Q5 which is −0.36. The scatter diagram in figure 9.7 confirms there is a problem with this question 'I am incentivised to work hard and well'. Participants who scored *high* for Q5 scored *low* in the total for the other five questions, and participants who scored *low* for Q5 scored *high* in the total for the other five questions. In the context of this study, this is not a reliable question; it is not doing its job. It is not measuring leadership style.

If you have piloted the questionnaire, you may change this question before the main study is administered. If you have not piloted it, you should write under a section heading 'Limitations and criticisms of the study' about how this unreliable question may have skewed results, or alternatively you may prefer to conduct the main analysis, with answers to this question excluded.

If you have used a gold standard questionnaire, on the one hand it may be argued that such tests are not necessary since they have already been done. However, it is appropriate that you still perform them. Perhaps the questionnaire has not been tested in the context of a study similar to yours; also doing the tests is an opportunity for you to demonstrate your analytical skills and understanding of reliability. That should get you extra marks.

9.11 Which test?

This chapter has covered (a) chi-square, (b) difference in means, (c) correlation tests. Which test to use is driven by the level of measurement of the data: (a) categorical or nominal, (b) ordinal, (c) interval or ratio. There are also important issues about whether the data is

Table 9.22 Correlations to calculate internal reliability of a multiple-item scale.

	A	B	C	D	E	F	G	H	I	J	K	L	M	N
1	Question number >	Q1	Q2	Q3	Q4	Q5	Q6							
2		Coding of possible responses: never 0, rarely 1, occassionally 2, often 3, mostly 4 and always 5												
3	Participant number v	My boss asks me politely to do things, gives me reasons why, and invites my suggestions.	I am encouraged to learn skills outside of my immediate area of responsibility.	I am left to work without interference from my boss, but help is available if I want it.	I am given credit and praise when I do good work or put in extra effort.	I am incentivised to work hard and well.	If I want extra responsibility, my boss will find a way to give it to me.	Sum Q1 to Q6	Total column H minus column B (Q1)	Total column H minus column C (Q2)	Total column H minus column D (Q3)	Total column H minus column E (Q4)	Total column H minus column F (Q5)	Total column H minus column G (Q6)
4	1	5	4	5	3	2	5	24	19	20	19	21	22	19
5	2	4	5	4	4	1	3	21	17	16	17	17	20	18
6	3	0	2	2.20	3	4	2	13.20	13.2	11.2	11	10.2	9.2	11.2
7	4	3	3	3	4	2	4	19	16	16	16	15	17	15
8	5	4	5	3	5	1	5	23	19	18	20	18	22	18
9	6	4	3	2	5	2	2	18	14	15	16	13	16	16
10	7	2	2	2	0	5	2	13	11	11	11	13	8	11
11	8	5	5	5	4	1	5	25	20	20	20	21	24	20
12	9	1	1	0	0	0.80	2	4.80	3.8	3.8	4.8	4.8	4	2.8
13	10	2	2	2	3	0	3	12	10	10	10	9	12	9
15	Correlation coefficents	Q1 = 0.78	Q2 = 0.89	Q3 = 0.87			Q3 = 0.87	Q4 = 0.56	Q5 = - 0.36		Q6 = 0.70			

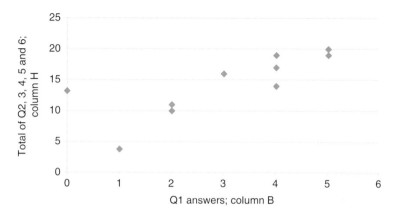

Figure 9.6 Scatter diagram showing the relationship between question 1 and the other five questions; correlation coefficient = 0.78.

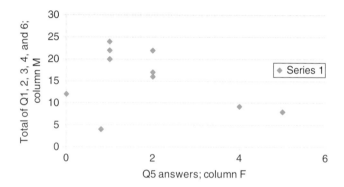

Figure 9.7 Scatter diagram showing the relationship between question 5 and the other five questions; correlation coefficient = −0.36.

non-parametric or parametric, and in difference-in-means tests whether the data is related or unrelated. Given the range of tests, how is it decided which test to undertake?

The lead issues in making a choice are the following:

- Be absolutely clear about your IV and DV.
- Be clear about the level of measurement of both variables (categorical/nominal, ordinal, interval or ratio). Interval and ratio can be considered as though the same level of data when selecting the test.

Stage 1 in making a decision is illustrated in table 9.23. Given three possible levels of measurement for each variable, there are six possible permutations when the two variables are paired. The table shows that for rows 1, 4, 5 and 6, you can go almost immediately to the appropriate test. If you are clear that you are chi-square or correlation, the options are few.

Table 9.23 Which test; chi-square, difference-in-means or correlation?

	A	B	C	D
1		Level of measurement for the two variables (IV and DV); 6 permutations	Which test category	Test name
2	1	Categorical and categorical	Chi-square	Pearson's chi-square
3	2	Categorical and ordinal	Difference in means	See table 9.22
4	3	Categorical and interval		
5	4	Ordinal and ordinal		Spearman's rho
6	5	Ordinal and interval	Correlation	
7	6	Interval and interval		Pearson's product moment

The only choice in the chi-square test is the size of the contingency table, e.g. 2×2, 2×3 etc. The only choice in correlation tests is the Spearman test for non-parametric data or the Pearson test for parametric data.

In difference-in-means tests, the choices are many. When you select your tests, the issue is complicated in the statistical textbooks since the same tests are sometimes referred to by many different names. Rows 2 and 3 in table 9.23 indicate that you must refer to table 9.24. In table 9.24 there are three factors shown in columns A, B and C that drive the choice of difference-in-means test:

There are two options in each of these three factors:

Column A: How many groups; classify as either two or more than two – more than two groups can also be labelled as three or more groups?

Column B: Are the datasets unrelated or related (see section 9.5.1)?

Column C: Are the datasets non-parametric (C1) or parametric?

It is absolutely essential that you are clear about columns A and B. It is desirable that you are clear about whether your datasets are non-parametric or parametric in column C.

Three factors (A, B and C), considered alongside each other with two values for each factor, give eight possible permutations and eight types of test. There are many other tests too for work at advanced levels. It is likely that the most often used difference-in-means test at undergraduate level is the Mann–Whitney. At this level, it is probably perfectly acceptable, if your study is group-based, that you restrict yourself to just two groups. This will lessen the need to consider rows 5 to 8 in table 9.24. You can substantiate just two groups in the context that your objectives should be modest, and there is no need, for example, to measure the sustainability performance (the DV) of construction materials in three groups, e.g. steel, concrete and timber, when better depth is possible and better knowledge may be gained by just examining two. If the articulation of your problem really drives one of your variables to be split into three or more groups, that is fine. But you can still lessen the need to do tests identified in rows 5 to 8, by doing analysis in pairs, e.g. three groups A, B and C, all with a list of scores on some concept, perhaps sustainability performance. First test scores of group A versus group B, then group A versus group C, then group B versus group C. While doing it this way, there are three tests instead of one,

Table 9.24 Which difference-in-means test?

	How many groups; 2 or more than 2?	Unrelated or related?	Non-parametric or parametric?	Type of test?	Recommended software?
	A	B	C		
2					
3	2 groups	Unrelated	Non-parametric	Mann-Whitney 'U' test	
4			Parametric	Unrelated 't' test	Excel
5		Related	Non-parametric	Wilcoxon matched pairs signed rank	
6			Parametric	Related 't' test	
7	More than 2 groups	Unrelated	Non-parametric	Kruskal-Wallis	
8			Parametric	Unrelated ANOVA	SPSS
9		Related	Non-parametric	Friedman	
10			Parametric	Related ANOVA	

the output for a singular test, e.g. the Kruskal–Wallis test, will tell you if there is a significant difference somewhere in the three groups. It may be that two groups are similar, and one group is different from one or both of the others. Such an outcome will warrant further interrogation of the data, to determine precisely where differences lie; advanced studies may use Tukey post-hoc analysis.

In the same sort of way that language used by statisticians can be unhelpful when different words or phrases are used to describe unrelated and related data, the language is also unhelpful when naming tests. For example, the Mann–Whitney U test is also called a t-test (a non-parametric unrelated version of a t-test). Also the Wilcoxon is called a T-test, but with a capital letter (a non-parametric related version). The unrelated t-test (row 2 of table 9.25), is also called, among other things, an independent t-test. An ANOVA, the name used for some tests involving three or more groups, is merely an abbreviation for 'analysis of variance'. This name can be used to loosely describe all inferential tests, since whether there are two or more groups, providing there are at least two variables, all tests are analysing variance. You need to try to cut your way through the language, as far as is possible.

9.12 Confidence intervals

Confidence intervals are potentially an added aspect of complexity. If you can manipulate the formula, and show understanding of the principles involved, that gives you the opportunity to gain extra marks. Its application is in the calculation of mean scores; and remember that mean scores are part of the descriptive analysis, if you are performing difference in mean tests or correlations.

Mean scores are often determined for samples: the sample size may be just 30, with a population size much larger than that. The whole sampling strategy hangs on the premise that results from the sample reflect the results that would have been achieved, had it been possible to survey the whole population. In table 8.3 cell 17K, the leadership style mean score is 57.67%; are we confident that if we were able to survey the whole of the population, we would get that same mean score? The answer is no.

Therefore, what confidence intervals will do is to set a range of scores, where we can state that we are 95% confident (or other specified significance level) that the mean score of the

Table 9.25 Summary of statistical test results in this chapter.

Test number	Chapter and table reference	Hypothesis with variables	Test type	Alpha level (sig)	Results	p value	Reject or cannot reject the null hypothesis	Manual or by Excel
1	9.3 and table 9.5	method of procuring work (IV) does not influence cost predictability (DV)	2 x 2 Chi-square	0.05	see contingency table 9.1	> 0.05	Cannot reject	Manual
2	9.3.1 and table 9.7	ditto	ditto	0.05	ditto	≤ 0.05	Reject	Excel
3	9.3.4 and table 9.11	the data in the sample for completion of projects on time is not different to industry norms	Single-row chi-square	0.10	see contingency table 9.11	≤ 0.10	Reject	Excel
4	9.3.4 and table 9.12	the actual number of accidents per age group of riders is not different the theoretical spread	ditto	0.05	see contingency table 9.12	≤ 0.05	Reject	Excel
5	9.6 and table 9.13	method of procuring work (IV) does not influence cost predictability (DV)	Mann-Whitney	0.05	negotiated projects complete 0.78% below budget; competitive projects complete 3.52% over budget	≤ 0.05	Reject	Manual
6	9.7 and table 9.17	the variance between stage 1 and 2 bids is not different to the variance between stage 2 and 3 bids	Wilcoxon	0.05	the variance between stage 1 and 2 bids is +1%, and the variance between stage 2 and 3 bids is +7.31%; these two variances are significantly different	≤ 0.05	Reject	Manual
7	9.8 and table 9.17	ditto	Related t test	0.05	the variance between stage 1 and 2 bids is +1%, and the variance between stage 2 and 3 bids is +7.31%; these two variances are not significantly different	> 0.05	Cannot reject	Manual
8	9.9.3 and table 9.20	there is no relationship between time and cost predictability	Spearman's correlation	0.05	there is a relationship between time and cost predictability; the correlation coefficient is 0.41. This can be classified as 'modest'. The mean percentage cost overrun is 7.31%, and the mean percentage time overrun is 3.81%	≤ 0.05	Reject	Manual

population will lie within a calculated range. If we need to be 95% confident, there is a 5% probability that the mean score will lie outside that range.

A narrow confidence interval range is welcome; that lends to the validity of the study. A wide confidence interval range severely limits validity; limited validity at undergraduate level is defensible, providing you recognise those limits. Confidence intervals help you to do that.

The key message with confidence intervals is that it is important that in your discussion and conclusions, and in interpreting the validity of your own study, you recognise that your mean scores may not accurately reflect the population mean scores. Confidence intervals are calculated from the formula:

$$\bar{\mu} = \bar{X} \pm t \frac{s}{\sqrt{n}}$$

where:

$\bar{\mu}$ = population mean to be calculated
t = critical value α = 0.05 with a value from appendix F7 n = 10, df = 9 = 2.262
$\bar{X}$ = sample mean = table 8.3 cell 17K = 57.67%
s = standard deviation = table 8.4 cell 14C = 21.47%
n = sample size = 10

It can be seen from the formula that small standard deviation and high n will reduce the range. The critical t value is obtained from the table in appendix F7, critical values of the t distribution. It is the size of t resulting from a t-test calculation that cuts off the tails of the normal distribution; t values equal to or greater than the stated values are significant. It is noted that t becomes smaller as n increases; this gives further weight to the principle that higher sample sizes will give narrower confidence interval ranges.

Using the data from table 8.3:

Step 1: $\sqrt{N} = \sqrt{10} = 3.162$

Step 2: $s / 3.162$
$= 21.47 / 3.162 = 6.79$

Step 3: Critical value of $t = 2.262 \times \text{step 2}$
$$= 2.262 \times 6.79$$
$$= \pm 15.35\%$$

Step 4: Statement of finding. We are 95% confident that the mean score of the population lies in the range of 57.67% ± 15.35. This gives a potential range of 42.32 to 73.02%; in practice this range is far too wide. The research shows with 95% certainty that the population mean could be anywhere between these figures. The calculation gives this wide range because $n = 10$ is far too small. The calculation repeated with a sample size of 30 gives a range of ± 8.01%.

9.13 Summarising results

If your analysis section includes a lot of tests, it may be helpful to readers if these are summarised in one table at the end of the chapter. The chapter in this text includes a lot of tests; they are summarised in table 9.25.

Summary of this chapter

Inferential statistics involves more complexity than descriptive statistics, but if you are able to execute such tests, that gives you the opportunity to gain more marks. Do not attempt these tests if you do not have a firm understanding of the underlying principles. The inferential tests can be put on the continuum of: chi-square, difference-in-means and correlation. The appropriate test is driven by the level of the data for the IV and the DV, and that is similarly on a continuum of categorical/nominal, ordinal and interval/ratio. You should identify whether your dataset is non-parametric or parametric. Correlation tests can be used to test internal reliability and validity. Since you may perform many statistical tests on your data, it is useful to summarise them in a table at the end of your chapter.

References

Bryman, A. and Cramer, D. (2005) *Quantitative data analysis with SPSS 12 and 13: a guide for social scientists.* Routledge, London.

Cohen, M. and Holliday, L. (1996) *Practical Statistics for Students.* Chapman Publishing, London.

Egan, J. (1998) *Rethinking Construction.* HMSO, London. http://constructingexcellence.org.uk/wp-content/uploads/2014/10/rethinking_construction_report.pdf Accessed 09.09.15.

Egan, J. (2002) *Accelerating Change.* London. http://constructingexcellence.org.uk/key-industry-publications/ Accessed 09.09.15.

Gray, C. and Kinnear, P. (2014) *IBM SPSS Statistics 21 made simple.* Psychology, Hove.

Glenigan (2015) *KPI zone.* UK Industry Performance Report. https://www.glenigan.com/construction-market-analysis/news/2015-construction-kpis Accessed 23.09.15.

Hinton, P.R. (2004) *Statistics explained.* 2nd edition. Routledge, London.

Latham, M. (1994) *Constructing the Team.* Final Report on the Joint Review of Procurement and Contractual Arrangements in the UK Construction Industry, July. HMSO, London. http://constructingexcellence.org.uk/wp-content/uploads/2014/10/Constructing-the-team-The-Latham-Report.pdf Accessed 09.09.15.

Matthews, R. (1998) The great health hoax. *Sunday Telegraph Review,* 13th September.

Miles, M.B. and Huberman, A.M. (1994) *Qualitative Data Analysis.* Sage, London.

Siegel, S. and Castellan, N.J. (1988) *Nonparametric Statistics for the Behavioural Sciences.* 2nd Ed. McGraw Hill, London, p. 9.

10 Discussion, conclusions, recommendations and appendices

The titles and objectives of the sections of this chapter are the following:

10.1 Introduction; to provide context for the final part of the study
10.2 Discussion; to explain the expected contents
10.3 Conclusions and recommendations; to explain the expected contents
10.4 Appendices; to identify the type of material to be included
10.5 The examiner's perspective; to give an overview of the dissertation or project expectations
10.6 Summary of the dissertation or project process; to give an overview of the dissertation or project process

10.1 Introduction

The final two chapters may be a discussion and, integrated into one chapter, the conclusions and recommendations. Together these two chapters may account for one-third of your dissertation or project. You should make sure that you keep these chapters focused on your objectives.

A discussion should bring together the literature, findings and results from all the data that you have collected and analysed, and develop the critical appraisal that you have undertaken in the literature review. Conclusions may state what needs to happen or explain cause and effect issues. Your recommendations should come from your data, and should not be too ambitious or suggest radical change in government or industry direction that is unlikely to materialise. To recommend another study may be appropriate. Appendices should include all the material that is not necessary to meet study objectives. At the end of your study you should undertake your own appraisal and recognise the limitations of what you have done, including judgements about reliability and validity. You will need to do a lot of early reading and make sure you write as you go.

Writing Built Environment Dissertations and Projects: Practical Guidance and Examples, Second Edition.
Peter Farrell, Fred Sherratt and Alan Richardson.
© 2017 John Wiley & Sons, Ltd. Published 2017 by John Wiley & Sons, Ltd.
Companion Website: www.wiley.com/go/Farrell/Built_Environment_Dissertations_and_Projects

10.2 Discussion

The discussion should be focused on aims and objectives. Perhaps it should be articulated under the heading of each. Discussion chapters in documents are sometimes weak or even non-existent; documents sometimes go straight from results to conclusions. It is as though students run out of time, and they are exhausted after completing the analysis. Others label the final chapter 'discussion, conclusions and recommendations', but it is only a title, and little real discussion takes place. Documents that appear to be destined for good marks in the early and middle parts, may weaken towards the end. You should view the discussion as an opportunity to take the dissertation or project destined for good marks, on to excellent marks. Though time management is an important issue (you need to complete the analysis several months before the final submission date), there is also a need to focus upon what good discussions should comprise. The Oxford dictionary definition of a discussion (Stephenson, 2010) is:

an examination by argument;
a debate.

The discussion requires that results and findings be evaluated against each other and against the theory and the literature. You should integrate what you have found in your data with what the literature states. It is not to do more reading; just go back to your literature chapter and bring forward the same sources. Cite them again, e.g. 'this study found that photovoltaic cells have a payback period of thirty years. This is supported by the work of Smith (2010) and Brown (2008), but it is contrary to the work of Baker (2006), who though using a different methodology found a shorter period of twenty years'.

It is in this chapter that you can take some imaginative and intelligent leaps in interpreting your findings. To the extent that this study is 'you', this is where you can demonstrate intellectual insight. But you should also get some help, to avoid the danger of viewing your findings too narrowly. There is huge potential to do some qualitative work towards the latter part of your study that could gain extra marks. The thrust of some informal interviews could be to report what you have found and ask participants what they think.

The findings are based on what the data tells you; they are 'fact'. Now that you have (or have not) found relationships between variables, the discussion gives you the opportunity to carefully speculate about important issues such as 'why' and 'how', which are important parts of theory building. If there is poor discussion, findings about relationships are effectively left isolated. There should be links into any theories identified in earlier parts of the document. To illustrate that your speculation is careful, use some caveats such as 'perhaps" or 'possibly'. Then start to develop in your discussion new hypotheses as recommendations for further study. In laboratory work, if you have found that material 'A' performs better than 'B', what are the scientific reasons for that – does the answer lie in understanding chemical behaviour of materials – why? If you find that construction professionals are less stressed working on projects procured using partnering methods, why is that? How can other methods of procurement learn from partnering to reduce stress levels?

Discussion should deduce explanations for similarities and differences and examine potential causes or potential variables. The discussion should start to open the door for what will come in the conclusion, perhaps in the form of stating what needs to happen. Conclusions should not just 'pop out' in the conclusion chapter. They should be developed in the discussion, and any arguments supporting them should be articulated. You should have the aspiration to do something original in your work, in a very modest way. The discussion should tease out what had been done that has been different, and how it may have implications for future studies. Provided your work is well founded and structured, however modest it is, it has its place in the knowledge base of your chosen subject area. Do not be afraid to emphasise that.

You should not have come into your research with intransigent preconceived ideas about what outcomes will be. You may have had perceptions and even 'hopes', but you must let the data give you answers. Some students use the discussion to embark on their own personal treatise about how to solve the problems of industry. Do not do this; there are analogies similar to the literature review. The discussion is a weighing exercise where you bring the evidence to bear, and you sensibly recognise the weight of your experience in the context of the experience and research of others detailed in literature. You must give your judgement, but make sure it is based on all the evidence, and not unreasonably biased by your own experience. The evidence before you is the literature, the new data collected and analysed in your study, and to a lesser extent your own experience. To the extent that you stray outside these evidential base points, that is only with careful speculation.

There are many issues in life where there are two sides to the debate. You must let the argument take place. The political world is full of issues where the decision making process is often controversial, with seemingly compelling evidence on each side of an argument. Judgements are often complex, and it may be the case that you have to articulate that in one set of market conditions the answer is 'abc', but perhaps in another 'xyz'. Many issues in life are not black or white; they are grey. A discussion should be written with sensitivity, recognising blurred boundaries that exist. It must try to have empathy with and respect for what others may judge. While you may make a judgement on one side, the discussion chapter should not leave readers to examine the arguments from the other side. In your writing you must recognise both sides, even if you are overwhelmingly convinced by the evidence in one direction. When making your judgements, it may be that others, including your examiner, disagree with you. That is fine, and you will not be marked down, providing you are able to defend your judgements based on correct interpretation of the evidence.

You should also consider the cynic who may be critical of the research. It is useful to articulate what the cynic may find, recognising limitations of what you have done. If your work is qualitative, you may choose not to have a separate discussion chapter. It may be embedded in the narrative that closes out your analysis, results and findings chapter. If this is your choice, fine, but do make sure that the narrative is of the appropriate weight, and includes all the features of a discussion as though it were a separate chapter. Alternatively, you may wish to take the narrative out of analysis, results and findings, and place that as a separate chapter with the discussion.

10.3 Conclusions and recommendations

The conclusion brings together the whole of the research. A conclusion should be a logical outcome of all that has gone before. Conclusions are often weak, sometimes comprising merely one side of writing not related to objectives. Students run out of time and are exhausted after completing the analysis. Students may write conclusions the day before submission is due; this must not be the case. The conclusions must be laboured over long and hard. They should be the result of many iterations, approached with insight and given a great deal of thought. Dissertations or projects that appear to be destined for good marks in the early and middle parts, fall or even worse, fail at this point. No matter how good the early work of documents, it is not possible to pass without substance in conclusions – absolutely impossible. Time management is important, but there is a need to focus upon what good conclusions should comprise. The Oxford dictionary definition of a conclusion (Stephenson, 2010) is 'a final result; a termination; a judgement reached by reasoning; the summing up of an argument, article or book; a proposition that is reached from given premises'.

The purpose of academic study is to come up with some really good conclusions. To reiterate, the whole study must hang around the objectives. In the introduction chapter you have stated your research questions and have moved through your document in a structured sort of way to come to some answers. The objectives imitate questions. In the same way that previous chapters have been focused on aims and objectives, so too should the conclusion. The conclusion closes the study out, but linking it back to the aim and objectives in the introduction. The strap line for objectives could be given as separate headings to make sure you do not lose sight of them. Number the conclusions to reflect the numbering of objectives.

There should be no new material and no quotations from the literature. As a chapter, the conclusion should stand alone, and readers should be able to understand concepts behind each conclusion without having to refer to earlier chapters. Therefore, at the risk of being monotonous, key parts of earlier chapters will be repeated. Be mindful that some readers will not take the full journey through your document. To help them, restate important parts of the literature and summarise the method, including populations and samples. Repeat your results and findings. This may include some key numbers such as mean scores or correlation coefficients. Important elements of the discussion should also be summarised succinctly. Judgements, which have started their formation in earlier chapters, can now be asserted firmly. Some words may be very similar to those used in the abstract.

What is a conclusion? Studies executed on the premise of establishing causes may be appropriate to a conclusion writing style which suggests that something needs to change or happen; that is to determine what the IVs are and to conclude that they need to be manipulated. However, some studies may not be appropriate to this style. Try to answer cynically put questions: 'Now that you have these results and findings, so what? Where do we go from here?' If it is found that A is better than B, you may draft conclusions on the assumption that something needs to change or something needs to happen. Either B needs to improve, or A should be used exclusively at the expense of B. The objective of some research may be to develop theories or explanations as to why things happen, and look at the cause and effect issues. The explanation may not be clear cut and definitive, and therefore a conclusion may be substantive in its own right, if it merely articulates why something may happen. Such an explanation needs to be academically robust and must stem from the data analysis, results and findings in the study. In the conclusion, also extend the 'why' and 'how' theories (or better labelled hypotheses) that are started in the discussion. If the evidence to support findings is overwhelming, the tone of conclusions can be assertive. If more work is needed, the tone of conclusions may be more tentative.

An opportunity to gain marks is available through a carefully written section under a subheading of 'Limitations and criticisms of the study'. You should not let readers spot weaknesses in your study. The undergraduate dissertation or project is only training for research, and to give you appreciation of the role of research in society. That appreciation is enhanced if you know where you went wrong. If you went wrong, that is understandable; but if you went wrong and do not know it, then you will not score high marks. Some weaknesses in your study may be where you made mistakes or misunderstood something. Some weaknesses may be due to failings in time management, which you reflect should not have occurred. Do not be afraid to mention these. You may, however, have completed a dissertation or project within the reasonable bounds of what universities can expect. Some fundamental weaknesses in the study may have been unavoidable, due to lack of resources. It is here where you should reflect on how valid your work is. Suppose it were appropriate to measure validity on a scale of 0–10, and in your undergraduate dissertation or project it was only reasonable to expect you to score 5/10. If you do indeed score 5/10, you should score excellent marks; but if you score 5/10 and you can articulate why you could not score 10/10 and how 10/10 could have been scored, you will get even more marks.

Recommendations follow conclusions. Some students do not write recommendations, while others use them to put right all the ills of industry, unrelated to study objectives. You should not write recommendations that are hunches or personal hobbyhorses; they must come from data in studies. Do not be too ambitious in writing recommendations, and do not write too many. Recommendations in authoritative reports written for governments sometimes fail at implementation. Your recommendations will similarly fail if action would cost substantial resources. Recommendations should reflect objectives and therefore implicitly reflect conclusions.

If you have compelling findings that would benefit industry, you can write your recommendations for that audience. However, it may be more likely the case that the appropriate audience is academia. On that basis, it may be better to write your recommendations as though for more study. That fits well with the scenario of theory building, testing and re-testing. Recommendations for more study should be well thought out and fully developed. Therefore, in one or two sides, describe the problem, state the research questions, objectives and hypotheses. Define the variables and state how they should be measured. Describe populations, sampling method and analytical method. This is potentially a lot of work, so it is probably sufficient that you do it for just one objective. It may be that as part of your literature review you noted an important, related topic area, but you decided not to formulate an objective for it since you were limited by resources. A recommendation for further study could be founded in such an area. There is the possibility that subsequent students could pick up one of your recommendations and carry it as a baton for their work. Having written the limitations and criticisms of your study, you may reflect with the benefit of hindsight that your study did not take you where you really wanted to go. That feeling is understandable, and it is not to get upset about it. The recommendations are a really good opportunity for you to write the full research proposal for the study you wish you had done.

10.4 Appendices

Appendix singular, appendices plural, provide supplementary information and necessary supporting evidence. They are not included in word counts. The following strap line may help, though not to be taken too literally:

If it should be, it should be in; if it should be out, it should be out. If in doubt, put it in the appendices.

The context of 'if it should in' is whether it is relevant to the objectives. All material that is necessary to meet study objectives should be in the main body of documents. Appendices should not be too long. If your study objective comprises a detailed appraisal of a British Standard, or a Standard Form of Building Contract, or a government report, do not include that full document. If readers really want to access those documents, they can do so by following leads in your citing and referencing. If there are one or two pages that are crucial to the objectives, by all means provide them for the convenience of readers. Appendices should be in the same order as they appear in the narrative, and they should be designated by a capital letter ('Appendix A', ' Appendix B' …) rather than by an Arabic number (1, 2, 3 …). It is useful to provide a header to appendices to label all pages A, B …, to allow readers to locate them easily. New appendices should be on separate pages and all page numbered following on from the references/bibliography sections. There should be a list of appendices in the preliminary pages at the front of the document.

Appendices often include such things as survey instruments, questionnaires, verbatim transcripts of interviews and photographs. Photographs may be of architectural features, defects in completed buildings or failure modes of materials tested to destruction in laboratories. Those that are important to the objectives should be in the main body. It is often sufficient to include only summaries of the raw data in the main body; the raw datasets themselves can usually be in an appendix. Results and findings should be in the main body. The analytical process may be assisted by computer. If you do your detailed analysis without software support, that may involve many calculations. A few calculations may be placed in the main body, but if you have reams of them, place them in the appendices. If you have used a questionnaire, do not include all those that are returned by participants. Only a blank copy is required. If you have asked closed questions and converted qualitative responses to numbers (e.g. excellent = 4, very good = 3 …), also include another copy, with the numerical coding marked on it. You should present in the main body a table that summarises the responses from any questionnaires.

During the proofreading process, you may reflect that some parts of the document are too long, and you have overstepped the word guide. Overstepping may occur in the literature review, or you may have included too much detail in your analytical chapter. The material may not be directly relevant to your objectives. Rather than delete it and lose the material completely, include it in an appendix. Examiners may give you some credit for it. Examiners will not necessarily read your appendices, but they may browse them, perhaps if they have some doubts that need to be explored. Subsequent researchers may want to use details in appendices to develop their work or to establish the validity of your work in their minds.

10.5 The examiner's perspective

Examiners are required to benchmark dissertations and projects against carefully written and designed criteria. The criteria help to ensure that the mark that you are allocated is as objective, reliable and valid as possible. A test of marking validity would be that two markers who read a document, independent of each other, allocate the same or nearly the same mark.

There is inevitably some subjectivity involved in marking most assessments; the criteria are in place to move judgements along the continuum from subjectivity to objectivity, as far as is possible. The following examiner's perspective was adapted by Rudd (2005) from work by Brown and Adkins. It could be summarised around two generic themes: originality and validity. You may find it useful to guide your work from the start, in making your own judgements about the validity of your study, and finally in writing a section in your conclusion chapter about limitations and criticisms:

Initial overview

Read title and abstract for thesis, key idea, question, topic.
- Turn to last chapter/conclusion to see how far this has been addressed.
- Note any methodological weaknesses.
- Turn to contents page to see if sufficient appropriate evidence has been covered.
- Read introduction for close definition of problem in context.
- General impression? Formulate questions, based on introduction, as to what you would hope to be answered/discussed – e.g. congruency of methods with problem, possible sources of bias.

Review of literature

- To what extent is it relevant to the research issue?
- To what extent is it 'this is all and everything I know about this'?
- Is the review just descriptive, or is it analytical, evaluative?
- How well is technical/theoretical literature mastered?
- Are links between literature review and design of this study explicit?
- Is there a summary of the essential features of other work as it relates to this study?

Design of study

- Precautions against likely sources of bias?
- Limitations of the design? Is candidate aware of them?
- Is the methodology for data collection appropriate?
- Are the techniques of analysis appropriate?
- Has the best design been chosen?
- Has the design been justified?

Presentation of results

- Does the design appear to have worked satisfactorily?
- Have the hypotheses been tested?
- Do the solutions relate to the questions posed?
- Is the level and form of analysis appropriate to the data?
- Could the results have been presented more clearly?
- Are patterns and trends in results identified and summarised?
- Is a picture built up?

Discussion and conclusions

- Is the candidate aware of possible limits in reliability/validity of the study?
- Have the main points in results been picked up and discussed?
- Are there links made with other literature?
- Is there an attempt to reconceptualise problems, to rethink/develop theory?
- Is there intelligent/imaginative speculation? Does it grow out of the results?

Summative overview

- Is the standard of presentation adequate (English, style, notes, bibliography)?
- Is the thesis generally the candidate's own work?
- Does the candidate have general understanding of the field and how the thesis relates to it?
- Has the candidate thought through implications of findings?
- Is there evidence of originality?
- Does the study add to existing knowledge of this area?
- Is there evidence of development in research skills?
- Is it worth publishing in some form?
- Are any qualms/reservations found in the material, or do they arise from ideological/epistemological differences?

10.6 Summary of the dissertation or project process

Good time management and focus upon objectives is essential. The process that you follow needs to be underpinned by reading about your specialist subject area and about research methodology. While you inevitably learn as you do so, you need to have a reasonable grasp of both at an early stage to be able to set out a plan; therefore early reading is paramount.

A possible scenario for a study is as follows:

- write as you go; keep communicating with your supervisor
- a well-founded description of the problem from which a provisional research question emerges
- the research question is translated into an objective
- agree a protocol for working with your supervisor
- complete your permission-to-do-research form and information sheet in accordance with your university ethical code
- exploratory work (networking, informal or formal interviews) to redefine the problem, and reshape the question/objective
- undertake the literature review
- more exploratory work to reshape the question/objective
- establish the objective firmly; identify the variables
- write the objective as a hypothesis
- open up the objective/hypothesis and its variables by defining them in the narrative in the context of your study
- open up the definition of the variables by designing a research instrument to measure them; the research instrument and its surrounding methodology must be the most appropriate instrument to meet the objective and measure the variables

- collect the data and do the analysis
- further networking and informal or formal interviews to help shape the discussion and conclusions
- extensive thought, interpretation and insight in writing discussion and conclusions
- final write up and submit draft submission for feedback in good time
- final proofreading and submission
- **revert back to figure 1.6 – the very last button you press before 'print' or 'submit' should be 'spelling and grammar check'**
- only after spelling and grammar check, submit ☺

Summary of this chapter

A discussion should bring together the literature, findings and results from all the data that you have collected and analysed, and develop the critical appraisal that you have undertaken in the literature review. Conclusions may state what needs to happen or explain cause and effect issues. Your recommendations should come from your data, and should not be too ambitious or suggest radical change in government or industry direction that is unlikely to materialise. To recommend another study may be appropriate. Appendices should include all material that is not necessary to meet study objectives. At the end of your study you should undertake your own appraisal, and recognise the limitations of what you have done, including judgements about reliability and validity. You will need to do a lot of early reading and make sure you write as you go. Proofread your document and use the spelling and grammar check facility often; your last task before printing should be to spell and grammar check again.

References

Rudd, D. (2005) Unpublished handout University of Bolton; adapted from Brown, G. and Atkins, M. (1988) Effective Teaching in Higher Education, Routledge, London.

Stephenson, A. (2010) Oxford Dictionary of English. 3rd edition. Oxford University Press. Available online

List of appendices

A   Glossary to demystify research terms, with examples
B   Research ethics and health and safety examples
C   An abstract, problem description and literature review
D   Eight research proposals
E   Raw data for a qualitative study
F   Statistical tables

Writing Built Environment Dissertations and Projects: Practical Guidance and Examples, Second Edition.
Peter Farrell, Fred Sherratt and Alan Richardson.
© 2017 John Wiley & Sons, Ltd. Published 2017 by John Wiley & Sons, Ltd.
Companion Website: www.wiley.com/go/Farrell/Built_Environment_Dissertations_and_Projects

Appendix A: Glossary to demystify research terms

This glossary is intended to provide a starting point for students encountering this new vocabulary of research terms in their work. Many of the terms below have much more complicated ideas behind them than can be included here, not to mention a whole history of philosophical development which for some even spans centuries. So here we have just aimed to provide simple definitions, to place these ideas within helpful contexts and to provide a basic understanding.

A priori: as opposed to a posteriori, this is knowledge that can be gained through reasoning, thinking and theory rather than experience.

A posteriori: as opposed to a priori, this is knowledge that has been gained from experience or the results of experiments.

Abstract: a concise summary of all your research placed at the front of the research document, best given as precisely one full whole page. It should give the aim, describe the problem, and the main research question, objective or the hypothesis. Also state the result, findings, conclusion and recommendation.

Action research: an approach in which the researcher, working closely with an organisation or project, develops a beneficial change to a system or process which is then implemented in practice and the resultant impacts monitored. This follows the Deming Cycle of plan-do-check-act, and the research can involve several action research cycles to tailor and enhance the developed change.

Aim: the overall goal of the research project and of a strategic nature. Since it is visionary and only part of a bigger picture in the topic area, it will not necessarily be met by your research.

Alternative hypothesis: as opposed to the null hypothesis; a provisional supposition that something is true, or there is an association or relationship between two variables.

Bias: the potential for external influence on the research making it lose objectivity, for example through the researcher's own beliefs, the way they ask questions or the selection of a biased sample which is not representative of the population, but some epistemological approaches would argue that all research is biased, and so it is accepted and made explicit through a reflexive approach by the researcher (Taylor, 2001).

Bibliography: all background reading, not cited in text, listed immediately after references.

Case study: a research approach that investigates a phenomenon in great detail within a specific context, using different data sources and methods of collection through triangulation to provide an in-depth analysis of the case (Proverbs and Gameson, 2008). In the construction industry, individual construction projects can easily provide a case for research.

Coding: the application of labels to the data to seek themes or categorisations – these labels can be prescribed and applied to the data in the form of a coding framework developed from the literature, or emerge from the data itself.

Conceptual model: a model developed to help demonstrate concepts or ideas that have been developed by the research, for example a health and safety project might seek to develop a flow diagram to show the relationships between decision making, the local environment and the various outcomes that might occur for construction operatives working at height.

Conclusion: the summary of the research project, what has been found out? This can be broken down into conclusions for the project objectives, which should in turn draw together to enable a conclusion to be made of the project aim.

Constructionist: an epistemological approach that sees the world as socially constructed by those within it, through ever-changing interactions and events. There are no facts to be discovered, just a good fit with reality as viewed by those who experience it. Only our experiences and perceptions can be accessed for research, and these cannot be judged against reality for their validity or accuracy (Burr, 2003).

Content analysis: an analysis carried out on qualitative data that involves the identification of key words or phrases within data to provide information on dominant themes. This can use counting and statistical methods of analysis or more interpretive approaches to the data.

Deductive approach: as opposed to the inductive approach, a deductive approach seeks to use data to test theories.

Delphi technique: a process by which a questionnaire survey is carried out to seek the opinions of an expert sample. On completion, these opinions are then analysed and collated and the average responses presented again to the sample in further rounds of the survey, giving individuals the opportunity to change their opinion given the response of others in the sample. Eventually the answers should converge on an agreed response.

Dependent variable: a variable that will move in an upwards or downwards direction as a result of the manipulation of an independent variable.

Discourse analysis: the study of talk and texts, it involves a set of methods and theories for investigating language-in-use and in social contexts (Wetherell et al., 2001). Emphasis is on how things are said, the analysis aiming to reveal how these linguistic interactions reveal and shape the way we understand the world.

Discussion: an examination by argument or debate, examining a problem from all sides.

Empirical: an approach in which data is collected through practical observations or experiments.

Epistemology: Flick (2009) defines epistemology as 'theories of knowledge and perception'; in looking at theories and knowledge, we are asking how can we know what is out there?

Ethics: research ethics are guidelines to ensure that people or animals are not negatively affected by research, either physically, emotionally or mentally (Gray, 2009). All academic institutions will have their own ethical management system in place, and all research proposals should be checked for institutional ethical compliance.

Ethnography: A research approach in which the researcher is immersed in the everyday life of the environment to be studied (Henn et al., 2006), seeing the world from that point of view.

This allows the collection of information about relationships, beliefs and values of that community (Angrosino, 2007).

Focus group: a group interview (Gray, 2009), where selected participants discuss and share their ideas around the research questions asked.

Generalisability: the extent to which the conclusions found by the research are relevant beyond the study itself (Angrosino, 2007). Within scientific research, the ultimate goal is to produce universal generalisations (Lincoln and Guba, 1985), but research is often limited by scope of sample and other limits to validity on this scale.

Grounded theory: developed by Glaser and Strauss, this method of research develops theory from the data without any preconceived ideas or hypotheses being used. It is an inductive approach to research which can be used to seek explanations and theories which 'may then be used to predict and explain phenomena' (Hunter and Kelly, 2008).

Hawthorne effect: where people behave in different ways when they know they are under observation or being researched (Kumar, 2005).

Homogeneity: data that hangs together well and consistently around a finding, not influenced by spurious variables.

Hypothesis: a provisional supposition. It may be about one variable (there is water on the planet Mars), or about an association, difference or relationship between two variables

Independent variable: a variable that is manipulated in an upwards or downwards direction with a view to moving a dependent variable.

Inductive approach: as opposed to the deductive approach, an inductive approach seeks to develop theory from the data.

Interpretivist: an epistemological approach based on researchers seeking to understand the world from the point of view of the people studied, rather than explaining actions through cause and effect (Henn et al., 2006). The researcher must interpret what they find, to explore the meanings behind what those being researched have said or done.

Literature review: a summary of the pertinent literature (such as previous research or government reports) around the research subject area; this should also include critical analysis of this literature and should draw towards the identification of a gap in current knowledge.

Method: a tool for data collection or analysis, such as a questionnaire. The results of the questionnaire can be analysed using a statistical method such a chi-square test.

Methodology: the underlying research framework, including ontology and epistemology. Be careful not to title your methods chapter a 'methodology' if it does not deal with such research paradigms.

Mixed methods: using various different methods to gather and analyse your data, for example collecting data from different sources to explore the same phenomenon.

Mnemonic: a pattern of letters that aids in recalling something; the letters may form a word e.g. NOIR: nominal, ordinal, interval, ratio.

Model: a visual representation of a phenomena, perhaps of a process or system, or the relationships between different factors and how they interrelate within the context under investigation.

Monte Carlo technique: a process of random number selection for use in simulations (Fellows and Liu, 2008).

Naturally occurring data: data that has not been created by the researcher specifically for the research (Silverman, 2001). For example, interview transcripts are not naturally occurring, but company webpages or tender documents are, as they have been created for reasons outside of the research.

Neural networks: statistical models that map the relationships between various elements in a network.

Null hypothesis: as opposed to the alternative hypothesis, the hypothesis of no association, difference or relationship.

Objectivism: an ontological position that sees the world as existing beyond social interactions, a tangible 'real world' that can have influence over social existences and about which we can determine facts.

Objective: a statement of what you will do in the main body of your dissertation or project. The success of your research will be measured against your objective(s).

Objectivity: as opposed to subjectivity; data that is fact based.

Ontology: often employed as the accepted concept of reality (Creswell, 2007), ontology relates to our ideas of the world, what is out there that can be known. The basis of research philosophy.

Paradigm: a distinct way of thinking about things. For example you could carry out your work using a positivist research paradigm.

Participant: a person who contributes data to the research.

Phenomenological: An approach that seeks to understand the world from the perspectives of those being studied and how they experience and give meaning to what is going on around them (Gray, 2008).

Pilot study: a small, limited version of the study carried out to check the suitability, reliability and validity of the planned research approach. And problems can then be changed (and the study re-piloted if necessary) before the main study is carried out, to ensure a robust research design.

Plagiarism: taking the work of others and passing it off as your own writing or idea.

Population: the total number of people or things that make up the whole group that meet specified inclusion criteria.

Positivism: an epistemological approach based on using scientific methods to measure things, including people, to determine facts and truths about the world.

Post-positivism: an epistemological approach that still seeks to measure things, including people, but allows for the possibility that people may be inaccurate in their representations of the world they live in (Creswell, 2007). It still applies the empirical methods of positivism, but within the framing of this caveat.

Primary data: new data, not already existing, that is collected by you for your research.

Qualitative: qualitative data refers to non-numerical data, qualitative analysis refers to the analysis of this data to explore *how* and *why* things are as they are.

Quantitative: quantitative data refers to numerical data, quantitative analysis refers to statistical analysis, either descriptive or inferential, which can establish *what* things are.

Quantitising: the numerical translation and conversion of qualitative data to quantitative through such methods as content analysis, or assignment of values as in a Likert Scale (Naoum, 2006).

Rationalism: an ontological position which states that the world out there can be known through deductive reasoning.

Realism: an ontological position that states that the world out there exists independently of us and can be tested and known without the need for people to perceive it.

Recommendation: a course of action to follow; in a dissertation that may often be a proposal for more research in the same subject area.

References: all citations in text to other authors, listed immediately after the conclusion chapter towards the end of the research document.

Reflexivity: the need for a researcher to acknowledge their own influence on the research work Gibbs (2007). This could be through their own bias, gender or other influence on the research being carried out.

Relativism: an ontological position that states that the world out there can only be known through our immediate perceptions, which are therefore subjective and relative and change with immediate space and time.

Reliability: whether a study can be easily replicated or repeated by others to produce consistent results (Seale, 2004); use of clear research methods will help you demonstrate this.

Research question: arising from the description of the problem; something that society does not know the answer to and which will be revealed by the data collection and analysis.

Researcher generated data: the opposite of naturally occurring data, this is data that would not have been produced had the interviewer not instigated it in some way (Potter and Hepburn, 2007).

Sample: a smaller part of the population. Inferences will be made that the results from samples replicate those that would have been obtained if it had been possible to survey the whole of the population.

Secondary data: data already existing, perhaps in the public domain or in company files.

Significance: associations, differences or relationships between variables that are genuine, and that statistically have not occurred owing to chance.

Subjectivity: as opposed to objectivity; data that is opinion based and liable to change over time.

Taxonomy: A taxonomy is a presentation of the categorisation and classification of data, and the relationships between these classifications, for example Garrett and Teizer (2009) provided a taxonomy for human error awareness in construction safety.

Triangulation: using one or more method, measure or type of data source to overcome problems of bias or to ensure validity in the research (Love et al., 2002).

Theory: Strauss and Corbin (1998, p. 15) define theory as 'a set of well-developed concepts related through statements of relationship, which together constitute an integrated framework that can be used to explain or predict phenomena'.

Theoretical framework: the ontology and epistemology underlying a piece of research.

Thematic analysis: the analysis of data, usually carried out by coding, to establish the dominant themes and ideas that are present.

Thesis: the final output of your research, the final document in which you set out your work for examination or assessment by others. Most often a whole qualification gained by one research document, such as a PhD thesis.

Validity: the truth of a research project: does it 'tell the truth' given the research paradigm in which it is operating? Validity is often examined from two perspectives: internal, the extent to which the conclusions are themselves supported by the study, and external, the extent to which these findings can be generalised (Seale, 2004).

Variable: as opposed to a constant; something with a value that is capable of moving and of having more than one value.

References

Angrosino, M. (2007) *Doing Ethnographic and Observational Research*. London: Sage Publications Limited
Burr, V. (2003) *Social Constructionism*. 2nd Ed. East Sussex: Routledge
Creswell, J.W. (2007) *Qualitative Inquiry and Research Design: Choosing Among Five Approaches*. 2nd Ed, Sage, London.
Fellows, R. and Liu, A. (2008) *Research Methods for Construction*. 3rd Ed., Wiley-Blackwell, Chichester.
Flick, U. (2009) *An Introduction to Qualitative Research*. 4th Ed. Sage, London.
Garrett, J. and Teizer, J. (2009). Human Factors Analysis Classification System Relating to Human Error Awareness Taxonomy in Construction Safety. *Journal of Construction Engineering and Management*, 135(8) 754–763.
Gibbs, G. (2007) *Analysing Qualitative Data:* The Sage Qualitative Research Kit. Sage, London.
Gray, D.E. (2009) *Doing Research in the Real World*, 2nd Ed., Sage, London.
Henn, M., Weinstein, M. and Foard, N. (2006) *A Short Introduction to Social Research*. Sage, London.
Hunter, K. and Kelly, J. (2008) 'Grounded theory', In Knight, A. and Ruddock, L. (2008) *Advanced Research methods in the Built Environment*. Wiley-Blackwell, Chichester, 86–98.

Kumar, R. (2005) *Research Methodology*. 2nd Ed. Sage, London.

Lincoln, Y.S. and Guba, E.G. (1985) *Naturalistic Inquiry*. Sage, London.

Love, P.E.D., Holt, G.D. and Heng, L. (2002) 'Triangulation in construction management research' in Engineering, Construction and Architectural Management 9(4) 294–303.

Naoum, S.G. (2006) *Dissertation Research and Writing for Construction Students*. 2nd Ed. Butterworth-Heinemann, Oxford.

Potter, J. and Hepburn, A. (2007) 'Life is out there: a comment on Griffin.' *Discourse Studies*. 9(2), pp. 276–282

Proverbs, D. and Gameson, R. (2008) Case study research In Knight, A. and Ruddock, L. (2008) *Advanced Research methods in the Built Environment*. Wiley-Blackwell, Chichester, pp. 99–110.

Seale, C. (2004) 'Validity, Reliability and the Quality of Research.' In: C. Seale (Ed.) *Researching Society & Culture*. 2nd edn. Sage, London pp. 71–83.

Silverman, D. (2001) *Interpreting Qualitative Data*. 2nd edn. Sage, London.

Strauss, A. and Corbin, J. (1998) *Basics of Qualitative Research*. 2nd Ed., Sage, London.

Taylor, S. (2001) 'Locating and Conducting Discourse Analytic Research.' In: M. Wetherell, S. Taylor and S.J. Yates (Eds) *Discourse as Data: A Guide for Analysis*. Sage, London, in association with the Open University, pp. 5–48.

Wetherell, M., Taylor, S. and Yates S.J. (2001) *Discourse as Data: A Guide for Analysis*. (Eds) Sage, London, in association with the Open University

Appendix B: Research ethics and health and safety examples

Appendix B1:   Research ethics checklist
Appendix B2:   Sample covering email to a questionnaire survey
Appendix B3:   Participant information sheet
Appendix B4:   Confidential Health Questionnaire for students and staff participating
                in Laboratory Work

Appendix B1

Research ethics checklist

Answer 'yes' or 'no' to each question:

Will the study involve participants who are particularly vulnerable or who may be unable to give informed consent (e.g. children, people with learning disabilities, emotional difficulties, problems with understanding and/or communication, your own students)?

Will the study require the cooperation of a gatekeeper for initial access to the groups or individuals to be recruited (e.g. students at school, members of self-help group, residents of nursing home)?

Will deception be necessary, i.e. will participants take part without knowing the true purpose of the study or without their knowledge/consent at the time (e.g. covert observation of people in non-public places)?

Will the study involve discussion of topics which the participants may find sensitive (e.g. sexual activity, own drug use)?

Could the study induce psychological stress or anxiety or cause harm or negative consequences beyond the risks encountered in normal life?

Will the study involve prolonged or repetitive testing?

Will financial inducements (other than reasonable expenses and compensation for time) be offered to participants?

Will participants' right to withdraw from the study at any time be withheld or not made explicit?

Will participants' anonymity be compromised or their right to anonymity be withheld, or information they give be identifiable as theirs?

Might permission for the study need to be sought from the researcher's or from participants' employer?

Will the study involve recruitment of patients or staff through the NHS?

If **any** of the above items are answered 'yes', you will need to describe more fully how you plan to deal with the ethical issues raised by your research. This does not mean that you cannot do the research, only that your proposal will need to be approved by a research ethics committee or subcommittee.

Source: adapted from Code of Practice for Ethical Standards in Research Involving Human Participants. University of Bolton. http://www.bolton.ac.uk/Students/PoliciesProcedures Regulations/AllStudents/ResearchEthics/Documents/CodeofPractice.pdf (accessed 07.07.2015).

Appendix B2

Sample covering email to a questionnaire survey

Dear Mr Smith,

Re: Electronic questionnaire survey on health and safety in construction.

My name is xxx. I am in the final year of study on a BSc(Hons) in Construction Management at the University of xxx. I am writing to you because I believe that you have some experience with contractors in the UK. I invite you to take part in an electronic survey on health and safety. If you are happy to do so, it should take no more than ten minutes of your time. There are thirty tick-box questions; I will use the answers in my dissertation and in an academic context only. Participation is voluntary. If you prefer not to take part, that is fine and there is no need to reply. If you start the questionnaire and wish to withdraw part way through, that is fine too. I am writing to 50 construction practitioners. I hope to receive at least 30 replies.

The aim of my research is to investigate opportunities to improve health and safety performance, and the objective is to determine the compliance of UK contractors with best practice in health and safety. I need to complete the study by mid-April 2016. To allow myself some time to process results and develop conclusions, I would like to receive your reply by xxx [suggest allow two weeks].

I will protect your confidentiality and anonymity; I do not require your name unless you would like to see some of my results and findings. In that case, you will be asked at the end of the questionnaire for contact details, which I will store securely.

If you do agree to participate, just click on the link below. The software is operated by xxx. There are six pages. The first eight questions are about your role in construction; subsequently there are 20 questions about compliance with best practice. There are finally two open text box questions; the first invites you to add any comments you may have about health and safety, and the second asks for your contact details if you would like to see my results. You will complete your answers by clicking on the 'submit' button.

If there are any problems or if you wish to contact my supervisor, her contact details are: University of xxx, Department of Civil Engineering and Construction, tele xxx. email xxx.

Thank you for taking the time to read this email.

Yours faithfully

Xxx

The link address: insert a real link

Appendix B3

Participant information sheet

Perceptions of owners of timber frame housing about the quality of their homes

My name is xxx. I am undertaking some research as part of my BSc/BEng degree at the University of xxx. I am looking at perceptions of owners of timber frame housing about the quality of their homes. I am talking to home owners on several estates in the area as part of my work.

What will I have to do if I take part?
If you agree to take part, I will ask you to answer some questions. There aren't any right or wrong answers – we just want to hear about your opinions. The discussion should take about an half an hour at the longest. Please note that some of the questions may relate to your personal experiences.

Do I have to take part?
No, **taking part is voluntary**. If you don't want to take part, you do not have to give a reason and no pressure will be put on you to try and change your mind. You can pull out of the discussion at any time.

If I agree to take part what happens to what I say?
All the information you give us **will be confidential** and used for the purposes of this study only. The data will be collected and stored in accordance with the Data Protection Act 1998 and will be disposed of in a secure manner. The information will be used in a way that will not allow you to be identified individually. **However, we must inform the University if, in the unlikely event, something you have said leads me to believe that either your health and safety or the health and safety of others around you is at immediate risk.**

   If you feel I have not acted appropriately after the discussion, and need help dealing with your feelings, it is very important that you talk to someone right away.

The contact details for the person to talk to are:
Name:Your supervisor
Tele:
Email:

Thank you very much for your help!
Adapted for the Faculty of Advanced Engineering and Sciences from a document approved by the University of Bolton Research Ethics Committee
Source: adapted from: www.bolton.ac.uk/Students/PoliciesProceduresRegulations/AllStudents/ResearchEthics/Home.aspxL:\AQAS\Common\Research\ResearchEthics\Example 1 Participant.doc

Appendix B4

Confidential health questionnaire for students and staff participating in laboratory work

Complete form, photocopy and submit to supervisor/tutor/manager/school office

Your personal health and medical information is requested for the purpose of potentially assisting in an emergency situation. This information will not be used for any other purpose.

Please keep this form with you at all times.

Full name:

Date of birth: _____ Course & Year: _____

Current home address (while at university):

_____

_____

Current emergency contact number: _____

Name and address of next of kin (to be contacted only in an emergency):

_____

_____

Telephone number of next of kin:

_____

Name and address of your doctor:

_____

Do you have any medical conditions that we should be aware of? (If so please give details)

_____

_____

_____

Have you received vaccination against tetanus in the last five years?

Yes   ☐   No   ☐

Are you receiving medical or surgical treatment of any kind from either your doctor or a hospital?

Yes   ☐   No   ☐

Have you been given specific medical advice to follow in emergencies?

Yes   ☐   No   ☐

If the answer to either of the last two questions is yes, please give the details here (including dosage of any medicines/tablets)

Signed (researcher/student): _____ Date: _____

Appendix C: An abstract, problem description and literature review

Integrating trust building mechanisms into partnering projects to achieve optimum success

By Jason Challender

Abstract

In recent times, government initiatives through public procurement frameworks have largely focused on increasing participation of partner-led consortium strategies for collaborative procurement of major capital projects in the further education (FE) sector. Such initiatives are heralded as a vehicle to obtain best value, improved levels of quality and optimum service delivery. Yet there is still evidence of low levels of client satisfaction, owing mostly to poor cost and time predictability. The study aim explores the extent to which trust is a necessary part of this process and a viable tool in collaboratively procuring more successful UK further education procurement strategies. It gives greater understanding of how trust building mechanisms and initiatives can be designed and implemented for improving project outcomes. The research population is restricted to those contracting, consulting and client organisations that have had experience of collaboratively procured further educational projects. A review of literature identifies a framework for measuring the degree of trust through established trust-related attributes and behaviours while trust building mechanisms will be used under three themed groups: motivational, ethical and organisational initiatives. A mixed method approach of quantitative and qualitative methodologies is adopted with the former using survey questionnaires and subjecting data to correlation analysis. The qualitative approach consists of nine semi-structured interviews where raw data is coded using content analysis and sorted into themes from transcribed recording for analysis.

Study findings provide an insight as to why organisations may feel vulnerable about vesting trust in their partners, and these include scepticism of realisable benefits, opportunism and inequitable working relationships. This lack of trust may be causing a lack of appetite for taking perceived unnecessary risks considering certain practices, attitudes and behaviours of partnering organisations. This is especially the case in project partnering, where relationships are perceived to be short term, as opposed to strategic partnering. Potential trust building measures to overcome such dilemmas are presented as informal networking, professional development and team workshops. Furthermore quantitative study findings have determined that there is a correlation coefficient of 0.87 between these trust building mechanisms/initiatives (as independent variables) and the degree of trust in collaborative working (as a dependent variable) suggesting a significant influence. Future research is recommended to further explore how trust building initiatives can be designed and implemented in developing a framework for increasing trust in partnering strategies.

Introduction and the problem

The UK Government's Construction 2025 report (HM Government, 2013) and the Construction Products Association (HM Government, 2010) both highlighted a growing need for increased collaboration, integration and trust across the industry in order to make greater contributions to the pursuit of efficiencies. Notwithstanding these measures and perceived benefits for construction clients, consultants and contractors, partnering and other collaborative strategies have not always achieved their expected outcomes. This may have resulted from stakeholders' poor commitment to partnering arrangements grounded in a lack of trust, which could be damaging the interests of the whole supply chain. This lack of trust, according Larson (1997) has emerged from the highly competitive nature of the UK construction industry where commercial considerations and opportunities have prevailed over partnering philosophies. Such examples include clients adopting strategies linked to bullying contractors to gain lowest price tenders and main contractors deliberately slowing construction progress to force clients into instructing costly acceleration programmes (Korczynski, 1996). This has led in some cases to deep rooted adversarial practices and behaviours. Wong et al. (2008) explained that this is largely a result of a general mistrust in the construction industry on both an individual and inter-organisational level. Such lack of trust could therefore explain the downward trend in collaborative working practices in recent years as identified by the RICS (2012), in favour of more market-based approaches to contractor procurement. Initiatives designed to encourage partnering have also suffered from 'collaborative inertia' due to lack of trust, guidance, support and understanding.

Review of theory and literature

Background to collaborative working and partnering in the UK construction industry; perfection through procurement

The choice of procurement strategies on projects has long been a contentious issue within the UK construction industry. Banwell (1964) and Emerson (1962) outlined deficiencies within traditionally procurement methods and made recommendations for change, which included bridging the gap between design and construction and encouraging early contractor involvement in areas such as value management and buildability. For these reasons partnering and collaboration have long been championed as the future of the UK construction industry. Latham (1994) sought to 'Construct the Team' and was heavily critical of traditional procurement and contractual routes, largely due to the lack of coordination between construction and design. He suggested a change in culture and a move to partnering to increase fairness, encourage teamwork and enhance performance through collaborative engagement of clients and design teams with contractors. Following this lead Egan (1998) saw early establishment of construction teams as an essential aspect of cooperative construction, with contractors able to contribute to management, buildability, health and safety, procurement and supply chain management of projects. It was thought that such early collaboration reduces disputes, reduces tender costs and improves team working practices. Other benefits of collaboration have been argued to include an increase in profits brought about by sharing expertise, knowledge, ideas, innovation, best practice and promoting efficiencies and improvements in decision-making (Hansen and Nohria 2004). More recently, the Government's Construction 2025 report 'Industry Strategy: Government and Industry in Partnership' (HM Government 2013) has emphasised the need for incentivising the extent and degree of collaboration on building projects, thus stimulating innovation and successful outcomes.

Definition of construction partnering and the importance of trust
as a collaborative necessity

Although there are many different definitions of partnering (National Audit Office, 2001), it can be defined as 'business relationships designed to achieve mutual objectives and benefits between contracting organisations' or alternatively as 'a structured management process to focus the attention of all parties on problem resolution' (Larson, 1997).

Although there is a general lack of consensus as to the meaning of trust (Bigley and Pearce, 1998) it can be defined as 'the willingness to become vulnerable to another whose behaviour is beyond his control'. Trust is considered to be a 'bonding agent' between collaborating partners and as an 'essential foundation for creating relational exchange' (Silva et al., 2012). Fawcett et al. (2012) presented a perspective that 'without trust collaborative alliances cannot be created or maintained'. Despite this, trust appears to be a stranger in construction contracting where confrontation remains the prevalent environment (Wong et al., 2008). One contributory factor for such lack of understanding may emanate from trust receiving only limited attention in construction project management literature (Maurer, 2010). This can be corroborated by recent findings from Strahorn et al. (2014) who report that in UK construction management 'trust repair skills appear to be rare' especially following disputes. These arguments appear to support the case that trust among construction project teams certainly needs to be significantly increased (Dainty et al., 2007) especially since it is 'central to every transaction that demands contributions from the parties involved' (Cheung et al., 2011). There has been much debate in academia as to how to achieve this in practice. Cheung et al. (2003), in this regard, stressed the importance for project teams to communicate well and operate within an environment leading to 'an upward cycle of trust'. Conversely some academics have argued that it is the creation of shared ethos based on equity and fairness embedded in aligned organisational strategies that best promotes trust between partners (Thurairajah et al., 2006). Notwithstanding these views there has been little written on trust building measures and mechanisms for construction relationships and even less for construction partnering.

Incentives and benefits for building trust relationships in partnering arrangements

Many academics have focused on theories relating to the creation and development of trust as a potential means to reduce opportunism especially when business environments are prone to hidden agendas and conflicting objectives (Silva et al., 2012). Other theories, conversely, advocate that trust within relationships can safeguard against excessive formal contractual relationships developing between partnering organisations which could be misinterpreted as signs of distrust (Li, 2008). This is supported by Colquitt et al. (2007) who found that the potential benefits of developing and nurturing trust in the workplace could have positive influences on job performance while allowing vital risk taking where there are no other safeguards to protect partners. Based on Latham (1994), Egan (1998) and Egan (2002), figure C.1 illustrates some of the perceived benefits, especially on complex projects that could emanate from the early involvement of contractors.

Challenges for developing trust in construction partnering; potential barriers and problems

While the potential benefits of partnering practices have been widely articulated collaborative working practices have, however, attracted their critics in some instances and there is evidence to suggest that that there can be barriers to the adoption of partnering. Egan (1998, p. 24),

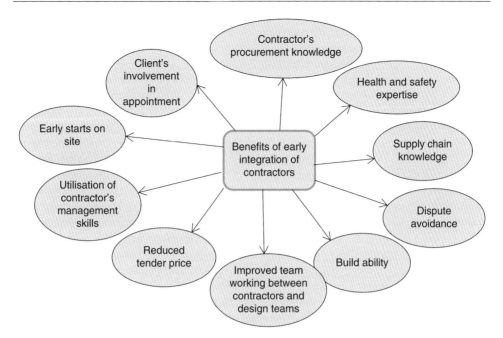

Figure C.1 Benefits of early integration of contractors (adapted from findings of Latham, 1994; Egan, 1998 and Egan, 2002).

for instance, referred to these potential difficulties in recognising that collaborative procurement is fast becoming a challenging option for contractors, subcontractors and suppliers and stated that:

> There is already some evidence that it is more demanding than conventional tendering, requiring recognition of interdependence between clients and constructors, open relationships, effective measurement of performance and an ongoing commitment to improvement.

This assertion can be corroborated by the RICS (2005, p. 2) who outlined that collaborative working practices and partnering 'may be difficult to integrate within a traditionally adversarial environment owing to its reliance on trust' with economic and cultural factors becoming the main barriers to greater participation. Furthermore some have debated whether such reliance on trust is appropriate where large sums of money are involved and opportunism could emerge (Lann et al., 2011). The other contentious factor is whether the fractious nature of the UK construction industry, based largely on one-off projects, facilitates the right environment and conditions for trust to prosper (Fawcett et al., 2012). This may explain why the UK construction industry has been slow to take on board the concept of trusting behaviours in supply chain management when compared to other sectors such as manufacturing (Akintoye et al., 2000).

The quality of collaboration can be reinforced or weakened, depending on the behaviour, approaches and attitudes of organisations and individual participants (Kaluarachi and Jones, 2007) and in practice the time that is needed to nurture key relationships is often lacking in

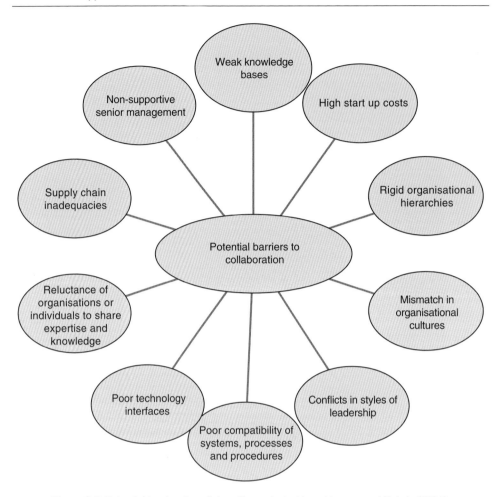

Figure C.2 Potential barriers to collaboration, adapted from Hansen and Nohria (2004).

construction management procurement systems (Walker, 2009). Also the project-based nature of much construction work can be seen as a fundamental barrier to the development of trust in practice, where relationships are often perceived to be short-term, and true collaborative working practices struggle to emerge (Walker, 2009). Furthermore reliance on the known and controllable has previously been identified within the UK construction industry, as a symptom of a lack of trust and 'negative culture', sceptical and suspicious of new initiatives.

Other problems for partnering have emerged on occasions where a perceived abuse of power has occurred (National Audit Office, 2001) or deployment of market leverage to disadvantage their 'partners' (RICS, 2005). Briscoe and Dainty (2005) supported this through development of their propensity to trust theory, and in practice this could manifest itself as 'buyers' dictating to 'sellers' the terms of their employment and what is required of them (Mathews et al., 2003). Korczynski (1994) referred to this type of practice and other forms of opportunism as the main source of mistrust in the UK construction industry.

According to (Hansen and Nohria, 2004, pp. 22–25) potential barriers to collaboration could result from a variety of factors and some examples of these are shown in Figure C.2.

Summary

Collaborative working through partnering arrangements has been spearheaded as the means by which such efficiencies can be introduced on capital projects to lower costs, achieve earlier completion dates and increase quality. Other benefits for clients, contractors and consultants include greater innovation and cooperation brought about through the integration of design and construction. Despite authoritative calls for such partnering practices there has been a growing trend over recent years for organisations to move back to traditional procurement routes. There have been reports that clients may be feeling that the only way to assure themselves that they are not paying too much is to market test their projects in a highly competitive environment. Organisations have been feeling vulnerable to partnering and reluctant to take unnecessary perceived risks. The main barriers and challenges for partnering have emerged from a general lack of trust between partners. These have been reportedly brought about through cases of inequitable working relationships, opportunism, lack of commitment and adversarial behaviours. There have also in recent years been reports of scepticism by some practitioners within the construction industry about realisable benefits of partnering arrangements. Furthermore the one-off and short-term project-based nature of the construction industry has hindered the development of trust through good working relationships and repeat business.

References

Akintoye, A., McIntosh, G. and Fitzgerald, E. (2000). A survey of supply chain collaboration and management in the UK construction industry. *European Journal of Purchasing & Supply Management*, 6 (2000) 159–168.

Banwell, H. (1964). *The Placing and Management of Contracts for Building and Civil Engineering Work.* HMSO, London.

Bigley, G.A. and Pearce, J.L. (1998). Straining for shared meaning in organisation science: Problems of trust and distrust. *Academy of Management. The Academy of Management Review*, 23(3): 405–421.

Briscoe, G and Dainty, A. (2005). Construction supply chain integration:an elusive goal? *Supply Chain Management: An International Journal*, 10(4): 319–326.

Cheung, S.O., Thomas, S.T., Wong, S.P. and Suena, C.H. (2003). Behavioural aspects in construction partnering. *International Journal of Project Management*, 21(5): 333–343.

Cheung, S.O., Wong, W.K., Yiu, T.W. and Pang, H.Y. (2011). Developing a trust inventory for construction contracting. *International Journal of Project Management*, 29: 184–196.

Colquitt, J.A., Scott, B.A. and LePine, J.A. (2007). Trust, trustworthiness, and trust propensity: A meta-analytic test of their unique relationships with risk taking and job performance. *Journal of Applied Psychology*, 92(4): 909–927.

Dainty, A., Green, S. and Bagihole, B. (2007). *People and Culture in Construction*, New York. Taylor and Francis.

Egan, J. (1998). *Rethinking Construction,* The Report of the Construction Task Force. London: DETR. TSO.

Egan, J. (2002). *Accelerating Change*, London: Rethinking Construction.

Emerson, H. (1962) *Survey of Problems Before the Construction Industries.* Ministry of Works. HMSO, London.

Fawcett, S.E., Jones, S.L. and Fawcett, A.M. (2012). Supply chain trust: the catalyst for collaboration. *Business Horizons*, 55(2012): 163–178.

Hansen, M.T. and Nohria N. (2004). *How to build a collaborative advantage.* 'MIT Sloan Management Review, 46 (1), 22–30.

HM Government (2010). *Low Carbon Construction Final Report* (November 2010). London: HM Government. 52–62, 196–199.

HM Government (2013). *Construction 2025. Industry Strategy: Government and Industry in Partnership* London: HM Government. 23–25.

Kaluarachi, DY and Jones, K (2007). Monitoring of a strategic partnering process: the amphion experience. *Construction Management and Economics*, 25(10): 1053–1061.

Korczynski M. (1994). Low trust and opportunism in action: evidence on inter-firm relations from the British Engineering Construction Industry. *Journal of Industry Studies*, 1(2): 43–64.

Korczynski M. (1996). The low-trust route to economic development: inter-firm relations in the UK construction industry in the 1980s and 1990s. *Journal of Management Studies*, University of Loughborough, 33(6): 787–808.

Lann, A., Voordijk, J. and Dewulf, G. (2011). Reducing opportunistic behaviour through a project alliance *International Journal of Managing Projects in Business*, 8(4): 660–679.

Larson, E. (1997). Partnering on construction projects: A study of the relationship between partnering activities and project success, *IEEE Transactions on Engineering Management.* 44(2): 188–95.

Latham, M. (1994). Constructing the Team: The Stationery Office, London.

Li, P.P. (2008). Toward a geocentric framework of trust: an application to organisational trust. *Management and Organisation Review*, 4(3): 413–439.

Mathews, J., Kumaraswamy, M. and Humphreys, P. (2003). Pre-construction project partnering: from adversarial to collaborative relationships. *Supply Chain Management*, 8(2): 166–178.

Maurer, I. (2010). How to build trust in inter-organisational projects: the impact of project staffing and project rewards on the formation of trust, knowledge acquisition and product innovation. *International Journal of Project Management*, 28 (2010): 629–637.

National Audit Office (2001). *Modernising Construction*, Report by the Controller and Auditor General HC 87 Session 2000–2001. The Stationary Office, London. 5–6.

RICS (2005). An exploration of partnering practice in the relationships between client and main contractors, *Findings in Built and Rural Environments.* London: RICS Research. 2–3

RICS (2012). *Contracts in Use. A Survey of Building Contracts in Use During 2010.* Royal Institution of Chartered Surveyors Publications.

Silva, S.C., Bradley, F. and Sousa, C.M.P. (2012). Empirical test of the trust performance link in an international alliances context. *International Business Review*, 21(2012): 293–306.

Strahorn, S., Gajendran, T. and Brewer, G. (2014). Experiences of trust in construction project management: The influence of procurement mechanisms *In:* Raiden, A B and Aboagye-Nimo, E (Eds) *Procs 30th Annual ARCOM Conference,* 1–3 September 2014, Portsmouth, UK, Association of Researchers in Construction Management, 463–472.

Thurairajah, N., Haigh, R. and Amaratunga, R.D.G. (2006). Cultural transformation in construction partnering projects. COBRA. *Proceedings of the Annual Research Conference of the Royal Institution of Chartered Surveyors* 7–8 September. University College London.

Walker, A. (2009). *Project Management in Construction*, Oxford: Blackwell Publishing Ltd.

Wong, W.K., Cheung, S.O., Yiu, T.W. and Pang, H.Y. (2008). A framework for trust in construction contracting. *International Journal of Project Management*, 26: 821–829.

Appendix D: Eight research proposals

Appendix D1: The behaviour of cyclist and motorists at signal controlled junctions
Appendix D2: The impact of community action group work on the regeneration of heritage buildings in north-west of England
Appendix D3: Investigating properties of timber
Appendix D4: Failure in bolt connections in temporary works systems
Appendix D5: An investigation on use of rainwater harvesting systems as sustainable measures and cost savings in the UK
Appendix D6: An empirical analysis of commitment to health and safety and its effect on the profitability of UK construction SMEs; by Andrew Arewa
Appendix D7: Determine the influence of temperature on synthetic fibre concrete
Appendix D8: Exploring the reasons behind cable strike incidents: A qualitative study

Appendix D1

The behaviour of cyclist and motorists at signal controlled junctions

Joanna Sammons

1. Introduction

The interaction of drivers and cyclists is an important aspect of road user behaviour with conflicts between the two well reported. These conflicts can often be seen at signalised and non-signalised junctions. In this instance, a general observational study will be undertaken of driver and cyclist behaviour at signalised junctions within the urban area of Bath.

2. Structure of dissertation

The dissertation will be structured in the standard way and include the following sections:

- abstract
- introduction
- literature review
- methodology
- results and analysis
- conclusion
- appendices
- references

3. Research context

The concepts of gap acceptance, red light running and stop line infringements are behaviours that are frequently discussed in literature. Ling and Wu (2004) undertook a study on cyclist behaviour at signalised intersections in Beijing, China and focused on crossing gap/lag acceptance behaviour. This study examined a signalised intersection which joined an east/west movement with a north/south movement. The intersection had a light sequence for each user with the lights all on a fixed cycle. The traffic light for the vehicles was three stage with a special green phase for left turning traffic in the east/west direction with a traffic light for cycles having two phases. As it could be expected, conflicts were reported between bicycles and motor traffic. In determining gap acceptance Ling and Wu (2004) referring to Miller (1971) defined the gap as the 'period between two successive vehicles beginning with the front of the leading vehicle and ending with the front of the following vehicle'. The lag was defined as 'the time from the moment the crossing vehicle reaches the point of conflict until the closing lag vehicle arrives'. The concept of the critical gap is analysed and is defined as the 'minimum gap duration which will be accepted'. As expected the gap acceptance and critical gap are variable across the population.

The data analysis for this study included statistical analysis of observational data collected which examined the characteristics such as crossing speeds, gap/lag analysis and group riding behaviour displayed at the study junction. Ling and Wu felt that the results of the study could be useful in building microscopic simulation models or theoretical models of cyclist behaviour to understand the behaviour of road users at signalised junctions. Regression modelling was also used in this instance to provide an estimation of the mean critical gap and lag.

Red light running or not obeying the red light during a traffic light sequence is a behaviour that is reported of both cyclists and drivers. The perception of red light running being a dangerous behaviour is reported by Wu et al. (2012). This was an observational study that looked at the rate, factors and characteristics of red light running by cyclists. In this case four armed signalised intersections were used. A number of variables were coded and observed relating to the rider's characteristics, movement information as well as situational characteristics. Collecting this data enables a picture to be developed of typical characteristics of those who run red lights. In analysing the data for this study a logistic regression analysis was undertaken to analyse the factors 'associated with red light running' and if there were any relationships between them.

Looking at behaviour during the crossing process enabled the authors of this study as well as other studies to classify cyclists into three types; this could also be applied to drivers who run red lights. The three categories are law obeying, risk taking and opportunistic and these relate to the point in the light cycle at which the rider will obey the traffic lights and cross the intersection.

Continuing under the same theme of red light running, but this time looking at drivers in urban settings, Porter and England (2000) reinforced the perception of red light running being a dangerous act. They noted that 'Reducing this risky behaviour depends on understanding its prevalence as well as the drivers involved.' Their study collected data on driver characteristics to test whether there were any 'predictors of red light running'. Logistic regression analysis was also used in this study and in this case it looked at red light versus yellow light running.

Johnson et al. (2013) examined the reasons for why cyclists infringe at red lights and the characteristics of those that do focusing on infringements by Australian cyclists. Among the reasons reported by the study for red light infringement were 'to turn left, when there were no other users present and at a pedestrian crossing'. An issue raised in their analysis of reasons for infringements is the potential consideration from a safety point of view of allowing cyclists to turn left on red. This proposal was put forwards as a result of the proportion of cyclists in the study who made a left turn against the red light. In commenting on the Australian media's portrayal of cyclists, Johnson et al. (2013) stated that 'red light infringement is frequently cited as the cyclist behaviour that most annoys drivers and is perceived as typical cyclist behaviour.' Red light infringement is often discussed in the UK media, particularly within social media settings.

A paper that offers a more qualitative view of cycling in the urban areas is McKenna and Whatling (2007). Their study examined the live experiences of urban commuter cycling. The findings of the study reported that influences on commuter cycling 'included: the weather, daily tasks, cycling infrastructure, driver behaviour and the value of cycling for physical and mental wellbeing'.

4. Research aim and objectives

The aim of this study is to investigate cyclist and driver behaviour at signal controlled junctions in the urban area. The objectives for this study are as follows:

(1) to survey red light running of cyclists and drivers
(2) to survey stop line infringement of cyclists and drivers

CCTV data will be used to observe cyclist and driver behaviour at a number of signalised junctions in Bath with the junctions used influenced by the availability of CCTV coverage. Potential junctions for this study are those at key points within the Bath urban network such as those along the A36 Lower Bristol Road and A4 London Road. Along the A4 one junction has recently been remodelled, and subject to data availability this junction will be included within the study.

The data collected is likely to be influenced by a number of factors including number of arms at the junction, whether there are advanced stop lines or other facilities for cyclists, whether there are any filters, the movement being made (i.e. left, right or straight on), the speed and volumes of traffic (this includes both motor vehicles and cyclists), the type of cyclist, the speed limit of the road, the time of day and the weather. Other information that will need to be considered in looking at gap acceptances and the critical gap includes the queuing behaviour of cyclists and this could feed into further investigation of how cyclist behaviour can be modelled in micro simulation.

5. Research methods

Following the final selection of junctions for the study, site plans will be made to ensure that there is a clear understanding of the possible movements at each junction for the road users. Information will also be gathered on traffic signal phasing, traffic flows and turning counts (where this data is held) to build up a picture of the operation of the junction.

To gather the behavioural information, a general observational study will be undertaken using CCTV footage of the selected junctions to analyse user behaviour and the extent to which cyclists and car drivers obey the rules. The CCTV will be examined to find evidence of red light running by cyclists and drivers as well as stop line infringements and may also include gathering information on road user characteristics such as those reported in previous studies. If possible, the study may be widened to include larger vehicles such as HGVs.

The data collection will take place at a neutral time, i.e. a period when there are no school or public holidays. Observations will be carried out at junctions throughout the day to ensure a fair representation of typical behaviour, this will include peak periods.

6. Ethics

Ethical considerations will need to be made prior to undertaking the research as well as throughout the duration of the project. The research methods involved in the study are primarily observations of driver and cyclist behaviour at signalised junctions. As this study will be a purely general, observational one it will not be necessary for a detailed ethics checklist to be completed. Consideration will also need to be made of the way in which data is collected and stored to ensure confidentiality and compliance with the Data Protection Act.

References

Ling, H. and Wu, J. (2004) A study on cyclist behaviour at signalised intersections. *IEE Transactions on Intelligent Transportation Systems*, 5 (4) pp. 293–299.

Wu, C., Yao, L. and Zhang, K. (2012) The red-light running behaviour of electric bike riders and cyclists at urban intersections in China: An observational study. *Accident Analysis and Prevention* 49 (2012) pp. 186–192.

Porter, B. and England, K.J. (2000) Predicting Red-Light Running Behaviour: A Traffic Safety Study in Three Urban Settings. *Journal of Safety Research*, 31 (1), pp. 1–8.

Johnson, M., Charlton, J., Oxley, J. and Newstead, S. (2013) Why do cyclists infringe at red lights? An investigation of Australian cyclists' reasons for red light infringement *Accident Analysis and Prevention* 50 (2013), pp. 840– 847.

McKenna, J. and Whatling, M. (2007) Qualitative accounts of urban commuter cycling, *Health Education*, Vol. 107 Iss: 5, pp. 448–462.

Appendix D2

A study of community action groups in developing sustainable heritage projects in the north-west of England

Clive Robinson

The conservation movement's philosophy is rooted in 19th century with the creation of the Society for the Protection of Ancient Building (SPAB) founded in 1877 by William Morris as an initial reaction to the industrial age, and the demolition and over-zealous restoration of parish churches. Other organisations emerged with similar objectives such as the National Trust, founded in 1894; its co-founders being Sir Robert Hunter, Octavia Hill and Canon Hardwicke Rawnsley (Cowell, 2013). The growth of amenity societies such as the Georgian Group, Victorian Society and 20th Century Society also reinforced the role of heritage in the 20th century. The basic conservation principles that we follow today through the work of SPAB and English Heritage (EH) ensure that some of our heritage buildings are well maintained, conserved and restored for future generations.

Our heritage continues to be put at risk, According to English Heritage (2008), they have 2433 listed buildings on their buildings at risk register, and we see many community groups across the country that have a heritage building in need of restoration. The groups usually comprise a loose body of people who are deeply concerned with their local historic environment and the need to save it for future generations. These groups commonly referred to as Community Action Groups (CAGs) are interested in saving local buildings that may be on a national list. The buildings could be a local church, pub, theatre, pavilion, pier, hotel, cottage or warehouse. The main theme that these building have in common is that they may have become functionally and economically obsolete. A current example of this is the trend of pub closures. According to Brandwood 2013:16:

> The British pub is going through tough times and during 2012 the closure rate varied between twelve and eighteen a week. Since 2008 no less than 4500 have closed and heritage pubs have suffered along with others.

The Campaign For Real Ale (CAMRA) are working in association with English Heritage to ensure that pubs of architectural and historic interest are listed to provide them with additional protection against demolition or significant changes that alter the exterior or interior of the building. CAMRA in association with English Heritage created a National Inventory of 270 pubs, 217 of which are listed (Brandwood, 2013).

There are typical examples of rural pubs being taken over by the local community to ensure that the pub continues to act as a hub for the community. Such initiatives are supported by CAMRA who publish a booklet *Saving Your Local Pub*. According to CAMRA a number have being saved by the new generation of micro-breweries (Brandwood, 2013).

The Architectural Heritage Fund (AHF) was also responsible for setting up the concept of Building Preservation Trusts (BPTs). Cambridgeshire Preservation Trust was one of the first founded in 1929. The full potential of BPT was recognised by the Civic Trust in the late 1960s. There were 30 BPTs in 1976, but this had increased to 165 by 1997. The concept of the BPT is a voluntary or non-profit organisation set up by likeminded people who are willing to devote time and effort without financial reward to a good cause. They exist in order to save high risk, low return historic buildings in which nobody else is prepared to invest (Wier, 1997).

An excellent example of this type of community-led regeneration was the formation of the Ancoats Building Preservation Trust in 1995, and the subsequent restoration and development of the huge complex of Murrays Mills at Ancoats Manchester in 2007 (Rose et al., 2011).

The CAGs are usually made up of volunteers from a varied spectrum of society with a common goal to save the historic building asset and ensure that it has a sustainable future. The road to achieving such a goal is difficult, complicated and full of potholes that need to be avoided in order that a group achieves their end goal. The barriers to success are numerous and very often each project can have a mixture of associated problems and challenges, including:

(a) location of the heritage asset, prosperous area, declining area
(b) deprived area or part of a regeneration area
(c) obsolete heritage asset
(d) planning issues
(e) building control issues
(f) funding issues
(g) loss of historic fabric due to water penetration and poor maintenance over time
(h) ownership issues and restricted covenants
(i) listed building consents
(j) factors affecting significance of the heritage asset
(k) sustainable future use of asset.

Patience, tolerance, determination, enthusiasm are all key traits in the character of a volunteer who becomes part of a CAG.

The UK Government's policy on heritage has been outlined in Heritage Counts (2011). This document is an overall snapshot of the heritage sector and its relationship with the wider agenda of the previous Coalition Government's Big Society initiative. This reveals how important the connection is between heritage and the voluntary sector organisations. It is a worthy starting point on how policy initiatives will shape the future landscape of the heritage sector and its association with local community groups (Heritage Environment Forum, 2011). These strategies are also supported by The Prince's Regeneration Trust under the Building Resources Investment and Community Knowledge (BRICK) UK-wide training and mentoring programme (The Prince's Regeneration Trust, 2014). The Heritage Lottery Fund (HLF) remains the largest funding organisation for heritage projects in England. The concept of community share offer is another method of raising money and was successful in raising £200,000 to finish a £4 million restoration project on a former Co-op building, Unity Works, in Wakefield Yorkshire (Big Issue in the North, 2014)

This research will focus on a project in the north-west of England that is currently listed by English Heritage, which is obsolete and in a poor state of repair. The concept of the community action group, and the desire to conserve or restore buildings for future generations will form a key part of the research.

Aim

To examine and apply best practice in a community-led approach to the regeneration and sustainability of a community heritage asset.

Objectives

  I.   Critically examine the role of community groups in the restoration of heritage assets.
 II.   Critically appraise good practice in community-led regeneration of historic assets.
III.   Translate best practice in community-led regeneration of historic assets into a case study project.
IV.   Evaluate the impact that a Community Action Group will have on the regeneration of a community heritage asset.

Outline methodology

This action research project will utilise qualitative research methods to investigate community-led projects.

M1. Review philosophy, and concepts of the meaning of community, groups, conservation and restoration of heritage assets. Access, relevant literature, contrasting different philosophical opinions of the subject matter, to inform the researcher of existing beliefs and any new shifts in attitude towards using heritage assets to regenerate communities (Naoum, 2007).

M2. Review literature related to the topic area. Current documents on funding, advice, guidance and associated reports. Literature – including academic journals, conference papers, publications, by various organisations such as English Heritage, The Prince's Regeneration

Trust, Heritage Lottery Fund, Architectural Heritage Fund. Books associated with the topic heritage regeneration and books on research methods such as (Farrell, 2011) and (Naoum, 2007).

M3. Review successful projects and interview key members who achieved their overall goal to establish best practice within the heritage sector using existing case studies that have had successful outcomes (Yin, 2009). The most appropriate approach to this study is therefore to seek opinions and attitudes of key members of CAGs that are involved in saving a building for future use as a sustainable venture. This is supported by Creswell (2009:18) who quotes (Morse 1991) 'If a concept or phenomenon needs to be understood because little research has been done on it, then it merits a qualitative approach.' The method of sampling for this kind of study will be purposive based on relevance and knowledge of existing members of CAGs within the north-west of England, including key members of each group and others working with the group in a professional capacity including funders, project managers, architects and surveyors (Denscombe, 2012).

M4. To network and access relevant community action groups within various locations in the north-west of England. Use interviews recorded for qualitative data collection to explore problems and identify barriers to successful projects. Observations, interviews and documents can all be used to inform the case study (Denscombe, 2012:54 and Yin, 2009). The interview transcripts will be analysed using content analysis of the text and sorting into coded data across a range of interviews. The main themes from the interviews will be combined in separate tables under key headings identifying consistencies and inconsistencies with the literature review. The data from the tables will be used to form a narrative as described by (Farrell, 2011). The narrative will be examined to form a detailed conclusion to the study along with key recommendations that can be implemented in the form of an action plan to be used by the case study CAG.

The research project will:

  I.   identify current problems and barriers to successful end project for case study CAG
 II.   systematic and rigorous enquiry of problems and barriers will be explored
III.   inform the current CAG of best practice is the sector to overcome barriers
 IV.   translate findings into action plan
  V.   instigate change within the CAG by implementing plan of action
 VI.   observe the impact of change initiatives on the project
VII.   review the process and make recommendations for future projects.

References

Brandwood G. (2013) *Britain's Best Real Heritage Pubs Pub Interiors of Outstanding Historic Interest*, St Albans: Campaign for Real Ale Ltd.

Cowell B. (2013) *Sir Robert Hunter Co-founder and Inventor of the National Trust*, Stroud: Pitkin Guides In association with the National Trust.

Creswell J.W. (2009) *Research Design Qualitative, Quantitative and Mixed Methods Approaches*, Third Edition, London: Sage Publications.

Denscombe M. (2012) *Open up Study Skills, The Good Research Guide, For Small-Scale Social Research Projects*, 4th Edition, Maidenhead: Open University Press.

English Heritage (2008) *Conservation Principles Policies and Guidance for the Sustainable Management of the Historic Environment*, Swindon: English Heritage.

Euston Arch Trust (2014) www.eustonarch.org (Accessed 28th August 2014).

Farrell P. (2011) *Writing a Built Environment Dissertation, practical guidance and examples*, Oxford: Wiley-Blackwell.

Heritage Environmental Forum English Heritage (2011) *Heritage Counts 2011 England*: Swindon: English Heritage.

Naoum S.G. (2013) *Dissertation Research & Writing for Construction Students*, Third Edition Oxon: Routledge Taylor & Francis Group.

Rose E., Falconer K. and Holder J. (2011) *English Heritage Ancoats Cradle of Industry*, Swindon: English Heritage.

The Big Issue in the North (4–10 August 2014): The Big issue in the North

Wier H. (1997) How *To Rescue A Ruin, by setting up a local Building Preservation Trust*, London: The Architectural Heritage Fund.

Yin R.K. (2009) Case Study Research Design and Methods Fourth Edition Applied Social Research Methods Series Volume 5, London: Sage Publications.

Appendix D3

An investigation of the assessment of the mechanical properties of timber using visual inspection and non-destructive testing in combination

Mike Bather

Background information

In order to assess the safety of existing timber structures an evaluation of the mechanical properties of the insitu timber is essential. Currently, there is no codified guidance in the UK on this topic, although some countries such as Italy (UNI, 2004) and America (Anthony et al., 2009) have limited guidance. Additionally, the mechanical properties of timber are difficult to measure without damaging or destroying the timber itself and are difficult to predict with accuracy. It is a very variable material, being highly anisotropic and heterogeneous (Blass et al., 1995). Also, it may be found in widely differing structures; from hardwood framing in prestigious historical properties to softwood roof members in Victorian terraced housing.

Visual stress grading rules for new timber have been developed which are not appropriate for assessing insitu timber in existing structures for several reasons. Just a few are presented below. Stress grading is carried out to sort new timber sections so that their mechanical properties can be relied upon to exceed certain minimum values and this leads necessarily to the undervaluing of almost all timbers. Additionally, in grading new timber, it is important to set the grade boundaries in order to make good use of the raw material. A simple system has been adopted in the UK which, for instance, categorises softwood as Special Structural, General Structural or reject. This subdivision of the entire range of the mechanical properties of softwood timber into just three groups is imprecise. Finally, the grading and associated design rules (for permissible stress design) are intended to be used before construction and so incorporate factors of safety that allow for potential construction variations in dimensions (affecting load widths, spans and cross-sectional areas of members) and load build-up. However, in existing structures, dimensions and loads can generally be measured with no risk of variation.

Around about a half of the money spent on construction is spent on maintenance and refurbishment and, within the building sector, most properties involved contain at least some structural timber. So, the issues with the assessment of insitu timber affect a substantial part of the construction industry. The unnecessary replacement of structural timber is not sustainable and brings pain to many people: the client (as it increases costs), the structural engineer (as it is inefficient) and finally to all of us (as it damages the environment and destroys our heritage). Improvements in the assessment of insitu timber would help to relieve some of this pain.

Area of research

What research has been carried out on insitu historical timber generally suffers from a restriction on sample size and the inability to test the timbers in question to destruction. Additionally, there is little published literature describing the efficacy of combining the results of visual inspection with those of non-destructive testing in predicting the mechanical properties of timber.

The testing of new timber allows for adequately large sample sizes, large specimens (sawn timber of structural size in place of clear specimens) and the ability to test specimens to destruction (as each specimen has little inherent worth, unlike many historical timbers). There is an element of convenience also, as the work currently carried out at the University, already includes destructive and non-destructive testing of sawn timber of structural size. It is hoped that the proposed work could be combined with this ongoing work, relatively straightforwardly.

As well as addressing the limitations of the current grading rules, the research must consider: the size effect of members, the wide range of species used in construction, the physical features affecting old timber (e.g. nails, bolts, dowels, notches, fungal and insect attack) as well as the relevant physical features affecting new timber (e.g. knots, slope of grain) and their known positions within the member, the effects of creep and possible previous overloading, moisture content and other factors such as density or the known orientation of growth rings. Finally, it would need to accommodate the likely inaccessibility of at least some insitu timber.

It is hoped that the linking of several pieces of information (gained from visual inspection and from non-destructive testing) will improve the accuracy and reliability of the prediction of the mechanical properties of timber. In doing this, it is essential that a statistically robust approach be developed.

Aims and objectives

In short, the aim of the research is to investigate the prediction of the mechanical properties of timber using a combination of visual and non-destructive techniques.

A literature review has started on this wide-ranging area of research and the author feels that at present, it is too early to define a complete set of objectives. However, from a preliminary review it can be seen that there are many journal and conference papers based on small pieces of research which do not directly link together. This suggests that a systematic review of different studies could be useful (even if this would be difficult to do). Bearing in mind the limitations of research carried out by one person in a short time frame, a single overarching research approach will not be appropriate, but instead, some smaller targeted pieces of research could be useful in addressing some of the key factors that determine the mechanical properties of insitu timber.

Thus the provisional objectives are:

(1) Complete a literature review of the topic area, investigating ways of combining the results of visual inspections with those of non-destructive testing to predict the mechanical properties of timber
(2) Coordinate with the testing programme at the University and combine visual inspections with the destructive and non-destructive testing regime already in progress
(3) Refine the predictive model in order to produce guidelines which will be of use to structural engineers assessing historical timber insitu (these guidelines may also potentially inform the work carried out grading new timber for the construction industry).

Schedule of work

It is proposed to carry out the research on a part-time basis. Broadly, the early months will be spent on the literature review. By the end of this phase a number of predictive models will be drafted. The next phase will be to liaise with the University to coordinate testing of suitable samples of timbers. The results from the testing will fed back iteratively into the theoretical models being assessed. The final phase will be the writing up of the project. It is not possible to attempt to draw up a more precise programme than this at this stage.

The testing facilities at the University are well suited to the scope of intended research. There are the usual testing facilities found within a civil engineering department; we currently have the equipment for testing small clear timber specimens in three-point bending and in shear.

References and bibliography

This limited bibliography gives an indication of some of the literature in the field.

Anthony, R.W. Dugan, K.D. and Anthony, D.J. (2009) A Grading Protocol for Structural Lumber and Timber in Historic Structures. Natchitoches, LA, USA: National Centre for Preservation Technology and Training.

Benham, C., Holland, C. and Enjily, V. (2003) Guide to machine strength grading of timber. BRE Digest 476. London: BRE Bookshop.

Blass, H. J. et al. (1995) STEP 1. Almere, The Netherlands: Centrum Hout.

BSI (1957) BS 373:1957 Method of testing small clear specimens of timber. London: British Standards Institution.

BSI (2007) BS 4978:2007 Specification for visual strength grading of softwood. London: British Standards Institution.

BSI (2002) BS 5268-2:2002 Structural use of timber. Code of practice for permissible stress design, materials and workmanship. London: British Standards Institution.

BSI (1971) CP 112:1971 The structural use of timber. London: British Standards Institution.

BSI (2003) BS EN 338:2003 Structural timber. Strength classes. London: British Standards Institution.

BSI (2010) BS EN 384:2010 Structural timber – Determination of characteristic values of mechanical properties and density. London: British Standards Institution.

BSI (2010) BS EN 408:2010 Timber structures – Structural timber and glued laminated timber – Determination of some physical and mechanical properties London: British Standards Institution.

BSI (2004) BS EN 1912:2004 Structural timber – Strength classes – Assignment of visual grades and species. London: British Standards Institution.

BSI (2004) BS EN 1995-1-1:2004 Design of timber structures – General – Common rules for buildings. London: British Standards Institution.

BSI (2005) BS EN 14081-1:2005 Timber structures – Strength graded structural timber with rectangular cross section – Part 1: General requirements. London: British Standards Institution.

BSI (2010) BS EN 14081-2:2010 Timber structures – Strength graded structural timber with rectangular cross section – Part 2: Machine grading; additional requirements for initial type testing. London: British Standards Institution.

Dinwoodie, J.M. (2000). Timber: Its nature and behaviour. London, E & FN Spon.

Forest Products Laboratory (2010) Wood handbook – Wood as an engineering material. General Technical Report FPL-GTR-190. Madison, Wisconsin, USA: U.S. Department of Agriculture, Forest Service.

Lavers, G.M. (1983) The strength properties of timber. London: Building Research Establishment Report, HMSO.

Thelandersson, S., Larsen, H.J. (Eds) (2003) Timber Engineering. Chichester, UK: John Wiley & Sons Ltd.

UNI (2004) UNI 11119 Wooden artefacts, Load-bearing structures – On site inspections for the diagnosis of timber members. Milan, Italy: Ente Nazionale Italiano di Unificazione.

Appendix D4

Mode of failure of a 90° bolted prop to prop connection for an existing cold formed steel temporary works system

Richard Forshaw

Background

The author is currently employed by a well-established temporary works company within the construction industry that offers a vast amount of proprietary equipment used for super- and substructure engineering problems.

One of the products they offer is a versatile modular propping system called 'Mass 50', which is most commonly used to construct supporting towers for various propping and jacking applications.

The main component of this propping system is the cold-formed steel prop bodies, which are formed from two back-to-back C-shaped sections and are available in various lengths ranging from 0.25 m to 3.0 m. Each prop body has a uniform hole pattern in both its webs, flanges and end plates. This allows various bracing and other ancillary items to be bolted together to form the desired configuration required.

This system was designed and tested by in-house engineers and was originally developed to provide a lightweight tower system. Over time, and mainly due to the system's versatility, bespoke configurations have been designed to suit client requirements. One of these bespoke configurations is to cantilever a horizontal prop, with an applied load, from a vertical prop without the form of any supporting bracing member, i.e. no knee brace. This 90° bolted connection is often frowned upon by in-house project engineers as the capacity and mode of failure for this detail is not fully understood as no form of testing or calculation has been previously done. It is thought that the mode of failure will be one of three possibilities:

(1)   local crushing of section webs due to a concentrated line load
(2)   local buckling of section flanges due to bolt pull out tension
(3)   a combination of both

It is this topic area that the author wishes to investigate. Understanding the structural capabilities and limitations of this bespoke application will not only enable the in-house engineers to design effective and efficient solutions, but will, most importantly, ensure that safe and practical schemes are produced in the future.

The data collection methods to be used are as follows.:

- Collate and review any existing company calculations and previous test data to understand what the Mass 50 system has been designed and tested for previously.
- Review relevant design codes, guidance documents, institution publications and any previous papers regards local crushing/bolt pull-out information for cold-formed steel C sections.
- Collate 2–3 typical company project schemes where this bespoke application has been used or considered, focusing on how knowing more about this connection would have benefited the project engineer at the time.

The main objective is to determine what load can be safely applied to a cantilevered prop before the point of failure is reached. The definition for the 'point of failure' is still to be decided, e.g. slight deformation or major deformation. Following this investigation, and subject to the validation of the findings, safe working joint load capacities are to be added to the company's Mass 50 technical data sheet, which can then be used by the project engineers in their future designs.

To achieve this goal, theoretical calculations are to be produced to any relevant design codes (Eurocode/BS5950) and practical testing is to be done in a controlled laboratory. Destructive practical tests are to be done on the company premises, with the use of a bespoke test rig inclusive of hydraulic jacks to apply the required loads and extensive use of calibrated monitoring equipment to measure displacements, i.e. strain gauges. Various configurations and conditions will be tested to help fully understand the mode of failure; these are still to be defined at present.

Appendix D5

An investigation on use of rainwater harvesting systems as sustainable measures and cost savings in the UK

Ayodele Rafiu and Yassin Osman

Background information

Of all the water on the planet, just 3% is fresh. Less than 33% of 1% of this fresh water is accessible for human utilization. The rest is solidified in ice sheets or polar ice tops, or is profound inside of the earth, beyond our range. To put it another way, if 100 litres speaks to the world's water, about a large portion of a tablespoon of it is fresh water accessible for our utilisation (Waterwise, 2015).

The UK has less accessible water per individual than most other European nations. London is drier than Istanbul, and the south-east of England has less water accessible per individual than the Sudan and Syria. In the UK, the human population keeps growing and the water demand in comparison to water availability suggests that water is scarce. The Environment

Agency (2008) suggested that meeting the future demands of society, while protecting and improving the environment in the face of the impact of climate change, will be a daunting challenge. This water scarcity is reflecting on the prices charged by water companies in the UK.

Rainwater harvesting systems might be able to help in the process of utilising a water resource alternative as well as saving UK commercial buildings money on water bills. It simply is the process of collecting and storing rainwater that falls on a catchment surface, typically on a roof of a building or other structure. The collected water can be used to replace mains water in non-potable applications, such as toilets, urinals, washing machines, irrigation systems or other general uses where water is not used for human consumption. Rainwater harvesting systems could be costly to install and maintain and this is one of the reasons why it not very popular in the UK. Most people do not even consider it because of the cost possibly being more than the cost of current water bills and other possible inefficiencies of the system. In this instance, an investigation will be carried out to analyse how efficient the systems are in different locations in the UK and how much money could be saved on water bills in these different locations.

Research aim and objectives

The main aim of this investigation is to test the rainwater harvesting system in commercial buildings as a means of saving money and helping the environment as a sustainable tool. This will lead to the following objectives:

- To test the viability of rainwater harvesting systems in four different locations with different climates
- To estimate how much money could be saved from water bills
- To estimate the payback period of the capital investment

Research methodology

The investigation will be based on four commercial buildings in different locations with the same demands and same size and shape of roof where rainwater harvesting systems will be installed in the four commercial buildings. Although the commercial buildings will be the same, the storage capacity will be estimated differently depending on the average rainfall in the four locations. This investigation will use three different methods to estimate the storage capacities needed so as to save money on storage tanks not being fully utilised or buying storage tanks that would be too small for the needs of the commercial building.

The first method, known as the simplified approach, will use the storage capacity charts alongside the average annual rainfall provided in the British standard rainwater harvesting system Code of Practice BS 8515:2009+A1:2013. The second method, known as the intermediate approach, will be using equations that are provided to allow a more flexible and accurate facility for calculating the storage needed. The advantage of this is that the variables can be modified to reflect the situation being considered. The third, more detailed, method will be conducted using a computer application called RainCycle Advanced. The function of the application is the hydraulic simulation and the whole-life costing (construction, operation/maintenance and decommissioning costs) for the rainwater harvesting systems in the four commercial buildings to be investigated. Annual rainfall data for the present year and years to come will be gathered from the Environment Agency or Met Office to help get a better picture.

Prices to install the rainwater harvesting systems will be obtained from companies to estimate how much it will cost. Prices to maintain the rainwater harvesting systems will also be gained from companies to estimate how much it will cost to maintain the systems annually. Other economic data will also be acquired in this investigation to consider how owners will get the capital investment to install the systems in their buildings. Assuming that money is borrowed, interest rates will be taken into account to give a more practical analysis of the rainwater harvesting systems. Calculations will be carried out to find how much money could be saved and how long it would take to pay back the capital borrowed.

Schedule of work

The first month will be used for the literature review and learning to use the RainCycle Advanced computer application. At the end of this month, the simplified approach method will be done and the hand calculations will begin to estimate the sizes of storage tanks needed at the four locations. Rainfall data will then be acquired from the Environment Agency or Met Office and prices will also be collected from RWH companies. After a couple of months, the data will start to go into the RainCycle Advanced computer application. After a couple more months, results will be analysed and conclusion will be drawn from it. The investigation will then be put into writing and submitted.

Bibliography

BSI (2013). Rainwater harvesting systems – British Standard Institution. Code of Practice. Available: Last accessed 27th Sept 2015.

Environment Agency (2008). Water resources in England and Wales – current state and future pressures. Available: http://webarchive.nationalarchives.gov.uk/20140328084622/http:/cdn.environment-agency.gov.uk/geho1208bpas-e-e.pdf. Last accessed 25th Sept 2015.

Fewkes, A. (1999). The use of rainwater for WC flushing: the field testing of a collection system. Building and Environment. 34 (6), 765–772.

SUD Solutions (2005) RainCycle Advanced User Manual.

Waterwise Project (2015) Why save water. www.waterwise.org.uk/pages/why-we-need-to-save-water.html. Accessed 26th Sept 2015.

Appendix D6

An empirical analysis of commitment to health and safety and its effect on the profitability of UK construction SMEs

Andrew Arewa (adapted by Peter Farrell)

1.1 Problem statement and research goals

While health and safety in the UK construction industry has improved significantly in the last decade, there are still too many fatalities, injuries and incidents. Not all people may agree, but some assert that while not having reached zero accidents, large companies have got health and safety 'just about as good as it can get'. There are relatively few serious cases in this sector. Small and medium-size enterprises (SMEs) account for a disproportionately high number of health and safety related statistics.

It is doubtful that poor health and safety records would be tolerated in other industries such as nuclear or aerospace. Indeed, it is arguable that construction can distance itself from poor health and safety performance, but only when undertaken under the watchful eye of clients such as these.

Many companies push for productivity, since simultaneously that may improve profitability and ensure completion of projects on time, thus avoiding or minimising financial penalties. Individual tradespeople also push their own productivity, since they are often paid on piecework; the more units of production completed, the higher the weekly wage.

Spending money to ensure health and safety is not always 'ring-fenced' expenditure. For example, maintaining mechanical equipment keeps it productive as well as safe; expenditure to keep sites clean is for both productivity and health and safety. However, there are some instances where money is spent solely for health and safety, such as personal protection equipment (PPE) or taking workers away from productivity for health and safety training. The human cost of a fatality is immeasurable.

There is perhaps a feeling among many practitioners that expenditure on health and safety is money taken out of profits. While not spending money may lead to an accident, statistically speaking the 'odds' are that accidents will not occur, so the attitude may be 'we will risk it and get away with it'. Some companies will only do the absolute statutory minimum (and in some cases even less) to fulfil their health and safety responsibilities.

The alternative view is that spending money on health and safety is an investment. It brings with it spin-offs into many parts of businesses. It is part of company corporate social responsibility (CSR). It reduces accidents with all the consequential costs that arise, and it improves public image, which may be helpful in securing more turnover in the marketplace. Doing business well is common sense; for example, keep sites clean, plan all activities carefully, employ sufficient supervisors, use proper and well-maintained plant and equipment, do not rush, employ qualified people and train, train and train. All have production benefits and contribute to keeping sites safe. However, in an industry which is extremely competitive, it is extremely difficult to convince business leaders to invest in these areas without being able to see tangible rewards. While captains of industry may argue that investing in health and safety improves profitability, it must be a questionable assertion, since there is no evidential base to support it.

As a consequence of the above, there are three imperatives: (1) to reduce the number of health and safety accidents and incidents, (2) to improve the profitability of the construction industry, (3) to determine if these two variables are compatible and related to each other. The research goals of this study are therefore:

- Aim: to investigate the feasibility of investing money to improve the health and safety performance of the UK construction industry
- Research question: does investment in health and safety influence profitability?
- Objective: to determine if investment in health and safety influences profitability.
- Hypothesis: investment in health and safety influences profitability.

It is noted that the aim is written using 'directional' or 'one-tail' terminology; that is 'improve'. Since there is no evidence to support a supposition that investment in health and safety 'improves' profitability, the word 'influence' is used in subsequent research goals, and analysis will be conducted against the two-tail hypothesis.

1.2 Key parts of the literature

There were 42 fatalities in the UK construction industry in the year to March 2014; that is an increase from 39 fatalities in the year 2012/13. Construction has the worst record of all industries in the UK. Rita Donaghy in a landmark inquiry submitted to the UK Secretary of State for Work and Pensions, used the title of her report to state that 'One death *is* too many' (Donaghy, 2009).

Two seminal documents illustrate historic data. In 1988, the report 'Blackspot Construction' identified that in the five-year period 1981–1986, an average of 148 people died in construction each year (HSE, 1988). As recently as 2001/02, there were 108 fatalities. This was at the same time as a government initiative, led by the then Deputy Prime Minister John Prescott, culminating in a report entitled 'Revitalising Health and Safety' (HSE, 2002).

In 2011/12 major injuries (now defined under RIDDOR 2013 as 'specified injuries' such as factures, amputations, loss of sight, crush injuries, serious burns or unconsciousness) were 2230 and over-3-day injuries (those requiring the victim to be absent for 3 days or more) were 5391. These statistics do not include people who die or have long-term illnesses resulting from exposure to inappropriate substances such as asbestos and dusts, or have other illnesses that do not become apparent for long periods such as vibration white finger and mental health issues.

While the record of UK construction is not acceptable, in comparison to many of our European partners, the number of fatalities is among the lowest. The HSE (2013) report that construction in 'Britain had one of the lowest rates of fatal injury published by Eurostat for 2010 (0.71 per 100,000 workers), and compares favourably with other large economies such as France (2.49 per 100,000), Germany (0.81), Italy (1.57) and Spain (1.76)'. Similarly, Britain continues to have one of the lowest levels of work-related ill health in the world (HSE, 2011). SMEs account for 90% of construction fatalities at work. Yet, a record number of unsafe acts remains in this category of firms, and a single adverse safety incident has the potential to substantially destabilise or liquidate an SME (Arocena and Nuez, 2010). The former UK Department for Business Enterprise and Regulation Reform (BERR, 2008) now known as the Department for Business, Innovation and Skills (BIS) claimed that many businesses have failed as a result of heavy costs incurred from penalties for not complying with health and safety best practice.

Smallman and John (2001) and HSE (2003) classify spending on health and safety as preventative and remedial. Each then has a classification of direct and indirect. Examples of direct costs are: cost of providing PPE, cost of health and safety supervision, compensation cost, insurance, cost of providing first aid/other health services and training costs, while indirect costs, often imbedded in production costs, are usually recognised as: costs of providing safe working environments, costs of providing remedial works in the event of adverse safety incidents and cost of carrying out risk analysis/regulatory supervision.

The European Agency for Safety and Health at Work (EASHW, 2001) argues that, 'the endless occurrences of adverse health and safety incidents in the construction industry suggests that practitioners tolerate hazards and accidents in the belief that, adverse safety incidents are either non-preventable or that a certain number are unavoidable.' This is arguably not acceptable, and Engasser (2010) put forward an alternative position that 'every accident is preventable … if there is genuine will to uphold best practice of working conditions'.

Leading authors frequently advocate that companies should invest in health and safety. For example McKinney (2002) argued that 'safety is, without doubt, the most crucial investment that can be made, and the question is not what it costs, but what safety best practice saves for organizations.' Waterman (2012) claims that 'managing health and safety well is not a cost, it is an investment.' However, the financial returns on investment in health and safety are immeasurable or intangible. The EASHW (2001) states that 'success is no accident' and that reducing work-related health and safety risk is not just a moral imperative, there is strong business case for doing so as well.

1.3 Methodology

The methodology is designed to meet the objective of this study thus:

Objective: to determine if investment in health and safety improves profitability.

This objective contains two variables: (1) investment in health and safety, and (2) profitability. The former is classified as the independent variable (IV), and the latter as the dependent variable (DV). That is, in a time frame an investment in health and safety will at a later stage lead to better profitability. It is, however, recognised that in practice the reverse could be the position. Companies that have good profitability are able to use that profit to invest in health and safety. Whichever is the IV and which the DV is not the important issue in this study; it is more to determine if there is indeed a relationship between the two variables. For convenience and clarity only, to reiterate, investment in health and safety is labelled the IV and profitability the DV.

As a part-time student, the author will use secondary data from her own company files to measure both variables. Care has been taken to follow the correct ethical procedures, including obtaining the permission of the author's line manager. The company is a medium-sized main contractor operating mostly in the north-west of England. It completes new build and refurbishment projects, mostly for commercial clients and residential social landlords, in the range of £100k to £3M. Its turnover in the last three years has been £17.1M, £16.3M and £19.4M. The 'population' will include all projects completed in the last three years, and current projects that have been in progress for at least four months. The four months period is an arbitrary figure, but stipulated to avoid just a snapshot of a project being taken over a short period. It is anticipated that data will be available for about 50 projects. The 'population' and the 'sample' are the same; that is, data is available for all projects. A theoretical population would embrace all UK construction projects completed in the last three years. However, since the data required is sensitive and indeed confidential to companies, it is simply not available. The sample may therefore be considered a 'sample of convenience' or analogous to a case study. It is recognised that findings cannot be used to infer replication across the whole of the UK construction industry.

To measure the IV 'investment in health and safety' in financial terms is difficult, since it has direct and indirect costs. The author's company costing system does not have 'cost codes' from which direct and indirect costs can be isolated. For example, it cannot be determined that on project A that direct safety costs were £x or indirect safety costs £y. Therefore the study assumes that investment in health and safety will be reflected in better health and safety performance. The author recognises that this assumption is fundamental and open to criticism. Preceding work by a colleague, had asked site workers to rate the health and safety operations on their sites. That was by questionnaire; for information, the multiple-item scale used is included as an appendix to this proposal. In this study, the health and safety performance will be measured using the reports of the company's health and safety officer. This officer visits every site each month, provides a report and gives a summative numerical score in the range of 0 to 10. A zero score is extremely bad and would require immediate site closure (hitherto not necessary within the author's company) and a score of 10 would indicate outstanding health and safety performance. It is proposed that one health and safety score will be used for each project; that will be the mean average of all monthly reports.

In order that the importance of the above assumption is not lost, the objective will be hitherto referred to as:

to determine if investment in health and safety (measured by health and safety performance) (the IV) improves profitability (the DV)

To measure the DV, 'profitability', the company internal management accounts will be used. For those projects that are completed, the company's final profit figures will be taken in percentage terms. For some projects, these will be definitive figures, if final accounts will have been completed between the main contractor and clients, and between the main contractor and subcontractors. In other cases, final account negotiations may be continuing, with agreements still to be reached. Final profit forecasts will be used in these cases. For projects that are still running, the percentage profit figures at the last valuation will be used. Profit percentage will be based on net cost, not gross excluding company overheads. For example: project final account £100k, net cost £90k, company head office overhead charge 7.5% of net cost = £6.75k, profit = £3.25k, profit percentage = 3.25k/90k = 3.61%. In cases where projects show a loss, these will be recorded as negative figures, thus: project final account £100k, net cost £95k, company head office overhead charge 7.5% of net cost = £7.125k, loss = 2.125k, loss percentage = 2.125k/95k = −2.23%. The range of scores for profitability may be from say −5% to +10%, but there may be exceptional projects outside this range.

Hypothetical datasets are presented in table D.1 and figure D.1.

Similar data to be collected from 50 projects

Based on a typical dataset in figure 1.1, both a Spearman's and Pearson's correlation coefficient will be calculated. A correlation coefficient above 0.35 or 0.40 will be interpreted such that there is a relationship between the variables.

Data on other miscellaneous variables (MVs), that are in the files and easily to hand, will also be collated. A hypothetical dataset for the whole study, to include these miscellaneous variables, is illustrated in table D.2. Correlations or paired t-tests will be executed to determine if any of these variables (to be labelled 'ancillary variables', AVs) impact upon both the IV and the DV. A rationale for conducting tests on this data will be developed; for example, it may be interesting to determine if projects with a high percentage of subcontract labour give better or worse health and safety performance. If worse, is there a problem in the company controlling subcontractors' health and safety? Do processes need to be improved? Or are subcontractors better than the main contractors' own labour?

A series of sub-objectives will be developed by pairing each of these AVs with the IV and the DV, as follows:

Size of project by actual or forecast final account (AV1); £
    Sub-objectives:
        SubOb 1a: To determine if size of project influences investment in health and safety (measured by health and safety performance)
        SubOb 1b: To determine if size of project influences profitability

Table D.1 Proposed company compliance with health and safety report against internal management account profitability figure.

Year 2013	Project 'A's monthly health and safety compliance reports complied by internal Health and safety officer The independent variable (IV)	Project 'A's profitability; %. The dependent variable (DV)
Jan	8/10	
Feb	7/10	
Mar	7/10	
Apr	6/10	
May	8/10	
June	5/10	
July	9/10	
Aug	8/10	
Sept	9/10	
Oct	7/10	
Nov	8/10	
Dec	9/10	
Total	91/120	
Mean	7.58/10	
Final account figure	–	0.763

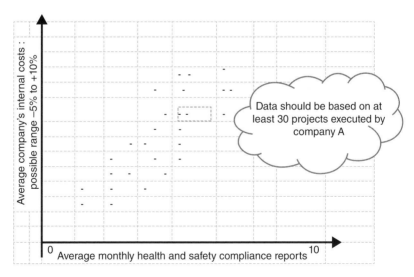

Figure D.1 Proposed correlation between average company's compliance with health and safety and profitability.

Actual or forecast completion date (AV2); early, on-time or late: %

Sub-objectives:

SubOb 2a: To determine if completion on-time influences investment in health and safety (measured by health and safety performance)

SubOb 2b: To determine if completion on-time influences profitability

Table D.2 Hypothetical dataset for 50 Projects. Data for the IV, DV and MVs 1 to 6 inclusive.

Project number ↓	The IV: Mean health and safety performance	Final or most recent net cost	Company head office overhead charge 7.5%	Actual or forecast final account, or most recent valuation (also MV1)	Profit	The DV: profitability percentage	Contract duration
			C * 0.075		E - (C+D)	F/C * 100	
1	7.8	90000	6750	100000	3250	3.61	15
2	6.22	95000	7125	100000	-2125	-2.24	18
3	8.88	253000	18975	273222	1247	0.49	26
Data set for projects 4 to 47 excluded excluded for clarity; in final dissertation, a full data set will be provided on a pen drive.							
	9.1	1950868	146315	2123666	26482.9	1.36	60
49	8.92	780000	58500	865212	26712	3.42	48
50	7.22	393222	29492	456123	33409.35	8.5	30
Total or sum (n = 50)	375	13000000	975000	14500000	525000		2000
Mean	7.5	260000	19500	290000	10500	4.04	40

Percentage of subcontract value at tender stage (AV3): %
   Sub-objectives:
      SubOb 3a: To determine if percentage of subcontract value influences investment in
         health and safety (measured by health and safety performance)
      SubOb 3b: To determine if percentage of subcontract value influences profitability
Background of site manager (AV4): 1 = ex-craftsman, 2 = graduate, 3 = experience of both.
   Sub-objectives:
      SubOb 4a: To determine if background of site manager influences investment in health
         and safety (measured by health and safety performance)
      SubOb 4b: To determine if background of site manager influences profitability
Client type (AV5): 1 = commercial, 2 = residential social landlord.
   Sub-objectives:
      SubOb 5a: To determine if client type influences investment in health and safety (meas-
         ured by health and safety performance)
      SubOb 5b: To determine if client type influences profitability
Procurement method (AV6): 1 = traditional, 2 = design and build.
   Sub-objectives:
      SubOb 6a: To determine if procurement method influences investment in health and
         safety (measured by health and safety performance)
      SubOb 6b: To determine if procurement method influences profitability
   To determine the values of the IV and DV are written as sub-objectives also, since to deter-
      mine their descriptive values may warrant further research and investigation. Values
      may lie outside (higher or lower) expected values or lie outside industry benchmark
      figures/key performance indicators. The sub-objectives are therefore:
      SubOb 7: To determine the value of investment in health and safety (measured by health
         and safety performance)
      SubOb 8: To determine profitability

Actual or forecast duration	Actual or forecast completion date (MV2); early, on-time or late: %	Tender figure	Value of subcontract work at tender	Percentage of subcontract value at tender stage (MV3): %	Background of site manager (MV4): 1 = ex-craftsman, 2 = graduate, 3 = experience of both	Client type (MV5): 1 = commercial, 2 = residential social landlord.	Procurement method (MV6): 1 = traditional, 2 = design and build.
	I-H/I * 100			L/K * 100			
16	6.67	101222	62333	61.58	1	2	1
16	−11.11	98333	42000	42.71	1	1	1
26	0	278359	200325	71.97	3	2	1
62	3.33	2123666	1796888	84.61	2	1	1
46	−4.17	875200	602212	68.81	3	1	1
38	26.67	402389	201010	49.95	3	2	2
2050		14550000	10000000	3436	n1 = 20, n2 = 12, n3 = 18	n1 = 30, n2 = 20	n1 = 33, n2 = 17
41	2.5	291000	200000	68.73			

References

Arocena, P. and Nuez, I. (2010) An empirical analysis of the effectiveness of occupational health and safety management systems in SMEs. *International Small Business Journal*, Sage publications, Spain, pp. 1–3.

BERR (2008) Improving outcomes from health and safety. A report to Government by the *Better Regulation Executive. Department for Business Enterprises and Regulatory Reform*, London, August, pp. 6–29.

Donaghy, R. (2009) One death is too many. A report to the Secretary of State for Work and Pensions. Inquiry into the underlying causes of construction fatal accidents, pp. 7–12.

EASHW (2001) Preventing accidents at work. New trends, Good practice European week 2011. *European Agency for Safety and Health at Work Magazine.* Luxembourg: Office for Official Publications of the European Communities, pp. 1–2.

Engasser, R. (2010) Safety is no accident. *Bechtel safety magazine brief.* Retrieved 21 February, 2011 from www.bechtel.com/assets/files/PDF/brief1298.

HSE (1988) Blackspot construction: a study of five years of fatal accidents in the building and civil engineering industries. Discussion Document. London. HMSO.

HSE (2002) *Revitalising Health and Safety in Construction.* Discussion Document. London The Health and Safety Executive Publication, December, pp. 3–6.

HSE (2003) Costs of compliance with health and safety regulations in SMEs. Research report 174. Prepared by *Entec UK Limited for the Health and Safety Executive.*

HSE (2011) New figures published of fatally-injured construction workers. Health and Safety Executive publications. Retrieved 30 July, 2011 from www.hse.gov.uk/press/2011/hse-fatalstatscon.htm.

HSE (2013) European Comparisons; Summary of GB Performance. Retrieved 24 December, 2013 from www.hse.gov.uk/copyright.htm.

McKinney, P. (2002) Expanding HSE's ability to communicate with small firms: a targeted approach (No. 420/2002): AEA Technology plc, for Health and Safety Executive, pp. 13–17.

Smallman, C. and John, G. (2001) 'British directors perspectives on the impact of health and safety on corporate performance', *Safety Science Journal*, Vol. 38, pp. 227–239.

Waterman, L. (2012) Game changer: how safety wins in 2012. *Construction News Magazine.* March, pp. 12–15.

Appendix to proposal D6

Coded copy of health and safety questionnaire with one hypothetical response

30 sample questions for a Likert scale: objective = 'to determine best practice with health and safety in construction'.
Based on the work of Arewa (2013) which includes 62 questions

Please read each question and place a tick (✓) in the column that best describes your judgement of health and safety practice in the UK construction industry. Please try to answer every question honestly and do not answer unduly to show your company or the construction industry off in a good light. Your confidentiality and anonymity is assured.

Hypothetical score for one participant; one missing score therefore maximum = 145. Insert this single response in a spreadsheet similar to table 8.1

Numbers in cells not to be shown to participants

In the context of site activities, please indicate the extent to which you agree or disagree with the following statements:	Very strongly agree	Strongly agree	Agree	Disagree	Strongly disagree	Very strongly disagree	Unsure	Hypothetical score
1 managers always insist that workers hold CSCS cards	5	4	3	2	1	0	—	4
2 workers are always focused on doing their jobs safely	5	4	3	2	1	0	—	3
3 welfare conditions are of a high standard	5	4	3	2	1	0	—	2
4 the standard of health and safety training could be improved	0	1	2	3	4	5	—	2
5 promotion of health and safety situational awareness on site is satisfactory	5	4	3	2	1	0	—	3
6 there is strong respect for safety abilities within teams	5	4	3	2	1	0	—	5
7 there are appropriate controls in place to limit working hours to avoid fatigue	5	4	3	2	1	0	—	4
8 regular sight and hearing tests of site workers are carried out	5	4	3	2	1	0	—	4
9 there is effective promotion of good canteen diets on site	5	4	3	2	1	0	—	4

10	lone workers are always provided with communication devices e.g. mobile phones	5	4	3	2	1	0	–	4
11	there are often communication barriers to health and safety practice on sites	0	1	2	3	4	5	–	2
12	relevant health and safety job-specific information is readily available	5	4	3	2	1	0	–	2
13	safety information attached to equipment can sometimes be inadequate	0	1	2	3	4	5	–	4
14	risk assessments are not adequately communicated to operatives	0	1	2	3	4	5	–	3
15	there is no strict adherence to health and safety work method statements	0	1	2	3	4	5	–	4
16	enforcement of safety rules that remove workers from site (for breaches) are not evident	0	1	2	3	4	5	–	4
17	there is insufficient attention paid to the tidiness of the internal working environment	0	1	2	3	4	5	–	3
18	guidance to prevent working in adverse weather conditions is not evident	0	1	2	3	4	5	–	4
19	there is careful consideration given to the selection of right equipment for work	5	4	3	2	1	0	–	1
20	site equipment is sometimes used inappropriately by workers	0	1	2	3	4	5	–	missing
21	retrofitting of safety aids on working tools most often do not meet safety standard	0	1	2	3	4	5	–	2
22	compliance with requirements to wear or use PPE is very good	5	4	3	2	1	0	–	5
23	PPE is always replaced after an appropriate period of use	5	4	3	2	1	0	–	5
24	there is regular safety training concerning the correct selection, care and use of PPE	5	4	3	2	1	0	–	3
25	company health and safety trainers are very competent and knowledgeable	5	4	3	2	1	0	–	2

(Continued)

(Continued)

Coded copy of health and safety questionnaire with one hypothetical response

30 sample questions for a Likert scale: objective = 'to determine best practice with health and safety in construction.' Based on the work of Arewa (2013) which includes 62 questions

Please read each question and place a tick (✓) in the column that best describes your judgement of health and safety practice in the UK construction industry. Please try to answer every question honestly and do not answer unduly to show your company or the construction industry off in a good light. Your confidentiality and anonymity is assured.

Hypothetical score for one participant; one missing score therefore maximum = 145. Insert this single response in a spreadsheet similar to table 8.1

Numbers in cells not to be shown to participants

In the context of site activities, please indicate the extent to which you agree or disagree with the following statements:	Very strongly agree	Strongly agree	Agree	Disagree	Strongly disagree	Very strongly disagree	Unsure	
26 health and safety is core to all staff training programmes	5	4	3	2	1	0	–	0
27 safety training is relevant to the context of a particular site conditions	5	4	3	2	1	0	–	5
28 there is a need to improve ways of reviewing and update safety procedures	0	1	2	3	4	5	–	2
29 safety procedures are developed with the help of people who actually do the job	5	4	3	2	1	0	–	4
30 safety procedures are concise and clear	5	4	3	2	1	0	–	4
								94

Maximum score - 30 x 5 = 150. 75/150 would equal 50%.
Adjust the percentage score if there are any 'unsure' responses; e.g. if only 28 responses the maximum score is 140; then 70/140 = 50%.

94/145 * 100 = 64.83%

Appendix D7

Determine the influence of temperature on synthetic fibre concrete

Alan Richardson

1.1 Problem statement and research goals

Synthetic fibres can provide many attributes to the mechanical performance of concrete. They can also control production problems such as bleed water control, plastic shrinkage and aggregate separation.

This research will investigate the effects of temperature upon the performance of concrete between the parameters of −20°C and 60°C. The data to be collected is production data and mechanical as-built data.

A control sample will be used at 20°C (room temperature) to determine the performance of the cubes and beams being tested. Samples will be tested at a ±40°C differential.

1.2 Key area for investigation – literature

See the reference list as provided that informs this work.

Particular regard should be given to the mechanical properties of polypropylene fibres at varying temperatures, as this is thought to be instrumental in determining the mechanical properties of the fibre concrete.

Polypropylene has low thermal transmittance, high resistance to electron flow, flexible qualities at normal temperatures and brittle qualities at freezing temperatures. The brittleness of polypropylene is particularly marked at low temperatures and is a serious disadvantage in certain applications.

The thermal coefficient for polypropylene is 1.1×10^{-4} at 20°C and the thermal coefficient for concrete at the same temperature is 2.2×10^{-4}, but polypropylene does not have a linear coefficient of thermal expansion as the value at 80°C is 1.7×10^{-4}.

As can be seen, the coefficient values for concrete are much higher than for polypropylene, which would indicate a greater potential for movement. Concrete has a bulk density of approximately $2400 \, kg/m^3$, whereas polypropylene has a bulk density of $525 \, kg/m^3$: there is a density difference in excess of four to one.

When concrete is manufactured and the initial set takes place, the concrete heats up as the tunnels surrounding the fibres expand, creating a void. When the initial heat subsides, the concrete starts to cure, shrinkage takes place and the fibres regain some of their original contact, to form a light bond between fibre and cement, within the concrete matrix.

It is clear from the aforementioned analysis that empirical data is required to substantiate the effects of temperature upon the bond between concrete and polypropylene fibres.

1.3 Methodology

Finite element analysis should be used to predict the performance, and the initial data will inform the sample size for production and can be compared against the final results.

The test will include compressive strength and flexural strength evaluation with a series of samples of sufficient number to obtain meaningful and statistically significant values.

The flexural strength tests are to be carried out in accordance to BS EN 12390-5:2009 using the centre-point loading method on the Lloyd LR100K Plus machine. The compression tests will be carried out to BS EN 12390-3:2009 using a calibrated ELE compression test machine. An optional test of temperature-related pull-out is recommended.

Outliers are to be identified and a standard deviation determined.

A paired comparison T or ANOVA test is to be used to determine whether or not the results are significant.

References

Journals

Richardson, A.E., Coventry, K.A. and Wilkinson, S. (2012), 'Freeze/thaw durability of concrete with synthetic fibre additions', *Cold Regions Science and Technology*, Volumes 83–84, December 2012, Pages 49–56 Elsevier. doi: http://dx.doi.org /10.1016/j.coldregions.2012.06.006.

Richardson, A.E. and Dave, U.V. (2008) 'The effect of polypropylene fibres within concrete with regard to fire performance in structures', *Structural Survey*, Vol. 26 No. 5, Emerald Group Publishing Limited, pp. 435–444.

Richardson, A.E. (2006) Compressive Strength of Concrete with Polypropylene Fibre Additions, *Structural Survey*, Vol. 24, No 2, August, MCB UP Ltd, UK, pp 138–153.

Richardson, A.E. (2003) Freeze/thaw durability in concrete with fibre additions, *Structural Survey*, Vol. 21, No 5, August, MCB UP Ltd, UK, pp. 225–233.

Conference papers

Richardson, A. and Batey, D. (2015) 'Impact resistance of concrete using Type 2 synthetic fibres', ConMat 15, *5th International Conference on Construction Materials*. Performance, Innovation and Structural Implications, August 19–21, Whistler, BC Canada. pp. 1771–1785.

Richardson, A.E. and Jackson, P. (2011) 'Equating steel and synthetic fibre concrete post crack performance', *2nd International Conference on Current Trends in Technology*, NUiCONE, December 8–10, Nirma University, Ahmedabad, India.

Richardson, A.E. Coventry, K. and Morgan M. (2011) 'Elevated temperature concrete curing – using polypropylene fibres', Fibre Concrete 2011 – 6th International Conference, September 8/9, Czech Technical University, Prague, Czech Republic.

Richardson, A.E. and Wilkinson, S.M. (2009) Early life freeze/thaw tests on concrete with varying types of fibre content, *9th International Symposium on Fiber Reinforced Polymer Reinforcement for Concrete Structures*, Sydney, Editors Oehlers D.J. and Seracino R., Australia, 13–15 July.

Appendix D8

Exploring the reasons behind cable strike incidents: A qualitative study

Mumtaz Patel (adapted by Fred Sherratt)

Context

An estimated 4 million holes are dug by utilities companies annually in the UK, a figure which does not include the many excavations carried out as part of ongoing construction projects (Stancliffe 2008). The precise number of cable strikes and subsequent injuries as a result of these excavations is more difficult to source due to parameters within the Reporting of Injuries, Diseases and Dangerous Occurrences Regulations 1995 (HSE 2012a) and the Electrical Safety, Quality and Continuity Regulations 2002 (as amended) (HSE, 2012). However the utility distribution industry does make its own estimates. For example, Industry Today (2011) puts the figure at 60,000 strikes annually that are reported, while industry insurer Zurich (2007) claims an average of 12 deaths and 600 serious injuries per year are attributed to cable strikes.

Despite a lack of clarity in the figures, the potential severity of the consequences of cable strikes is not in dispute. Damage to underground services can cause fatal or severe injuries (HSE, 2005) as well as the potential for fire or explosions, and these risks are not just to the workforce but also to the general public. Even non-fatal shocks can cause severe and permanent injury (HSE, 2010). Flash burns may occur as a result of arc formation; burns may be extensive and lower the resistance of the skin so that electric shock may add to the ill effects. Temporary blindness can also occur due to burning the retina of the eye (Hughes and Ferrett, 2007). Alongside the human costs, the financial repercussions can be severe. A recent case examined an accident in which an employee received life-threatening 60% burns after striking a high voltage cable with a hydraulic breaker on London's Crossrail project in 2008. In addition to the disabling injuries received by the operative, the employing company was fined £55,000 with £30,000 costs (Prior, 2012).

With regard to human safety, the HSE has published detailed guidance for best practice in the form of Guidance Booklet HSG47 'Avoiding Danger from Underground Services', which outlines the dangers that can arise from work near underground services and gives advice on how to reduce the risk (HSE, 2005). HSG47 requires detailed inspections prior to any excavation to ensure that all utilities are identified and safe systems of work employed to detect the presence of underground plant and services. Safe systems of work are a fundamental feature of HSG47 and include all aspects of the work: planning, utility drawings, cable locating devices and safe digging practice.

However, cases brought to court reveal that the HSE guidance and company procedures are not necessarily followed in practice. For example, HSE inspectors examining the Crossrail incident found that no effective lines of communication had been established, appropriate training in digging techniques had not been provided, key safety documentation showing the cable was not to hand at the work location, and although the site had been scanned, no markings had been made to show the locations of buried cables. As this incident happened in a busy London street, the HSE inspectors felt it was 'completely foreseeable that cables would be present' (Prior, 2012). Following an incident in 2002, the HSE (2002) voiced a warning to the construction industry to ensure that safe working practices are followed when working near buried electrical cables. This followed an incident where an employee suffered burns to the face and neck as a result of striking a live 11 kV electricity cable, and the employer was fined £10,000, yet the investigation concluded that had the method statement actually been followed in practice 'this was a preventable accident'.

Theoretical explanations for these behavioural challenges to procedures can be found within the management structures and payment systems of the utilities sector. In keeping with the practices of the construction industry as a whole, a large proportion of utilities work is subcontracted both by the operating companies to contractors, and also from contractor to contractor due to the fluctuating workload (Lingard and Rowlinson, 2005), resulting in potentially elongated supply chains and highly fragmented delivery systems (Loosemore et al., 2003). The utility industry subcontractors are paid on a price per metre basis as an incentive to increase productivity, facilitated by the ease with which outputs can be measured and rewarded (Harris et al., 2006). However, this practice has been found to encourage operatives, who are also paid on price per metre, to work as fast as possible to make the most money in a day or shift. As speed often means cutting corners and taking risks, safety is often sacrificed (Spanswick, 2007). In a work scenario where painstaking preparation through the use of cable avoidance tools, followed by careful and precise excavation using hand tools or mini diggers is essential for cable avoidance, a payment structure based on speed of installation appears somewhat incongruous.

Indeed, recent developments in training have shifted in focus from technical to behavioural. The online 'cable avoidance evaluation' assessment for operatives has been developed by a utilities industry training provider, to establish knowledge, confidence and attitudes rather than just technical knowledge to enable evaluation of skills gaps and training needs (Industry Today, 2011). Such an approach acknowledges the people in the process, and their influence and participation within cable strike incidents; indeed, a human influence can be identified in the case studies noted above. Consideration of the human element as a causal factor in accidents and incidents is not uncommon within the construction industry as a whole (HSE, 2009) and associated behaviours such as inaccurate assessments, bad decisions and poor judgements are often seen to be the root cause of incidents (Perrow, 1999; Dekker, 2006).

However, there has been a paradigm shift in the overall positioning of human error within the accident context, and the view that work-related accidents and injuries are a direct result of carelessness and unsafe behaviours has become outdated (HSE, 2007). The systems theory of accident causation has challenged this approach. This theory states that 'Human error is the effect, or symptom of deeper trouble … that it is systematically connected to features of peoples' tools, tasks and operating environment' (Dekker 2006, p. 15). It suggests people make incorrect assessments or take incorrect action as a result of failures in the systems which have created situations which dictate a certain course of action (Perrow 1999; Dekker 2006). It is no longer accepted that the system would work correctly if it were not for the behaviour of some 'bad apples'. Rather, there is a need for safety to be instilled at all levels of the organisation (Dekker 2006), including management, who may unwittingly create latent failures within the system through the choices they make in boardrooms (Reason 1990; Kletz 2001). Cultural influences have also been suggested to affect people working within complex systems, and therefore can influence safety in terms of acceptance of authority, need to conform to the social groups within organisations as well as organisational culture itself (Strauch 2004).

Aim

To explore the reasons for cable strike incidents among utilities installation operatives.

Proposed methodology

This study employed a qualitative and interpretivist approach (Creswell, 2003; Flick, 2009) in its desire to seek out the subjective experiences, understandings and attitudes of operatives. Semi-structured interviews will be used as the method of data collection (Gillham, 2005), employing open questions to enable probing where appropriate (Fellows and Liu, 2008) and facilitate development of talk around safety and cable strike incidents. This approach will enable the theoretical contexts of human error, systems theory and cultural influences to be explored in detail, and enable their consideration through practice.

A purposive sample will be made of 20 members of a utilities distribution operational workforce. While this is a relatively small sample size it will be able to provide insights as to the perspectives of operatives with regard to safety, cable strikes and the potential causes that lie behind them. It can also be argued that given the peripatetic nature of utilities distribution, the operatives' experiences, perceptions and attitudes are likely to be common within the industry as a whole.

The interviews will be digitally recorded, transcribed verbatim and subsequently coded, to highlight themes, consistencies, inconsistencies, patterns and irregularities (Silverman, 2001; Langdridge, 2005) when the data is viewed through the lens of the literature. Attention will be given to the causal factors as developed through the operatives' talk, and the connections and interactions described between them. These factors can then be further analysed to enable the development of causal chains, sparked by the initial thematic associations from the data, and developed within the context and understanding of industry practices.

Ethical considerations

As with all research projects, university ethical procedures will be following, including the use of participation information sheets – to explain the reasons for the study to the participants – and participant consent forms – to ensure that participants are happy to be part of the study and know that they are able to opt out at any time.

This particular study will involve human participants talking about a sensitive subject, with the potential for issues of self-incrimination. Therefore, anonymity of the participants will be ensured at all times, and data will be kept secure and encrypted in password-protected documents, in compliance with the Data Protection Act.

References

Creswell, J.W. (2003) *Research Design: Qualitative, Quantitative and Mixed Methods Approaches*. 2nd the ed. London: Sage Publications Limited.

Dekker, S. (2006) *The Field Guide to Understanding Human Error*. Surrey: Ashgate Publishing Limited.

Fellows, R. and Liu, A. (2008) *Research Methods for Construction*. 3rd Ed. Oxford: Blackwell Publishing.

Flick, U. (2009) *An Introduction to Qualitative Research*. 4th Ed. London: Sage Publications Limited.

Gillham, B. (2005) *Research interviewing: the range of techniques*. Maidenhead: Open University Press.

Harris, F., McCaffer, R. and Edum-Fotwe, F. (2006) *Modern Construction Management*. 6th Ed. Blackwell Publishing, Oxford.

HSE (2002) *HSE warns over buried electrical services following £10,000 fine for unsafe system of work*. Health and Safety Executive Online www.hse.gov.uk/press/2002/e02008.htm [31 March 2012].

HSE (2005) HSG47 *Avoiding Danger From Underground Services*. Sudbury: HSE Books.

HSE (2007) HSG48 *Reducing error and influencing behaviour*. Sudbury: HSE Books.

HSE (2009) *Phase 2 Report: Underlying Causes of Construction Fatal Accidents* – External Research. Norwich: HMSO.

HSE (2010) Electrical Safety and You. Health and Safety Executive Online www.hse.gov.uk/electricity [9 September 2010].

HSE (2012) ESQCR. Health and Safety Executive Online www.hse.gov.uk/statistics/tables/index.htm [31 March 2012].

Hughes, P. and Ferrett, E. (2007) *Introduction to Health and Safety in Construction.* 2nd Ed. Oxford: Butterworth-Heinemann.

Industry Today (2011) Reducing Cable Strikes – A Radical New Approach from Develop and Cognisco. *Industry Today Online* www.industrytoday.co.uk/utilities-industry-today/reducing-cable-strikes---a-radical-new-approach-from-develop-and-cognisco/8403 [31 March 2012].

Kletz, T. (2001) *Learning from Accidents.* 3rd Ed. Oxford: Butterworth-Heinemann.

Langdridge, D. (2005) *Research Methods and Data Analysis in Psychology.* Pearson Education, Harlow.

Lingard, H. and Rowlinson, S. (2005) *Occupational Health and Safety in Construction Project Management.* London: Spon Press.

Loosemore, M., Dainty, A.R.J and Lingard, H. (2003) *Human Resource Management in Construction Projects.* London: Spon Press.

Perrow, C. (1999) *Normal Accidents – Living with High Risk Technologies.* West Sussex: Princeton University Press.

Prior G. (2012) Worker suffers 60% burns in Crossrail cable strike. Construction Enquirer Online www.constructionenquirer.com/2012/02/27/worker-suffers-60-burns-in-crossrail-cable-strike/ [26 March 2012].

Reason, J. (1990) *Human Error.* Cambridge: Cambridge University Press.

Silverman, D. (2001) *Interpreting Qualitative Data: Methods for analysing talk, text and interaction.* 2nd Ed. London: Sage Publications Limited.

Spanswick, J. (2007) As Near As Dammit. '*Building*', 13.

Stancliffe, J. (2008) Third party damage to Major Accident Hazard pipelines. Health and Safety Executive Online www.hse.gov.uk/pipelines/ukopa.htm [16 March 2012].

Strauch, B. (2004) *Investigating Human Error: Incidents, Accidents and Complex Systems.* Aldershot: Ashgate Publishing Ltd.

Zurich (2007) *Underground service strikes: best practice guidelines for construction companies.* Zurich Insurance plc Online www.cablejoints.co.uk/upload/Underground_Cable_Strikes,_Best_Practice_Guidance___Zurich.pdf [26 March 2012].

Acknowledgement

Developed from work carried out and first published at:

Patel, M., Sherratt, F. and Farrell. P. (2012) Exploring human error through safety talk of utilities distribution operatives. *ARCOM*, Heriot Watt University, Edinburgh. 3–5 September.

Appendix E: Raw data for a qualitative study

Appendix E1:   Research objectives
Appendix E2:   Interview questions and prompts
Appendix E3:   Verbatim transcript of interviews with two site managers A and B
Appendix E4:   Coded version of verbatim transcript with site manager A
Appendix E5:   Analytical tables
Appendix E6:   The final narrative

Appendix E1

Research objectives

Objective 1: To determine whether the propensity to put completions before quality influence profit within the PHS

Objective 2: To determine whether the propensity to put completions before quality is recognised by commentators within the current body of literature

Objective 3: To determine whether the propensity to put completions before quality is recognised by practitioners within the PHS

Objective 4: To determine whether the propensity to put completions before quality varies at different times of the year

Appendix E2

Interview questions and prompts

Q1   What is the approximate proportion of subcontractors and directly employed labour currently working on this development?

Q1a   What are the differences between these two types of employee?

Q2   How would you describe your working relationship with your workforce?

Q2a   Under what circumstances are tradespeople pressured to work within unrealistic timescales?

Q2b   What are the implications for quality?

Q2c   How do tradespeople react to such pressure?

Q2d   How do subcontractors ensure that satisfactory quality is achieved?

Q2e   What is your policy regarding subcontractors' meetings?

Q3   How would you describe your working relationship with your contracts manager and directors?

Q3a   In what ways do these relationships change?

Q3b   What conflicts of interests are there when achieving completions?

Q3c   How does this issue affect your bonus payments?

Q4   In what way do you believe your build programme influences your ability to build houses to an adequate standard of quality?

Q4a   What additional help is available to you if needed?

Q5   Based on your experience, what would you say are the main reasons for sometimes being unable to build houses to an adequate standard of quality?

Q5a   Explain any reasons for feeling compelled to comply?

Q6   Describe how the build quality is controlled by this company?

Q6a   What causes some inspections to be missed out?

Q6b   How do you ensure there is sufficient time to rectify defects before the final check by the building inspector?

Q6c   Describe how thoroughly the building inspector checks a property during the final inspection?

Q6d   Explain your company's formal procedure for resolving defects?

Q6e   For what reasons is the system not always followed?

Q6f   What causes some defects to be rectified after occupation?

Q7   How would you compare customer satisfaction within the private housing sector to, say, five years ago?

Q7a   What effect do you think customer expectations and their perceptions of housing quality have on customer satisfaction?

Q7b   Do you feel customers tend to report more defects than, say, five years ago?

Q7c   At what stages can customers view their property?

Q8   To what extent do customers exhibit a willingness to recommend this company to family or friends?

Q8a   What about negative word-of-mouth referrals?

Q8b   How would you say this issue compares to, say, five years ago?

Q9   What awards for quality have you or any other site managers at this company been nominated for or received?

Q9a   What differences are there in how award-winning sites are run compared to the rest?

Q10   Finally, would you say that the opinions you have voiced in answer to these questions are commonly held by other site managers currently working for this company?

End of interview.

Appendix E3

Verbatim transcripts of interviews with two site managers A and B

Site manager A

Q1   What is the approximate proportion of subcontractors and directly employed labour currently working on this development?
A-Ans1. Well the forklift driver's cards-in and I also have two bricklaying gangs who are cards-in but apart from that everyone else is a subbie. Even the labourers we use are all from an agency.

Q1a   What are the differences between these two types of employee?
A-Ans1a. I have to say that the standard of work I get off the cards-in brickies is noticeably better than the subcontracting brickies and they're also a lot tidier and more cooperative. You always find that directly employed forklift drivers are harder working, more reliable and more conscientious than agency drivers as well.

Q2   How would you describe your working relationship with your workforce?
A -Ans2. With most of them it's good, very good. There are a couple of the trades on here that I seem to have constant running battles with but I always manage to get what I want in the end so on the whole I'd say the working relationship's fine.

Q2a   Under what circumstances are tradespeople pressured to work within unrealistic timescales?
A-Ans2a. Sometimes, when you're really busy such as at year-end or half-year-end, you do pressure the subbies to work late or come in early and we definitely mither them to work weekends and that means full weekends too, all day Saturday and all day Sunday, not just a couple of hours on a Saturday morning.

Q2b   What are the consequences of this?
A-Ans2b. It becomes an absolute nuisance. You end up with things like the joiners first-fixing before the windows are fitted, which means the plots not watertight; the plumbers and sparks in together and pinching each others pre-drilled service holes in the joists; and the plasterers boarding the upstairs ceilings and one side of the stud walls while the building inspector's walking around the plot doing his pre-plaster inspection, which isn't ideal really. Then you might double up on your plasterers to get it plastered quicker but still have to put joiners second-fixing downstairs while they're finishing off skimming upstairs while the plumber's fitting his heating and sanitary wherever he can – it becomes madness at times, shear madness.

Q2c   What are the implications for quality?
A-Ans2c. The implications for quality are the worst thing for me. I can deal with the long hours and trades having a 'barny' with one another but it's the bad affect on quality that really ****** me off at those times. A lot of the time I do think these situations occur simply because the construction director won't say no to the MD. I wish he'd just once say 'no, you can't have that plot in three weeks, it's impossible. You'll have to have it next month instead or it'll be a load of **** or something like that.

Q2d   How do tradespeople react to such pressure?
A-Ans2d The subbies themselves all basically do as they're told. The lads on site may moan about working weekends but they do get a bonus of about £25 per man, per day, for coming in, on top of whatever price-work they do so at the end of the day they are earning more money for a few weeks. Ehm, as for their bosses, well, they moan too but at the end of the day they daren't refuse because they're **** scared of not getting any more work off us. I don't think any of them are bothered about quality really because if it is bad they can just blame us for making them rush it.

I'd have to say at times there are problems with health and safety too. Ehm, it's unavoidable really if you've got men working outside normal working hours. There's just bound to be times when someone ends up working on site without the benefit of a first aider present or even access to the first aid kit or toilets or washing facilities if the cabins are all locked up. Your painter can often end up as a lone worker too because they'll just lock themselves in a plot and stay to all hours to get it finished. I suppose it's quite dangerous practice when you stop and think about it.

Q2e   How do subcontractors ensure that satisfactory quality is achieved?
A-Ans2e. Well none of them use supervisors or foremen or anything like that, not really, no. I mean, if I have any issues with any of the lads on site regarding quality I'll ring their bosses and get them out to look at it with me and they always 'play ball', to be fair. Anyway, all you have to do is threaten not to pay 'em until they do come out and they'll soon turn up to sort it.

Q2f   What is your policy regarding subcontractors' meetings?

A-Ans2f Well according to our company policy we are supposed to do them once a month and we have a pro forma on which to record the meeting and report it back to head office but no manager does them on a regular basis as far as I'm aware. I will always do one at the beginning of a new site as a way of introducing myself face-to-face to the subcontractors who I don't know or haven't used before. I find this initial meeting useful as it puts a face to a name and makes it easier in the future when you're dealing with them over the phone. Doing them each and every month though just isn't practical especially at crucial times of the year when you're mad busy because you just don't have the time and neither have they really. I mean you're talking to all the subbies on a daily basis anyway because the job's moving so fast so most of the managers don't see the point of them really and the contracts manager and construction director only seem to mention them when things are quite and they have time to 'nit pick' over things – just like they do with the health and safety policy.

Q3   How would you describe your working relationship with your contracts manager and directors?

A-Ans3. Good, very good. We have a good understanding. I know what they expect and I've got high expectations, as have they, so, ehm, as long as we work together, we do a good job.

Q3a   In what ways do these relationships change?

A-Ans3a. I wouldn't say the relationship ever becomes tense or bitter as such, ehm, not really. I mean it can be difficult when you've got a lot of plots, a lot of completions in one particular month, ehm, but as long as we all work together and pull together, the help is already there so all you've got to do is ask for it.

Q3b   What conflicts of interests are there when achieving completions?

A-Ans3b. I think that naturally there is some, there would be a small amount of conflict there wouldn't there? Ehm, with them pushing to get profit by achieving the sales completions, which obviously increases the chances of them getting full bonuses. I wouldn't say it was unhealthy though, to have that.

Q3c   How does this issue affect your bonus payments?

A-Ans3c. Well my bonus is based on my NHBC items, reportable items that you get when the building inspector records defects he's spotted so it can affect how much bonus I get if they force me to hand over a plot with loads of items on it instead of giving me extra time to get them sorted.

Q4   In what way do you believe your build programme influences your ability to build houses to an adequate standard of quality?

A-Ans4. Overall I would have to say we are given enough time in the build programme to build a good house because quality seems to come before productivity at this company. Occasionally you are given less time than others depending on the plot and upon how and when it's been sold during a particular month and when the completion needs to be in for. Time pressures can be more severe on one plot than it could be for another but, you know, they're nearly always achievable. It's often last minute sales and stuff like that the cause the problems with time constraints. If we sell a plot and it has to be in within four to five weeks and it's achievable, we will certainly do our best to achieve it. They don't usually give you impossible scenarios. We can usually achieve what were asked to achieve.

I mean I have had situations in the past where I've been extremely rushed and where, say, a sixteen-week build, we've had to try and achieve it in twelve weeks. Well naturally, you know, I've not been happy with that, with having to do that, because with them extra three or four weeks, you can get things more pristine. The funny thing is though, when they do let you stick to the programme the pressures are probably greater because the expectations are higher because you've been given more time. Motivation is a key in this. You need to be motivated and ensure that you stay motivated, even though you might have a bit more time to complete it. You see, the overall expectation is far greater because you've had more time and once you've handed over the plot complete, they expect it to be to a really good standard, with greater emphasis on quality.

Q4a   What additional help is available to you if needed?
A-Ans4a. Probably the biggest negative aspect of working here, compared to other house builders I've worked for, is that cost controls are extremely tight. Everything you order is scrutinised and anything extra you ask for is questioned; even something like an additional broom for sweeping the welfare cabins out with. It becomes really frustrating and annoying at times, the way they 'penny-pinch', especially if they're asking you to build a plot in 30% less time than normal but they don't want to pay for temporary lighting even though we haven't got electric to the plot yet. They expect miracles on a shoestring budget but we usually get there in the end, which of course means that the situation never changes. The more often you do it that way, the more often they want it that way.

Q5   **Based on your experience, what would you say are the main reasons for sometimes being unable to build houses to an adequate standard of quality?**
A-Ans5. I haven't really experienced that within this company, no. With other companies I've worked for, yeah, but not here really, not with the current structure we've got. I think my finished units are always very good. We do have various stages and various viewing by both the contracts manager and construction director. We all look at it, you know, at various stages of the actual finishing stages and by the time we come to actually handing it over, I would say my standard's quite good.

You do get times when you want another day or two to finish a plot off but you're told no because they need it by that date to get the money in on time for the year-end deadline and sometimes it is difficult but the way I look at it, if it has to be in for that month, it has to be in. If it means putting a few extra hours in to achieve that, well we have to do it. If, you know, nine times out of ten it can be done, then yes, we certainly try our best to achieve it, although sometimes it's extremely difficult.

Q5a   Explain any reasons for feeling compelled to comply?
A-Ans5a. You do sort of feel compelled to do it, basically because it's the worry about losing your job and we've had that in the recent past. The demand can be so high that, ehm, that we would be rushed because of demand. Besides that, if we didn't cooperate

and we didn't cope with demand, then we would either not get our bonus or our bonus would be extremely reduced.

Q6   Describe how the build quality is controlled by this company?
A-Ans6. It's down to the site manager mainly. The contracts manager will have a bob round about once a week when he comes to site but apart from that it's down to me. I do all the inspections really. We don't have a quality control manager or anything like that.

Q6a   What causes some inspections to be missed out?
A-Ans6a. I don't believe I ever miss any of the required inspections out. Like I mentioned earlier, our bonuses, the site managers' bonus that is, are only based on the amount of reportable items you get so it's in your own interest to thoroughly check plots before key-stage inspections by the NHBC inspector, otherwise you may end up with no bonus.

Q6b   How do you ensure there is sufficient time to rectify defects before the final check by the building inspector?
A-Ans6b. During busy periods like at year-end when you've got some tight plots because they've been thrown on you at the last minute, you might not get round all the plots as often as you'd like or spend as much time in each one as you normally would but you still make sure you do your checks, even if it means you being on site at six o'clock in the morning. Having said that, I've had many occasions where the NHBC inspector has turned up to do a CML while I'm half way through doing my final checks. You just have to keep your fingers crossed that you haven't overlooked any major defects when that happens.

Q6c   Describe how thoroughly the building inspector checks a property during the final inspection?
A-Ans6c It really depends on which inspector you get as to how thorough he is. Some can be real pernickety ******** and others quite lax. My inspector on here is a bit of both; he can be okay one day and a complete 'jobsworth' the next. At times like year-end it also depends on how many visits they have on. With a lot of builders having they're year-end around the same time, if they're all wanting CMLs on the same day then he obviously won't have as much time to spend in each one so that can work to your advantage.

Q6d   Explain your company's formal procedure for resolving defects?
A-Ans6d. We have a procedure set out in the quality management books. What's supposed to happen is that a few days before the CML is done, the site manager and the sales negotiator 'snag' the plot and the site manager sorts the snags before the CML. After the CML is done, the sales people make an appointment with the client to come and view the property in what we call a 'home demo' but in the book it's called a 'Pre-Handover Viewing' and if the client picks up any other snags, the site manager is supposed to try his best to get them done before the client moves in. At the actual handover of the property to the client, which the book calls a 'Key Release', they're again asked to note down any defects but also to sign off where issues have been resolved. Then seven days after they have physically moved in, the site manager does another inspection with

the client and once they've signed off the book to say he's done them the book goes into the office and the customers then have to contact head office direct with any problems.

Q6e   For what reasons is the system not always followed?

A-Ans6e. I say that's what's supposed to happen because it very rarely pans out that way. Firstly, the sales people – I won't say sales women because we do have a man as a sales negotiator – the sales people are all 'bone idle', especially when it comes to filling out the 'Quality Management' books. You see, it's their responsibility to do them. They have the books from day one and it stays with them until I do my seven-day call. I can't remember the last time I did a pre-CML check with a sales negotiator. I do, do them of course but I do them on my own. The other thing is that, especially at year-end, you often have the scenario where clients are moving in the same day you get the CML or the house was only actually completed on the day of the CML so you obviously can't follow the procedures in the book. It's all a good idea in principle but I think it's mainly just a token gesture to try and make the company look efficient.

Q6f   What causes some defects to be rectified after occupation?

A-Ans6f. Some defects often have to be rectified after completion but it's nearly always due to time limitations. I don't think it's ever about the site manager just not trying hard enough. The other thing is that all house builders work within tolerances so there are certain things we don't have to attend to but the clients often think we do. If it's something that's only half-an-hour of a job, such as a painter touching up after they've had carpets fitted, then you'll often just do it as sort of a gesture of good will to try and get the customer on your side and establish a good relationship with them from the start. I would say we probably do around 30% of the snags after completion.

Q7   How would you compare customer satisfaction within the private housing sector to, say, five years ago?

A-Ans7. I think that house buyers are more demanding of good quality than in the past, yes, very much so, especially with the internet and programmes on TV and in the general media, its ehm, it is becoming more difficult as time goes by. I mean, the media in general over the past few years has promoted a bad image and people these days are expecting far more for what they're actually paying out, whereas in the past they were just, you know, happy to purchase a plot for instance and just go along with it but nowadays, they just, they want the best for what they're paying for.

Q7a   What effect do you think customer expectations and their perceptions of housing quality have on customer satisfaction?

A-Ans7a. I think the quality of the actual build has remained the same, if not improved; it's the expectations of the actual client, the purchaser that has risen. Their expectations have got far greater because they want value for money and they want what they want and they seem to know a lot more about what they want through the use of the internet and the media. Ehm, I think our attitudes have remained the same and if the job, if the build is actually built to the way it should be along the process of the NHBC inspector looking at it at the various key stages and we all know as site managers what the company expects, then I'd have to say that quality has probably improved.

Q7b   Do you feel customers tend to report more defects than, say, five years ago?
A-Ans7b. If the defect's there to be reported, then the customer reports it, so yes, I suppose they do tend to report more defects now than a few years ago.

Q7c   At what stages can customers view their property?
A-Ans7c. The home demo is usually the first chance the customers get to see their property. It's the best time really; to see it when it's completely finished because it gives them a better impression of what they've actually bought. You sometimes get the odd one wanting to view the plot at first-fix or when it's plastered but site managers discourage this because clients end up walking round the house pointing at all sorts saying 'this isn't finished' or 'that isn't finished' and you feel like saying 'I know it's not finished you ****** fool, we're still building the ******* thing'.

Q8   **To what extent do customers exhibit a willingness to recommend this company to family or friends?**
A-Ans8. Ehm, not really sure about that one. We sometimes get clients giving us bottles of wine and stuff as a thank you if they're really happy with their new home and some do send letters in to head office praising the site staff but apart from that I don't really know. We've actually had relatives of people who work in our head office buy houses off us but that's because they think they'll get a good deal more than anything.

Q8a   What about negative word-of-mouth referrals?
A-Ans 8a. You only really get to hear from the ones who threaten to go to the press or threaten to tell everyone they know not to buy a *********** house but I don't think that's changed much over the years.

Q8b   How would you say this issue compares to, say, five years ago?
A-Ans 8b. Ehm, they used to reckon one in every ten home buyer would be a 'customer from hell' so judging from that I don't think we get that many really bad ones but you never know what people say to their family and friends, do you?

Q9   **What awards for quality have you or any other site managers at this company been nominated for or received?**
A-Ans9. I was nominated this year but I haven't heard anything of that yet and as far as I know at least two other site managers have been nominated for 'Pride in the Job' awards.

Q9a   What differences are there in how award-winning sites are run compared to the rest?
A-Ans9a. For some companies I definitely know they are run differently but at this company, no, and that's mainly due to the cost constraints. We don't have a great deal of money to spend on the likes of fancy sales and site compound areas and other presentation ********like other companies do. We do still try to achieve the standard of quality that would enable us to win an award but it's not really possible to achieve the overall standard that the NHBC is looking for when it comes to awards without spending loads of money 'tarting' the site up. Besides, everyone in the game knows that the biggest house builders take the NHBC bosses on golfing weekends and trips abroad and what

have you so a lot of it's plain bribery really and little to do with the way they actually build the majority of their houses.

Q10   **Finally, would you say that the opinions you have voiced in answer to these questions are commonly held by other site managers currently working for this company?**
A-Ans10. Yes, definitely. I would say all of them. I know all the managers here person-ally and I can't think of any one of them, of the top of my head, who would disagree with anything I've said.

End of interview.

Site manager B

Q1   **What would you say is the approximate proportion of subcontractors and directly employed labour at this company?**
B-Ans1. They're all subcontractors, all 'subbies'.

Q2   **How would you describe your working relationship with your workforce?**
B-Ans2. In the main, I would say it was quite good.

Q2a   Under what circumstances are tradespeople pressured to work within unrealistic timescales?
B-Ans2a. It happens when they shorten the build programmes, that kind of thing. It just becomes more vertical all the time, the build programme and they end up having to work Saturdays, Sundays, Bank Holidays, seven or eight o'clock at night; all that busi-ness. It's basically during the run up to year-end from the summer up to September. The half-year-end is a similar scenario but not quite as much pressure. In between half-year-end and year-end it's more or less a steady pace.

Q2b   What are the consequences of this?
B-Ans2b. Well you feel under tremendous pressure because they want so many units in so you end up flooding it with men and you'll have every trade working on top of each other and coming up to September, year-end, you know that you'll start having to get the handovers coming in at the end of July and through August and if you're looking after thirty or forty plots yourself and you're handing over six or seven a week, you're expected to go in and 'snag' them as well and you can't physically do everything. The amount they ask you for at year end, you can't physically do it with the amount of hours in the day; there's never enough hours in the day. It's not that you can't do your job it's that you don't have enough hours to do it in.

On this site, when you've got people coming in to start work really early, you've got to come in yourself to open up because you can't trust the labourers to do it for you. You see, we are under obligation as part of the planning permission to employ the locals – well they are useless. So you're working with useless labourers and working in ****
crowded conditions all the time.

To cope with being under that sort of pressure you have to work long hours and weekends as well. They never used to pay us for weekends but they've started paying us to soften the blow about having to keep giving up our weekends. I have times when I'm doing handovers and trying to 'snag' plots as well as do all my other jobs and I'll have some weeks where I'll turn in at half six in the morning, before everyone's got there, to start doing my work early, to be there to intercept subbies on their way in and what have you and then I'm going home and waking up in the middle of the night thinking about plots.

I find that it starts affecting my performance as a site manager as well. I won't be taking any breaks during the day and I suppose as the years go by you learn to manage the pressure better yourself but I'll have days when I won't have a lunch and you'll have the hierarchy or directors that come round and expect you to walk all the plots with them and that takes hours out of your day, keeps you from the work you should be doing.

Come the afternoon, sometimes your brain goes into 'meltdown' and your concentration levels just go, basically. In a morning you're 'on the ball' and whatnot but in the afternoon you'll be running out of steam, especially if you've been running up and down stairs in an apartment block. Then when I go home the missus is telling me I look gaunt and loosing too much weight because you're just on the go so much all the time and you do start making mistakes yourself.

Q2c   What are the implications for quality?

B-Ans2c. Well the subbies, it kills the subbies and they end up pulling practically anyone of the street because they end up short of men. For example, they could have plasterers in they don't know or joiners they've never used before because they're under that much pressure to supply the men and they could mess the plot up. You don't have your regular subbie in all the time so the standard of workmanship ends up suffering.

If you've got plasterers still working in a house and you have to send painters in because of the lack of time, how's he going to have a chance of doing a decent job when he goes in? The joiners can't always keep up with the programme because he can't supply enough men and you end up with poor workmanship running all the way through. You can't physically, I can't get, there comes to a point where you can't throw a subbie off a job. If the work is poor and you aren't rushed you could just say 'I don't want that man anymore' and the job could hang on a day or two 'til you got the right man in but when you're rushed you just have to put up with it because otherwise you know you won't get the handovers. So, if I had a set of boarders who were absolutely useless, I'd just have to let them carry on boarding, what else could I do? If I throw them out I just wouldn't get the handover. All in all it becomes a nightmare. They all end up working on top of each other, over crowded, out of sequence, working late, often unsupervised, poor lighting because you often don't get services on 'til the death – a nightmare. I mean you'll have all the support to set up temporary lighting but just trying to maintain the lighting…if you've got 'festoon' lighting in and people are banging and bulbs are going… you can spend all day ****** around with festoon lighting in an apartment block and you're not getting anything else done. The bulbs gone in this box and these bulbs have gone here and is health and safety gonna come on and see there's a bulb missing and the cage has fell off and someone's just pulled a lead out of the transformer – you're into all that.

I mean there is always someone there who's a first aider. I'd have to say that on that one they're insistent that there will be a first aider there but supervision wise, I couldn't.

If I have that many units and it's that busy and I have to supervise what all the trades are up to so, you know, I'll go into where someone has messed up a plot and everyone's blaming each other and there are so many trades in there that it just gets left a tip, which can obviously be dangerous health and safety wise.

When it's booming, like before this recession, the trades know they can go down the road and get another job so once you put them under a certain amount of pressure they'll just leave the job and **** off because they know there's another job to go to down the road. Now all this 'credit crunch' has come into being and work's become scarcer but we're still quite busy so I can hold the high ground and say 'well if you don't do it my way, go and I'll get someone else to do it'.

Q2d How do tradespeople react to such pressure?
B-Ans2d. You do end up putting the subcontractors under a lot of pressure, i.e. their supervisors and I'll be, like, kicking his head in of a day and once the pressure starts getting to him, he starts forgetting things as well and making mistakes. I mean it definitely reduces morale and causes friction between you and your subbies because I'll end up talking to subbies like, ehm, well, putting them under pressure. Everyone is under massive pressure from, I suppose, the MD of the subcontractor down to his contracts manager and then he passes it onto the supervisor who takes it out on his lads.

They might put up some resistance at first but not in the end because they're obviously being told by their superiors and me just to get on with it. Their bosses might try to be awkward sometimes but you have them by the *****, the company has them by the ***** because the subcontractor is so reliant on us because they'll have so much work with us that they can't afford to say no. You can see it in some of their bosses' eyes when they come out to site; they can't say no.

Q2e How do subcontractors ensure that satisfactory quality is achieved?
B-Ans2e. You will be looking at the quality as well yes but not so much to stop you throwing men at it. I mean, I might put a joiner in to second-fix a plot that will take him three days and then have to send the painter in after two days and then the joiner's final fixing while the painter's glossing out. Then, if I go to the painter before handover and tell him that his paintwork's a bit rough he'll just say 'what do you expect when I'm painting while the joiner's in and there's sawdust and all kinds of **** flying around' but you'll still sometimes force him to go back in and redo bits and threaten not to pay him if he doesn't.

Part of the subcontractors' package includes the responsibility to provide supervisors to monitor their lads on site. Now, every subbie is supposed to do that but it's a running battle to get them to stick to it. They will come out sometimes to check their lads' work but it's only 'once every blue moon'.

Q2f What is your policy regarding subcontractors' meetings?
B-Ans2f. We do hold subcontractor meetings very regularly, depending on how busy we are. Sometimes during the really busy periods we might have one every day and all the subbies' bosses will have to attend – we even make the cleaner's boss come out (laughs **IntOb1**). During less busy times we'll probably only have them once a month or something but when you're busy it definitely helps to have them more regular because you can hammer out exactly what you need from each of them and sometimes I'll deliberately embarrass them by giving them a ********** in front of the other bosses if they've let me down in any way.

Q3   **How would you describe your working relationship with your contracts manager and directors?**
B-Anw3.Good and bad.

Q3a   In what ways do these relationships change?
B-Anw3a. In the times leading up to year-end and half-year-end, they'll be constantly on top of you all the time, screaming 'when's this happening, when's that happening?' They'll say 'you'd better go out there and grow a set of ********. Ring him up now, I want him here now. I want this many people in tomorrow' so there's a lot more pressure being caused by the handovers; by them wanting the amount of units they're looking for. The pressure goes from the MD to the construction director to the contracts manager and down to the site manager. Like some weeks I could be handing over ten units a week; in a week.

Q3b   What conflicts of interests are there when achieving completions?
B-Anw3b. The senior managers and directors bonuses are all profit-related, based on achieving profit forecasts, which obviously has a lot to do with why they push you so hard to get your completions in. Yes, I'd have to say that's why they want to get so many units over because that's what's going to affect their bonuses.

On a personal level I don't like having to hand over a plot to customers who are all exited about moving into their new home, sometimes supposedly their 'dream' home, when I know they're going to find all sorts wrong with it and be really disappointed with the standard. It's not a nice feeling.

Q3c   How does this issue affect your bonus payments
B-Anw3c. To me it doesn't really make any difference or not whether a house goes over or not because I still get more or less the same bonus as long as I build it to the pro-gramme. My bonus is down to three factors: health and safety, building to programme and customer care. None of it is based on NHBC reportable items so as long as I get a CML it doesn't matter how many items I get. Obviously you don't want to get reportable items but as long as you don't get a 'red' item, which prevents the inspector from issuing you with the CML, then you can have 'green' items all over the place. It's just the 'red' items you mustn't get because that's a 'show-stopper'. So I have to say that we do get our bonuses, they do pay us our bonuses but they get their pound of flesh out of us; you know, the amount of work they put you through, the long hours and the sleepless nights but I'll always get my bonus, yeah.

Q3d   What additional help is available to you if needed?
B-Anw3d. More managers will sometimes come on board as the job goes on. On one phase you could have three assistants, two site managers, a project manager and a build manager so you could have five or six managers in total but you'd need that many. To be honest, a lot of the time when it's busy it's just as difficult running a smaller site with only one manager on it because there's still never enough time. More time is what you really need; not more men or equipment but they never give it to you.

Q4   **In what way do you believe your build programme influences your ability to build houses to an adequate standard of quality?**
B-Anw4. Everyone is looking at the programme and timescales all the time. The build programme when you look at it, the amount of weeks, yes at first looks enough but

because it's so vertical it only takes one trade to start 'flagging' and going by the wayside to put the whole programme out of 'sync'. Then, you know, as happens, if the plumber starts to struggle on a number of plots and then the joiner, he's under pressure and he's got to try and get joiners to go back in there to do pipe-boxings after the painter's already been in there but you've had to do things out of sequence because the sanitary was late going in.

The trades are all on price so when they have to keep going back in again and again because of the order of works being mixed up, they're losing money and in their eyes doing work for nothing so they stop being bothered about doing a good job. Then you've got to get your painter back in again but then if you don't get the same painter back, you're getting some fuddy-duddy who just goes around snagging here and there, he's not the quickest, he's not the best and quality is going to suffer again.

There aren't any allowances made on your programmes for things like bad weather either? I mean, the past two years and all the rainfall we've had during the summers, I know no one could have foreseen that but there's absolutely no leeway whatsoever. I mean, there's not even anything in the programme for a scaffold strip and some strips can be massive, like on apartment blocks and they can hold you up by as much as a week. The programmes change all the time; you'll get issued with a new programme but it'll keep changing all the time. It starts off with a nice gradient and goes to looking almost vertical as they keep shortening it.

Thing is, you get to September, year-end, finish and then the programme is often delayed for the start of the next phase. Instead of starting back up again in October and starting digging footings you could have problems with land remediation and the QS' arguing over prices and so the programme starts late again. Instead of starting in October or November you wouldn't be starting proper again until January or February and in the meantime you'd have nothing much to do; it could be very quite. So you can be weeks behind programme before you've even got on site so you know it's going to be unrealistic beforehand; you know what's coming.

Q5   Based on your experience, what would you say are the main reasons for sometimes being unable to build houses to an adequate standard of quality?
B-Anw5. A lot of time it's because of the workload. I know it's gonna come back and bite me because I haven't had enough time to do everything I need to do to get the quality to the right finish inside. You know that instead of just handing a house over and that's it and it's a nice job done and you can just walk away, you know that you're gonna have to go back in it; that someone from customer care is gonna go in and slaughter it and you're going to have to go back in.

I mean there's nothing wrong structurally wise, it's just the finishing quality that suffers. The defects are usually mainly in the finishing, in the finishing quality. For some plots you don't even have time to put a plaster patcher through them before you get them painted and then you'll have to go back in after handover and it often gets to the point where the painter is more or less doing a 're-dec' (redecoration) because they've had to gloss it without services on so there are runs all over the place because they were painted without any heating on, when the plot was still cold and damp and with other trades still working in there so you also end up with gritty paintwork; all because we were rushing the house through. Another common issue is having to send plaster patchers back in to cut holes in walls to find buried wires for the electricians because the plaster boarders have just boarded over them, you know, because they've been thrown in and told to do the job double-quick and they'll have been working at such a rate that they've screwed into loads of pipes too and then when you're testing the plumbing installations out, once you've got the services on, there'll be leaks everywhere but you can't be there all the time to monitor it because you're that busy. I suppose it's just music to their ears because they know they can just take the p**s and instead of taking the care to measure and cut holes and feed wires through for lights or alarms and such like, they just board them in.

Q5a Explain any reasons for feeling compelled to comply?
B-Anw5a. You feel compelled to do these things because, well, there's an atmosphere of intimidation from the hierarchy who definitely try and intimidate site managers and pressurise them a lot, you know, to get the handovers, so you end up being scared about losing your job because if they want to get rid of you they can always find a way. Yeah, I do, I do worry about my job or worry about it affecting my chances of being promoted within the company.

Q6 Describe how the build quality is controlled by this company?
B-Anw6. We have a customer care department with a customer care manager but he only looks at the finished product. He doesn't inspect a house as its being built, that's my department, he only looks after the finished product.

The procedure is, or it's supposed to be, you get the CML, you hand it over to sales and once they've handed the paperwork into the office, within three days customer care will come out and go through the plot and 'snag' it. Then you have five working days to put right anything they've picked up and then you have to invite them back for a revisit to check it.

Q6a What causes some inspections to be missed out?
B-Anw6a. Let's say on a decent standard house you can expect fifteen to twenty snags, which is mainly bits of paintwork and plasterwork because that's what they mostly look at but when it's really busy you'll get some of them with, like, 150 snags on it and then you're having to go back in and you've got all these other units you've still got to knock out. You're having to go through a snag list that's done by someone else, cut it down into trades, ring the people up and fax them and it's taking up so much of your time that your not looking through the next plots that are coming up to completion; your eye's off the ball. Really, they should get other people in to do that but they never do. You're expected to sort all that out so basically you're supposed to snag it yourself but sometimes the workload's so great that you can't and you know you're handing over **** and customer care are going to go in and shred it to bits. You know that's going to happen and that he's going to come back with a list of 100 or 200 items and yet you know that you've had to turn the house out like that because of the programme.
   If you've got ten plots with all them items it can take you hours and hours in a day to sort the jobs out from the snag lists from customer care. The way I do it is to photocopy loads of them for all the various trades because the quality is that poor that you'll have, like, ten different subcontractors involved. Then I'll highlight the different jobs for each of them because you can't just give them a copy of the sheet and trust them to go through it conscientiously, you have to give it to them in, like, big highlighted things so they can see it and then you'll hand it to them but you won't have time to go back and check

all that again. Then if customer care goes back in again and doesn't get it down to, say, five items, they'll fail it again and then there's a stewards enquiry with the hierarchy; 'why has it failed again?', 'why haven't you got it done?', 'well, I'm trying to do all this as well' and it just causes conflict and a bad atmosphere throughout the whole company.

So for some houses that you do, the procedure rarely works because you often have people moving in the same day as the CML and so the plot has to go over by default. So they just move in and then you have your fingers crossed as to how much they'll find. I mean, some punters will move in and they haven't got a clue and so they'll just go in and think its fine. Others would be all over it like a rash and then the emails to head office start. I'd say the proportion of houses that go over by default is probably between 5–10%, which isn't a massive amount as a percentage but still a lot of actual plots per year for the company as a whole.

Q6b How do you ensure there is sufficient time to rectify defects before the final check by the building inspector?

B-Anw6b. Well the short answer to that is you can't. You just can't ensure there's enough time because there seldom is. No, not at all, nothing like sufficient time.

Q6c Describe how thoroughly the building inspector checks a property during the final inspection?

B-Anw6c. To be honest we usually get away with murder with him although it depends on the inspector; some can be quite severe. It just depends but you can tell that they're under pressure to give you the handovers as well from our contracts managers and directors. I mean, one building inspector I had was just a thoroughly nice bloke and hardly ever gave me any items but his replacement was a complete pest. I suppose a lot depends on their individual personalities when it comes to how thorough they do their inspections.

Q6d Explain your company's formal procedure for completion defects?

B-Anw6d. Before the client moves in they have a 'demo' tour of their home then after their demo they have up to two weeks to contact customer care with their own list of snags and then customer care go in and agree which items they'll rectify and which they won't because, you know, some clients might give you a never-ending list but you're only obliged to deal with the ones that fall outside of the tolerances that the NHBC allow because, you know, house construction is subject to tolerances. Then after they've done that list with the customer care manager, that then gets handed back down to site. I never really understand this properly though because once we've done all customer care snags, as far as I'm concerned we should be out of the equation because they're supposed to use their budget to go back in but it always comes back to site again, always ends up being at the site's cost.

Some of the things they pick up are just ****** ridiculous but it's all about job justification isn't it? I mean anyone can find fault with anything and when you go through so much s**t when you've got completions and then you get some ******** from customer care going in pointing out piddling little flaws that you need a magnifying glass to see; it just gets your back up.

Q6e For what reasons is the system not always followed?

B-Anw6e. Well often a plot will go over by default. You can have a plot go over by default before customer care even go in to do their pre-handover inspection, if you get me, because if it's been right on the brink of year-end when you got it finished and then as soon as you got the CML the person will be moving in so customer care won't have a chance to go in, so, you know, it will be of a very poor quality. You never walk into one and say 'yeah, this is how I wanted to get it', you're always handing it over, looking at it, knowing that the quality is as poor as it is because you rushed it.

If you've got the CML and your customer care manager comes out and says 'right, you've got 200 defects here' but the customer's moving in the next day or even the same day, then the procedure goes out of the window and the customer just has to go in.

I'll give you a good example. We had a customer care guy who was a **** and I remember him going in this particular plot and recording 164 ******* items but it had to go over by default because the client needed to move in. When the client eventually did her snag list she only picked eight ******* snags up so I took the list back to the customer care manager and said 'go and ***** that ** your * ****** ****'. I know it doesn't sound good but I end up wanting them to go over by default because I know the client will never pick up as many defects as what customer care will pick up. Plus, if it goes over by default then it's not my fault.

Q6g   What causes some defects to be rectified after occupation?
B-Anw6g. It doesn't seem to matter how many customer complaints you get, although you get a ********** for having too many, the company never stops and says 'we've got to revise the way we do things', no, it's all about the units because that's bonus-based for them you see. They're the guys sitting up in the office with the MD agreeing to get all these units because they know they're gonna get such a big bonus out of it. They're attitude is get the CML and sort the defects later. It's a tricky one to put a figure on but I'd say roughly about 10% of defects have to be rectified after completion, yeah, about 10%.

Q7   How would you compare customer satisfaction within the private housing sector to, say, five years ago?
B-Ans7. I'd say it was a lot less, yes, especially on the busy sites like this anyway, definitely less. I think the majority of houses are okay but quality is definitely getting worse overall. At the end of the day, the majority of the plots when the client moves in are reasonable but, well, you get your good ones and your bad ones. Sometimes you get bad plots and they'll all be condensed next to each other because they've come at the critical time leading up to year-end. That's where you're worst customers and complaints would come from.

Q7a   What effect do you think customer expectations and their perceptions of housing quality have on customer satisfaction?
B-Ans7a. I think a lot of them are aware of the TV programmes you see a lot on the telly now, these fly-on-the-wall documentaries. I think they're more 'clued-up' nowadays as to what they're getting. I've had someone come into a private house and start crying as soon as they walked in because the quality was that bad, the painting was terrible and they were literally crying. That's because they're expectations were shattered the moment they walked in and I couldn't blame them because some of them will often move in the same day they get the CML so you don't even have time to rectify anything before they do move in. You tend to remember the bad ones more than the good ones because the good ones you hand over, you tend to forget about them.

Q7b   Do you feel that customers tend to report more defects today than, say, five years ago?
B-Ans7b. I'd say so yeah, yeah. I would say we spend about 50% more time than we

used to, rectifying defects to substandard houses than those we've been able to build to the quality we want.

Q7c   At what stages can customers view their property?
B-Ans7c. They can view it on completion of the build during the home demo with sales and customer care, once the CML is in.

Q8   **To what extent do customers exhibit a willingness to recommend this company to family or friends?**
B-Ans8. I don't know really. I never really deal with the customers apart from dealing with the snagging on the clients first list but even then they very often aren't in. A lot of the time you just get some keys off them and go in but generally it's the customer care department that has the most contact with the customers. They do the demo and they agree what we will attend to from the clients' first list.

Q9   **What awards for quality have you or any other site managers at this company been nominated for or received?**
B-Ans9. We actually won an award, a 'Pride in the Job' award, on the first phase of this site. I was only the assistant then so the site manager at the time was presented with the award but I also got a certificate as his assistant.

Q9a   What differences are there in how award-winning sites are run compared to the rest?
B-Ans9a. Well none really, not on here anyway. I mean we did have an advantage in the sense that it was only a small site then with twenty-two plots on it so with effectively two managers on it; it was pretty easy to run.

There was another site that won an award a few years ago though where they did throw money at it. You know, it had proper metal hoarding all the way round the perimeter instead of ***** old timber or ********* heras fence panels and a tarmac compound with flagged walkways and proper meeting rooms and all that. It also had some trees and nice landscaping around the compound too but that site did over 200 units in six months, which all sold so it made its money for the company in return.

Q10   **Finally, would you say that the opinions you have voiced in answer to these questions are commonly held by other site managers currently working for this company?**
B-Ans10. Definitely. All the site managers would say the same, all of them, all of the ones I've worked with here and that I know here. Yes.

Appendix E4

Coded version of verbatim transcript with site manager A

A-Ans1. Well the forklift driver's cards-in and I also have two bricklaying gangs who are cards-in but apart from that everyone else is a subbie. Even the labourers we use are all from an agency.

A-Ans1a. [I have to say that the standard of work I get off the cards-in brickies is noticeably better than the subcontracting brickies and they're also a lot tidier and more cooperative.] **QUAL3**

A-Ans1a. You always find that directly employed forklift drivers are harder working, more reliable and more conscientious than agency drivers as well.

A -Ans2. [With most of them it's good, very good. There are a couple of the trades on here that I seem to have constant running battles with but I always manage to get what I want in the end so on the whole I'd say the working relationship's fine.] **QUAL3**

A-Ans2a. [Sometimes, when you're really busy such as at year-end or half-year-end, you do pressure the subbies to work late or come in early and we definitely mither them to work weekends and that means full weekends too, all day Saturday and all day Sunday, not just a couple of hours on a Saturday morning.] **T1/ PROP3**

A-Ans2b. [It becomes an absolute nuisance. You end up with things like the joiners first-fixing before the windows are fitted, which means the plots not watertight; the plumbers and sparks in together and pinching each others predrilled service holes in the joists; and the plasterers boarding the upstairs ceilings and one side of the stud walls while the building inspector's walking around the plot doing his pre-plaster inspection, which isn't ideal really. Then you might double up on your plasterers to get it plastered quicker but still have to put joiners second-fixing downstairs while they're finishing off skimming upstairs while the plumber's fitting his heating and sanitary wherever he can – it becomes madness at times, shear madness.] **T2**

A-Ans2c. The implications for quality are the worst thing for me.

A-Ans2c. [I can deal with the long hours and trades having a 'barny' with one another but it's the bad affect on quality that really ****** me off at those times.] **COMP1/ QUAL2**

A-Ans2c. [A lot of the time I do think these situations occur simply because the construction director won't say no to the MD. I wish he'd just once say 'no, you can't have that plot in three weeks, it's impossible. You'll have to have it next month instead or it'll be a load of **** or something like that.] **PROP2**

A-Ans2d [The subbies themselves all basically do as they're told.] **QUAL3**

The lads on site may moan about working weekends but they do get a bonus of about £25 per man, per day, for coming in, on top of whatever price-work they do so at the end of the day they are earning more money for a few weeks. Ehm,

A-Ans2d [as for their bosses, well, they moan too but at the end of the day they daren't refuse because they're **** scared of not getting any more work off us. I don't think any of them are bothered about quality really because if it is bad they can just blame us for making them rush it.] **QUAL3**

A-Ans2d [I'd have to say at times there are problems with health and safety too. Ehm, it's unavoidable really if you've got men working outside normal working hours. There's just bound to be times when someone ends up working on site without the benefit of a first aider present or even access to the first aid kit or toilets or washing facilities if the cabins are all locked up. Your painter can often end up as a lone worker too because they'll just lock themselves in a plot and stay to all hours to get it finished.] **T3**

I suppose it's quite dangerous practice when you stop and think about it.

A-Ans2e. [Well none of them use supervisors or foremen or anything like that, not really, no.] **QUAL2**

I mean, if I have any issues with any of the lads on site regarding quality I'll ring their bosses and get them out to look at it with me and they always 'play ball', to be fair. Anyway, all you have to do is threaten not to pay 'em until they do come out and they'll soon turn up to sort it.

A-Ans2f [Well according to our company policy we are supposed to do them once a month and we have a pro forma on which to record the meeting and report it back to head office but no manager does them on a regular basis as far as I'm aware.] **COMP3**

I will always do one at the beginning of a new site as a way of introducing myself face-to-face to the subcontractors who I don't know or haven't used before. I find this initial meeting useful as it puts a face to a name and makes it easier in the future when you're dealing with them over the phone.

A-Ans2f [Doing them each and every month though just isn't practical especially at crucial times of the year when you're mad busy because you just don't have the time and neither have they really.] **COMP1/COMP4**

I mean you're talking to all the subbies on a daily basis anyway because the job's moving so fast so most of the managers don't see the point of them really and the contracts manager and construction director only seem to mention them when things are quite and they have time to 'nit pick' over things – just like they do with the health and safety policy.

A-Ans3. Good, very good.

A-Ans3. [We have a good understanding. I know what they expect and I've got high expectations, as have they, so, ehm, as long as we work together, we do a good job.] **PROP2**

A-Ans3a. [I wouldn't say the relationship ever becomes tense or bitter as such, ehm, not really.] **PROP2**

I mean it can be difficult when you've got a lot of plots, a lot of completions in one particular month, ehm, but as long as we all work together and pull together, the help is already there so all you've got to do is ask for it.

A-Ans3b. I think that naturally there is some,

A-Ans3b. [there would be a small amount of conflict there wouldn't there? Ehm, with them pushing to get profit by achieving the sales completions, which obviously increases the chances of them getting full bonuses. I wouldn't say it was unhealthy though, to have that.] **PROP2/ PROF3**

A-Ans3c. Well

A-Ans3c. [my bonus is based on my NHBC items, reportable items that you get when the building inspector records defects he's spotted so it can affect how much bonus I get if they force me to hand over a plot with loads of items on it instead of giving me extra time to get them sorted.] **PROF1/ T1**

A-Ans4. [Overall I would have to say we are given enough time in the build programme to build a good house because quality seems to come before productivity at this company.] **T1**

Occasionally you are given less time than others depending on the plot and upon how and when it's been sold during a particular month and when the completion needs to be in for.

A-Ans4. [Time pressures can be more severe on one plot than it could be for another but, you know, they're nearly always achievable.] **T1**

A-Ans4. [It's often last minute sales and stuff like that the cause the problems with time constraints.] **T4**

If we sell a plot and it has to be in within four to five weeks and it's achievable, we will certainly do our best to achieve it.

A-Ans4. [They don't usually give you impossible scenarios. We can usually achieve what were asked to achieve.] **PROP2**

I mean

A-Ans4. [I have had situations in the past where I've been extremely rushed and where, say, a sixteen-week build, we've had to try and achieve it in twelve weeks. Well naturally, you know, I've not been happy with that, with having to do that, because with them extra three or four weeks, you can get things more pristine.] **T1**

A-Ans4. [The funny thing is though, when they do let you stick to the programme the pressures are probably greater because the expectations are higher because you've been given more time.] **PROP1**

Motivation is a key in this. You need to be motivated and ensure that you stay motivated, even though you might have a bit more time to complete it.

A-Ans4. [You see, the overall expectation is far greater because you've had more time and once you've handed over the plot complete, they expect it to be to a really good standard, with greater emphasis on quality.] **PROP1/ T1**

A-Ans4a. [Probably the biggest negative aspect of working here, compared to other house builders I've worked for, is that cost controls are extremely tight.] **COMP2**

Everything you order is scrutinised and anything extra you ask for is questioned; even something like an additional broom for sweeping the welfare cabins out with.

A-Ans4a. [It becomes really frustrating and annoying at times, the way they 'penny-pinch', especially if they're asking you to build a plot in 30% less time than normal but they don't want to pay for temporary lighting even though we haven't got electric to the plot yet.] **COMP2**

They expect miracles on a shoestring budget but we usually get there in the end, which of course means that the situation never changes.

A-Ans4a. [The more often you do it that way, the more often they want it that way.] **PROP2**

A-Ans5. I haven't really experienced that within this company, no. With other companies I've worked for, yeah, but not here really, not with the current structure we've got.

A-Ans5. [I think my finished units are always very good. We do have various stages and various viewing by both the contracts manager and construction director. We all look at it, you know, at various stages of the actual finishing stages and by the time we come to actually handing it over, I would say my standard's quite good.] **QUAL2**

A-Ans5. [You do get times when you want another day or two to finish a plot off but you're told no because they need it by that date to get the money in on time for the year-end deadline] **PROF1**

and

A-Ans5. [sometimes it is difficult but the way I look at it, if it has to be in for that month, it has to be in. If it means putting a few extra hours in to achieve that, well we have to do it.] **COMP1**

If, you know, nine times out of ten it can be done, then yes, we certainly try our best to achieve it, although sometimes it's extremely difficult.

A-Ans5a. [You do sort of feel compelled to do it, basically because it's the worry about losing your job and we've had that in the recent past.] **PROP1**

The demand can be so high that, ehm, that we would be rushed because of demand.

A-Ans5a [Besides that, if we didn't cooperate and we didn't cope with demand, then we would either not get our bonus or our bonus would be extremely reduced.] **PROF3**

A-Ans6. [It's down to the site manager mainly. The contracts manager will have a bob round about once a week when he comes to site but apart from that it's down to me. I do all the inspections really. We don't have a quality control manager or anything like that.] **QUAL2**

A-Ans6a. [I don't believe I ever miss any of the required inspections out.] **QUAL2**

Like I mentioned earlier, our bonuses, the site managers' bonus that is, are only based on the amount of reportable items you get so

A-Ans6a [it's in your own interest to thoroughly check plots before key-stage inspections by the NHBC inspector, otherwise you may end up with no bonus.] **PROF3**

A-Ans6b. [During busy periods like at year-end when you've got some tight plots because they've been thrown on you at the last minute, you might not get round all the plots as often as you'd like or spend as much time in each one as you normally would but you still make sure you do your checks, even if it means you being on site at six o'clock in the morning.] **PROP3/ QUAL2**

Having said that,

A-Ans6b. [I've had many occasions where the NHBC inspector has turned up to do a CML while I'm half way through doing my final checks. You just have to keep your fingers crossed that you haven't overlooked any major defects when that happens.] **QUAL2**

A-Ans6c. [It really depends on which inspector you get as to how thorough he is. Some can be real pernickety ******** and others quite lax. My inspector on here is a bit of both; he can be okay one day and a complete 'jobsworth' the next. At times like year-end it also depends on how many visits they have on. With a lot of builders having they're year-end around the same time, if they're all wanting CMLs on the same day then he obviously won't have as much time to spend in each one] **QUAL2**

so that can work to your advantage.

A-Ans6d. [We have a procedure set out in the quality management books.] **QUAL2**

A-Ans6d. [What's supposed to happen is that a few days before the CML is done, the site manager and the sales negotiator 'snag' the plot and the site manager sorts the snags before the CML.] **QUAL1**

A-Ans6d. [After the CML is done, the sales people make an appointment with the client to come and view the property in what we call a 'home demo' but in the book it's called a 'Pre-Handover Viewing' and if the client picks up any other snags, the site manager is supposed to try his best to get them done before the client moves in.] **QUAL1/ PROF2**

A-Ans6d. [At the actual handover of the property to the client, which the book calls a 'Key Release', they're again asked to note down any defects but also to sign off where issues have been resolved.] **QUAL1**

Then

A-Ans6d. [seven days after they have physically moved in, the site manager does another inspection with the client and once they've signed off the book to say he's done them the book goes into the office and the customers then have to contact head office direct with any problems.] **QUAL1/ PROF2**

I say

A-Ans6e. [that's what's supposed to happen because it very rarely pans out that way. Firstly, the sales people – I won't say sales women because we do have a man as a sales negotiator – the sales people are all 'bone idle', especially when it comes to filling out the 'Quality Management' books. You see, it's their responsibility to do them. They have the books from day one and it stays with them until I do my seven-day call. I can't remember the last time I did a pre-CML check with a sales negotiator. I do, do them of course but I do them on my own.] **T4**

The other thing is that,

A-Ans6e. [especially at year-end, you often have the scenario where clients are moving in the same day you get the CML or the house was only actually completed on the day of the CML so you obviously can't follow the procedures in the book.] **PROP3**

A-Ans6e. [It's all a good idea in principle but I think it's mainly just a token gesture to try and make the company look efficient.] **QUAL2**

A-Ans6f. [Some defects often have to be rectified after completion but it's nearly always due to time limitations. I don't think it's ever about the site manager just not trying hard enough.] **QUAL1/ T1**

The other thing is that all house builders work within tolerances so there are certain things we don't have to attend to but the clients often think we do. If it's something that's only half-an-hour of a job, such as a painter touching up after they've had carpets fitted, then you'll often just do it as sort of a gesture of good will to try and get the customer on your side and establish a good relationship with them from the start.

[I would say we probably do around 30% of the snags after completion.] **QUAL1**

A-Ans7. I think that house buyers are more demanding of good quality than in the past, yes, very much so, especially with the internet and programmes on TV and in the general media, its ehm, it is becoming more difficult as time goes by. I mean, the media in general over the past few years has promoted a bad image and people these days are expecting far more for what they're actually paying out, whereas in the past they were just, you know, happy to purchase a plot for instance and just go along with it but nowadays, they just, they want the best for what they're paying for.

A-Ans7a. [I think the quality of the actual build has remained the same, if not improved; it's the expectations of the actual client, the purchaser that has risen. Their expectations have got far greater because they want value for money and they want what they want and they seem to know a lot more about what they want through the use of the internet and the media.] **PROF2**

Ehm, I think our attitudes have remained the same and if the job, if the build is actually built to the way it should be along the process of the NHBC inspector looking at it at the various key stages and we all know as site managers what the company expects, then I'd have to say that quality has probably improved.

A-Ans7b. [If the defect's there to be reported, then the customer reports it, so yes, I suppose they do tend to report more defects now than a few years ago.] **QUAL1**

A-Ans7c. [The home demo is usually the first chance the customers get to see their property. It's the best time really; to see it when it's completely finished because it gives them a better impression of what they've actually bought. You sometimes get the odd one wanting to view the plot at first-fix or when it's plastered but site managers discourage this because clients end up walking round the house pointing at all sorts saying 'this isn't finished' or 'that isn't finished' and you feel like saying 'I know it's not finished you ****** fool, we're still building the ******* thing'.] **PROF2**

A-Ans8. Ehm, not really sure about that one.

A-Ans8. [We sometimes get clients giving us bottles of wine and stuff as a thank you if they're really happy with their new home and some do send letters in to head office praising the site staff] **PROF2**

but apart from that I don't really know. We've actually had relatives of people who work in our head office buy houses off us but that's because they think they'll get a good deal more than anything.

A-Ans 8a. [You only really get to hear from the ones who threaten to go to the press or threaten to tell everyone they know not to buy a ************ house but I don't think that's changed much over the years.] **PROF2**

A-Ans 8b. Ehm,

A-Ans 8b. [they used to reckon one in every ten home buyer would be a 'customer from hell' so judging from that I don't think we get that many really bad ones but you never know what people say to their family and friends, do you?] **PROF2**

A-Ans9. [I was nominated this year but I haven't heard anything of that yet and as far as I know at least two other site managers have been nominated for 'Pride in the Job' awards.] **QUAL2**

A-Ans9a. [For some companies I definitely know they are run differently but at this company, no, and that's mainly due to the cost constraints. We don't have a great deal of money to spend on the likes of fancy sales and site compound areas and other presentation ********like other companies do.] **COMP1**

We do still try to achieve the standard of quality that would enable us to win an award but it's not really possible to achieve the overall standard that the NHBC is looking for when it comes to awards without spending loads of money 'tarting' the site up. Besides,

A-Ans9a. [everyone in the game knows that the biggest house builders take the NHBC bosses on golfing weekends and trips abroad and what have you so a lot of it's plain bribery really and little to do with the way they actually build the majority of their houses.] **QUAL2**

A-Ans10. Yes, definitely. I would say all of them. I know all the managers here personally and I can't think of any one of them, of the top of my head, who would disagree with anything I've said.

Appendix E5

Analytical tables

Data coding number	Main category headings	Sub-category headings	Frequency counts	Literature sources	Observations, implications or interpretations	Data consistencies	Data inconsistencies
1	Propensity	Culture	155	Atkinson (2002;1999), Tam et al. (2000), McCabe et al. (1998), Reason (1998)	If a propensity exists among the site managers then it would seem to stem from higher management targets. Various forms of intimidation are used to pressure the site managers.	Excessive pressure is placed on site managers to achieve company targets. Self-esteem is a prime motivator for site managers.	Job security is not the overriding factor regarding propensity. Some pressure is self-inflicted.
2		Hierarchy	88	Liu (2003), Reason (1995), Morris (1994)	Senior managers and directors operate from a position of self-interest motivated by personal monetary gain via bonus payments. A lack of respect for superiors prevails among site managers.	Most managers experience strained relations with senior managers during busy periods.	A few managers claimed their relationships with hierarchy is unaffected by pressure.
3		Specific periods	105		The PHS as a whole, experiences unusually busy periods in an annual and/or bi-annual cycle. There is a lack of support from higher management during busier times.	Adverse working practices are frequently adopted during such times.	Some become institutionalised by the casual attitudes of hierarchy towards product quality and H&S.

4	Completions	Workload	117	Langford et al. (2000), Barker & Naim (2008), Housing Forum (2001), Naim & Barlow (2003), Ozaki (2003)	Excessive workloads and unsociable hours are often imposed on site managers during year-end and half-year-end periods. This has detrimental psychological effects on many site managers on a bi-annual cycle.	Unfair expectations are placed upon site managers during crucial periods.	A few claimed to be comfortable with their workload during busy periods.
5		Resources	16		Additional resources are available when needed.	Extra resources cannot compensate for lack of time. Control of resources is difficult for site managers.	Some site managers feel that being given extra resources increases the pressure to achieve unrealistic targets.
6		Supply chain	44		The fragmented nature of the PHS is reflected in a dysfunctional supply chain. Lead times of manufacturers are inconsistent.	Problems with delayed or incorrect materials deliveries are rife.	Utility providers pose the most problematic external danger for house builders.
7		Communication	13		There is an inherent lack of formal communication between site managers and subcontractors.	Formal site meetings with subcontractors are not frequently conducted.	There are failures in upwards communication flows between site managers and superiors.

(Continued)

(Continued)

Data coding number	Main category headings	Sub-category headings	Frequency counts	Literature sources	Observations, implications or interpretations	Data consistencies	Data inconsistencies
8	Quality	Defects	142	Kim et al. (2008), Sommerville (2007), Garrand (2001), Kletz (2001), Reason (2000;1990), Minato (2003), Roy & Chocrane (1999), Atkinson (1998)	Increased defects are mainly the result of compressed construction duration and inadequate inspection times. Tradespeople can use reduced timescales as an excuse for shoddy work.	The site managers all displayed a conscientious attitude towards product quality.	Houses can always built to a reasonable standard.
9		Quality control	81	Barker (2004), Cheng et al. (2002), Tam et al. (2000)	Third party NHBC building inspections have limited significance for product quality. Doubt exists over the validity of NHBC quality awards. Very few subcontractors employ supervisors to ensure satisfactory quality is achieved.	There is an absence of any effective, standardised system of quality control among private house builders.	Even where formal quality control procedures exist they are usually compromised by completion rates.
10		Workforce	213	Power (2000), Tam et al. (2000)	Subcontractors are more concerned with securing future tenders than with producing quality work. Unreasonable working hours and timescales imposed from above, lowers moral and causes friction between site managers and tradespeople.	The use of subcontracted labour is prevalent within the PHS.	A preference for directly employed tradespeople exists among site managers; even with those who had not used this type of labour for some time.

11	Profit	Budget targets	72	Rosenfeld (2009)	There is an unstated assumption that profit must ultimately take priority over quality.	Senior managers and directors have a vested interest in securing overall budget forecasts for the company.	Site managers do not have a vested interest in securing overall budget forecasts for the company.
12		Repeat business	4	Mbachu & Nkado 2006, Power (2000), Zeithaml et al. (1996), Hoyer & MacInnis (2005), Pitcher (1999)	The negative perception of new build housing portrayed by the media is unfair. Site managers have only a vague awareness of customer satisfaction levels.	Most home buyers have unrealistic expectations.	The link between customer satisfaction and customer loyalty is given little or no credence within the PHS.
13		Bonus payments	69		The bonuses of senior managers and directors are mainly profit-related. The different criteria by which most site managers & apos; bonus payments are evaluated causes inevitable c	The bonuses of site managers are often dependant on not receiving reportable items (recorded defects) from the building inspector.	Site managers with bonuses scored according to similar criteria as senior managers experience less conflicts of interest.

(Continued)

(Continued)

Data coding number	Main category headings	Sub-category headings	Frequency counts	Literature sources	Observations, implications or interpretations	Data consistencies	Data inconsistencies
14	Time	Build programmes	20		Condensed build programmes is the main factor in the inability to produce good quality housing.	No allowances are made for delays.	The build programmes are adequate for most of the year.
15		Working conditions	22	Roy et al. (2003), Morris (1994),	During year-end and half-year-end periods, the working practices adopted by most private housing companies is not conducive to good quality.	Overcrowding and out-of-sequence works is commonplace during busy periods.	When houses are built to a reasonable programme, the expectations of good quality can then become unrealistic.
16		Health & safety	22	Reason (1997), Whittington et al. (1992)	Unsupervised out-of-hours working is often unavoidable but is a major health and safety concern among site managers.	Overcrowding and reduced supervision are the main causes of increased risk.	Effective forward planning combats health and safety concerns.
17		Sales issues	53	CIH (2008), Forsythe (2007), Livette (2006),	There is a philosophy among the hierarchy to complete the sales and worry about quality once the money is in the bank.	Last-minute sales are one of the biggest causes of rushed completions.	Sales staff are considered incompetent and contribute to delays in production during the busiest times.

Appendix E6

The final narrative

It is suggested that a culture of putting completions before quality exists, and the dataset illustrates that it originates with top-level management and passes down through the chain of command to site level. There is a perception that directors are aggressive in pushing for completions, and they remain silent about the consequences for quality, but realistically know that it will suffer. The systems are such that if problems arise after handover, SMs can be blamed. Contract managers, in their position between directors and SMs, are reluctant to say no to directors, since that may prejudice their positions. Directors seem motivated by 'brownie points' and bonus schemes that are based on payment for completions. Directors' remuneration packages may be substantially based on achieving targets for completions. There appears to be a common belief among SMs that the completion targets set by directors are motivated by their self-interest. Tam et al. (2000) lend weight to these grievances by positing that culture-related issues are the most important factors affecting construction quality.

Most SMs expressed a resistance to pressure placed upon their job security, at least in principle, but as one SM put it 'there's resistance there from the site managers but you have to do as you're told at the end of the day and get them through, so you just go for it'. Half of the managers voiced concerns over job security and most of them also hinted at issues of self-esteem as part of working within a team. One SM expressed a wish to be well thought of by superiors, and reflected this is just being part of human nature.

Adverse working practices are clearly commonplace at specific times of the accounting year, which are generically termed 'year-end' and 'half-year-end', when bonuses are calculated. As one SM commented 'it's just the completion periods that cause the problems really. I'd say on average there's a period of about a month really, twice a year, prior to year-end and half-year-end periods, when things get crazy'. One SM conceded that most of the time SMs are fighting losing battles if they try to delay completions, and they may be tempted to push completions through, and if there are problems later, blame them on directors and contracts managers; however, this can be quite demoralising. Interestingly, one manager believed some of the pressure was self-inflicted, stating that SMs should do more checking of quality and should not use time as an excuse for not doing so; SMs should work hard all the time and not 'slack-off' as is sometimes the case. Pressure to achieve completions gets passed on to tradespeople. Subcontractors are asked to start early, work late or all day Saturday and all day Sunday, not just a 'couple of hours' on a Saturday morning. Many different trades may all be working in one property at the same time, and this has consequences for productivity, quality and safety.

Rather alarmingly, a few of the interviewees considered their contracts managers to be ineffective at their jobs, as demonstrated through comments such as, 'my contracts manager isn't particularly good under pressure so he tends to stay away. He's pressurised on all the sites, everybody's chasing him'. Directors could also be said to be somewhat aloof from the realities of trying to build houses under severe time constraints and to an acceptable standard of quality. Some SMs think that contracts managers do not appreciate site problems or turn a blind eye to them. Even though they have site experience, they get 'removed' from reality after a few years. One SM reported that when things are tough contracts managers can 'sit back like the generals on the hill watching everybody else charge around and they won't even answer their phones half the time' and when SMs ask for help the response is 'get it sorted, everybody else is in the same boat, just shut up and get on with it'. Only one SM claimed to be able to successfully resist condensed build programmes and gave an example of telling head office that completions would not be achieved on Friday, but it will be on the following Tuesday or Wednesday. Only one manager claimed not to have any relationship problems with senior managers, 'I tend to be strong enough to get my own way with my contracts manager and even the director'. However, for most of those questioned, successful resistance was rare.

Bonus systems for some SMs may be set with two competing interests. One element of bonus may be based on having few defects on snag lists produced by building inspectors, while the other element is based on getting the completion itself. A bonus can comprise a significant proportion of SMs' remuneration packages, therefore, they can be as highly motivated as directors and contracts managers to achieve completions. Despite the obvious lure that significant amounts of money can present, the general impression from the data is that the main bone of contention for SMs is the inability to build houses to a satisfactory standard of quality.

Pressure applied by directors appears to have some disturbing psychological consequences for some SMs, such as negative impacts on working relationships as well as their personal and family lives. One SM reported 'you start to lose the team … and seem to lash out at them because you're frustrated, with the office, with the senior managers and the director'. Another SM recalled 'the amount they ask you for at year end, you can't physically do it; there's never enough hours in the day … I won't be taking any breaks during the day and I'll have days when I won't have a lunch and … I'll have some weeks where I'll turn in at half six in the morning, before everyone's got there … and then I'm going home and waking up in the middle of the night thinking about plots … come the afternoon, sometimes your brain goes into meltdown … then when I go home the missus is telling me I look gaunt and losing too much weight because you're just on the go so much all the time and you do start making mistakes yourself.'

This apparent transfer of disregard for product quality is perhaps reflected by Liu (2003, p. 149) who details how people become institutionalised through hierarchical power arrangements that affect their behaviour by guiding their thoughts and beliefs to create group values. Hence, it could be said that the attitudes and behaviours of senior managers and directors within the private house building sector towards quality will be mirrored among tradespeople (and some SMs) with resultant negative effects on quality.

Opinions vary concerning the additional workload imposed on managers when timescales are reduced. One SM claimed that although fast-track build completions were attainable, they did require the manager's full attention because 'if you're doing a house in two weeks from plaster to completion, you have to sit on that house and that's all that's on your mind and you kind of push everything else to one side a little bit to give that house the 100% attention it needs.' This suggests that while it may be possible to complete houses in this manner and to an adequate standard of quality, it could only be done on a limited number of plots before quality standards begin to be compromised.

Despite most of the managers indicating that additional resources such as assistant managers and labourers were readily available during busy periods if needed, none of them felt that this compensated for the lack of time in the build programme. In cases where there is initially adequate time, there can be problems with resources if they are taken off site to attend to urgent completions elsewhere. Trying to achieve completions can be frustrated by problems with material deliveries. If customers are allowed to pick their own kitchens, the delivery period may extend beyond the proposed completion date. Materials may be late or be delivered short or damaged, and it can be difficult trying to 'rob' items from other plots. Another widespread problem is the delay with service connections (gas, water and electricity) to properties, which can impact on production and quality in a variety of ways.

There appears to be an inherent lack of communication between SMs and subcontractors. Only three of the eight SMs held regular formal site meetings with subcontractors; they did claim the meetings to be worthwhile. The majority seem to view them as an unnecessary chore with little or no benefit, and argue that neither they nor the subcontractors have time. It would appear that most SMs prefer a more ad hoc approach to communication. It may be argued that the general fast pace of the private housing sector does not lend itself to such organisational tools as it does with the slower, more controlled process of the construction industry at large.

There is an appreciation that while money is the key driver, maximum profits are primarily derived from positive cashflow and not efficiency or lowest costs on site. This is particularly the case if high value properties are being completed in clusters, e.g. four properties at £250k each, total income £1M. This sum of money in the bank means far more than relatively minor problems with quality. It is perceived by some to be the lesser of two evils to have hassle with customers and work inefficiently to rectify defects after customers have moved into a house, rather than alternatively, not to have the cash. It is probably on this basis that owners of companies reward directors based upon housing completions rather than on quality.

Appendix F: Statistical tables

Appendix F1: The standard normal distribution tables
Appendix F2: F distribution
Appendix F3: Chi-square table χ^2
Appendix F4: Mann–Whitney U table, $p = 0.01$, two-tailed
Appendix F5: Mann–Whitney U table, $p = 0.05$, two-tailed
Appendix F6: Wilcoxon table, $p = 0.01$, 0.05 and 0.10; one and two-tailed
Appendix F7: Related t-test table, $p = 0.01$, 0.02, 0.05 and 0.10; one and two-tailed
Appendix F8: Spearman's rho table (ρ), $p = 0.01$, 0.05 and 0.10; one and two-tailed
Appendix F9: Pearsons's r table, $p = 0.01$, 0.05 and 0.10; one and two-tailed

Appendix F1

The standard normal distribution tables

To determine the probability of a z score, use the first column for the first decimal place, and the remaining columns for the second decimal place. Consider 1 standard deviation; the probability of scores being more than 1 standard deviation is 0.1587. The probability of scores being more than ± 1 standard deviation is twice that value = 0.1587 x 2 = 0.3174. The probability of score lying within ± 1 standard deviation = 1.0 − 0.3174 = 0.6826 or 68.26%

z ↓	0	1	2	3	4	5	6	7	8	9
0	0.5000	0.4960	0.4920	0.4880	0.4840	0.4801	0.4761	0.4721	0.4681	0.4641
0.1	0.4602	0.4562	0.4522	0.4483	0.4443	0.4404	0.4364	0.4325	0.4286	0.4287
0.2	0.4207	0.4168	0.4129	0.4090	0.4052	0.4013	0.3974	0.3936	0.3897	0.3859
0.3	0.3821	0.3783	0.3745	0.3707	0.3669	0.3632	0.3594	0.3556	0.3520	0.3483
0.4	0.3446	0.3409	0.3372	0.3336	0.3300	0.3264	0.3228	0.3192	0.3156	0.3121
0.5	0.3085	0.3050	0.3015	0.2981	0.2946	0.2912	0.2877	0.2843	0.2810	0.2776
0.6	0.2743	0.2709	0.2676	0.2643	0.2611	0.2578	0.2546	0.2514	0.2483	0.2451
0.7	0.2420	0.2389	0.2358	0.2327	0.2296	0.2266	0.2236	0.2206	0.2177	0.2148
0.8	0.2119	0.2090	0.2061	0.2033	0.2005	0.1977	0.1949	0.1922	0.1894	0.1867
0.9	0.1841	0.1814	0.1788	0.1762	0.1736	0.1711	0.1685	0.1660	0.1635	0.1611
1	0.1587	0.1562	0.1539	0.1515	0.1492	0.1469	0.1446	0.1423	0.1401	0.1379
1.1	0.1357	0.1335	0.1314	0.1294	0.1271	0.1251	0.1230	0.1210	0.1190	0.1170
1.2	0.1151	0.1131	0.1112	0.1093	0.1075	0.1056	0.1038	0.1020	0.1103	0.0985
1.3	0.0968	0.0951	0.0934	0.0918	0.0901	0.0885	0.0869	0.0853	0.8380	0.0823
1.4	0.0808	0.0793	0.0778	0.0764	0.0749	0.0735	0.0721	0.0708	0.0694	0.0681
1.5	0.0668	0.0655	0.0643	0.0630	0.0618	0.0606	0.0594	0.0582	0.0571	0.0559
1.6	0.0548	0.0537	0.0526	0.0516	0.0505	0.0495	0.0485	0.0475	0.0465	0.0455
1.7	0.0446	0.0436	0.0427	0.0418	0.0409	0.0401	0.0392	0.0384	0.0375	0.0367
1.8	0.0359	0.0351	0.0344	0.0336	0.0329	0.0322	0.0314	0.0307	0.0301	0.0294
1.9	0.0287	0.0218	0.0274	0.0268	0.0262	0.0256	0.0250	0.0244	0.0239	0.0233
2	0.0228	0.0222	0.0217	0.0212	0.0207	0.0202	0.0197	0.0192	0.0188	0.0183
2.1	0.0179	0.0174	0.0170	0.0166	0.0162	0.0158	0.0154	0.0150	0.0146	0.0143
2.2	0.0139	0.0136	0.0132	0.0129	0.0125	0.0122	0.0119	0.0166	0.0133	0.0110
2.3	0.0107	0.0104	0.0102	0.0099	0.0096	0.0094	0.0091	0.0089	0.0087	0.0087
2.4	0.0082	0.0080	0.0078	0.0075	0.0073	0.0071	0.0069	0.0068	0.0066	0.0064
2.5	0.0062	0.0060	0.0059	0.0057	0.0055	0.0054	0.0052	0.0051	0.0049	0.0048
2.6	0.0047	0.0045	0.0045	0.0043	0.0041	0.0040	0.0039	0.0038	0.0037	0.0037
2.7	0.0035	0.0034	0.0033	0.0032	0.0031	0.0030	0.0029	0.0028	0.0027	0.0026
2.8	0.0026	0.0025	0.0024	0.0023	0.0023	0.0022	0.0021	0.0021	0.0020	0.0019
2.9	0.0019	0.0018	0.0018	0.0017	0.0016	0.0016	0.0015	0.0015	0.0014	0.0014
3	0.0013	0.0013	0.0013	0.0012	0.0012	0.0011	0.0011	0.0011	0.0010	0.0010
3.1	0.0010	0.0009	0.0090	0.0009	0.0008	0.0008	0.0008	0.0008	0.0007	0.0007
3.2	0.0007	0.0007	0.0006	0.0006	0.0006	0.0006	0.0006	0.0005	0.0005	0.0005
3.3	0.0005	0.0005	0.0005	0.0004	0.0004	0.0004	0.0004	0.0004	0.0004	0.0003
3.4	0.0003	0.0003	0.0003	0.0003	0.0003	0.0003	0.0003	0.0003	0.0003	0.0002
3.5	0.0002	0.0002	0.0002	0.0002	0.0002	0.0002	0.0002	0.0002	0.0002	0.0002
3.6	0.0002	0.0002	0.0001	0.0001	0.0001	0.0001	0.0001	0.0001	0.0001	0.0001
3.7	0.0001	0.0001	0.0001	0.0001	0.0001	0.0001	0.0001	0.0001	0.0001	0.0001
3.8	0.0001	0.0001	0.0001	0.0001	0.0001	0.0001	0.0001	0.0001	0.0001	0.0001
3.9	0	0	0	0	0	0	0	0	0	0

Appendix F2

F distribution

Critical values of the F distribution

The calculated value of U must be larger than or equal to the table value for significance. p or $\alpha = 0.05$

Var1→ / Var2↓	1	2	3	4	5	6	7	8	9	10	12	15	20	24	30
1															
2	18.50	19.00	19.10	19.20	19.30	19.30	19.30	19.30	19.30	19.40	19.40	19.40	19.40	19.40	19.40
3	10.10	9.55	9.28	9.12	9.01	8.94	8.89	8.85	8.81	8.79	8.74	8.70	8.66	8.64	8.62
4	7.71	6.94	6.59	6.39	6.26	6.16	6.09	6.04	6.00	5.96	5.91	5.86	5.80	5.77	5.75
5	6.61	5.79	5.41	5.19	5.05	4.95	4.88	4.82	4.77	4.74	4.68	4.62	4.56	4.53	4.50
6	5.99	5.14	4.76	4.53	4.39	4.28	4.21	4.15	4.10	4.06	4.00	3.94	3.87	3.84	3.81
7	5.59	4.74	4.35	4.12	3.97	3.87	3.79	3.73	3.68	3.64	3.57	3.51	3.44	3.41	3.38
8	5.32	4.46	4.07	3.84	3.69	3.58	3.50	3.44	3.39	3.35	3.28	3.22	3.15	3.12	3.08
9	5.12	4.26	3.86	3.63	3.48	3.37	3.29	3.23	3.18	3.14	3.07	3.01	2.94	2.90	2.86
10	4.96	4.10	3.71	3.48	3.33	3.22	3.14	3.07	3.02	2.98	2.91	2.84	2.77	2.74	2.70
11	4.84	3.98	3.59	3.36	3.20	3.09	3.01	2.95	2.90	2.85	2.79	2.72	2.65	2.61	2.57
12	4.75	3.89	3.49	3.26	3.11	3.00	2.91	2.85	2.80	2.75	2.69	2.62	2.54	2.51	2.47
13	4.67	3.81	3.41	3.18	3.03	2.92	2.83	2.77	2.71	2.67	2.60	2.53	2.46	2.42	2.38
14	4.60	3.74	3.34	3.11	2.96	2.85	2.76	2.70	2.65	2.60	2.53	2.46	2.39	2.35	2.31
15	4.54	3.68	3.29	3.06	2.90	2.79	2.71	2.64	2.59	2.54	2.48	2.40	2.33	2.29	2.25
16	4.49	3.63	3.24	3.01	2.85	2.74	2.66	2.59	2.54	2.49	2.42	2.35	2.28	2.24	2.19
17	4.45	3.59	3.20	2.96	2.81	2.70	2.61	2.55	2.49	2.45	2.38	2.31	2.23	2.19	2.15

18	4.41	3.55	3.16	2.93	2.77	2.66	2.58	2.51	2.46	2.41	2.34	2.27	2.19	2.15	2.11
19	4.38	3.52	3.13	2.90	2.74	2.63	2.54	2.48	2.42	2.38	2.31	2.23	2.16	2.11	2.07
20	4.35	3.49	3.10	2.87	2.71	2.60	2.51	2.45	2.39	2.35	2.28	2.20	2.12	2.08	2.04
21	4.32	3.47	3.07	2.84	2.68	2.57	2.49	2.42	2.37	2.32	2.25	2.18	2.10	2.05	2.01
22	4.30	3.44	3.05	2.82	2.66	2.55	2.46	2.40	2.34	2.30	2.23	2.15	2.07	2.03	1.98
23	4.28	3.42	3.03	2.80	2.64	2.53	2.44	2.37	2.32	2.27	2.20	2.13	2.05	2.01	1.96
24	4.26	3.40	3.01	2.78	2.62	2.51	2.42	2.36	2.30	2.25	2.18	2.11	2.03	1.98	1.94
25	4.24	3.39	2.99	2.76	2.60	2.49	2.40	2.34	2.28	2.24	2.16	2.09	2.01	1.96	1.92
26	4.23	3.37	2.98	2.74	2.59	2.47	2.39	2.32	2.27	2.22	2.15	2.07	1.99	1.95	1.90
27	4.21	3.35	2.96	2.73	2.57	2.46	2.37	2.31	2.25	2.20	2.13	2.06	1.97	1.93	1.88
28	4.20	3.34	2.95	2.71	2.56	2.45	2.36	2.29	2.24	2.19	2.12	2.04	1.96	1.91	1.87
29	4.18	3.33	2.93	2.70	2.55	2.43	2.35	2.28	2.22	2.18	2.10	2.03	1.91	1.90	1.85
30	4.17	3.32	2.92	2.69	2.53	2.42	2.33	2.27	2.21	2.16	2.09	2.01	1.93	1.89	1.84
40	4.08	3.23	2.84	2.61	2.45	2.34	2.25	2.18	2.12	2.08	2.00	1.92	1.84	1.79	1.74
60	4.00	3.15	2.76	2.53	2.37	2.25	2.17	2.10	2.04	1.99	1.92	1.84	1.75	1.70	1.65
120	3.92	3.07	2.68	2.45	2.29	2.18	2.09	2.02	1.96	1.91	1.83	1.75	1.66	1.61	1.55
∞	3.84	3.00	2.60	2.37	2.21	2.10	2.01	1.94	1.88	1.83	1.75	1.67	1.57	1.52	1.46

Appendix F3

Chi-square table

Critical value of the chi-square (χ^2) distribution The calculated value of χ^2 must be larger than or equal to the table value for significance				
p value →	0.10	0.05	0.025	0.01	0.001
df↓					
1	2.706	3.841	5.024	6.635	10.828
2	4.605	5.991	7.378	9.210	13.816
3	6.251	7.815	9.348	11.345	16.266
4	7.779	9.488	11.143	13.277	18.467
5	9.236	11.070	12.833	15.086	20.515
6	10.645	12.592	14.449	16.812	22.458
7	12.017	14.067	16.013	18.475	24.322
8	13.362	15.507	17.535	20.090	26.125
9	14.684	16.919	19.023	21.666	27.877
10	15.987	18.307	20.483	23.209	29.588
11	17.275	19.675	21.920	24.725	31.264
12	18.549	21.026	23.337	26.217	32.910
13	19.812	22.362	24.736	27.688	34.528
14	21.064	23.685	26.119	29.141	36.123
15	22.307	24.996	27.488	30.578	37.697
16	23.542	26.296	28.845	32.000	39.252
17	24.769	27.587	30.191	33.409	40.790
18	25.989	28.869	31.526	34.805	42.312
19	27.204	30.144	32.852	36.191	43.820
20	28.412	31.410	34.170	37.566	45.315
21	29.615	32.671	35.479	38.932	46.797
22	30.813	33.924	36.781	40.289	48.268
23	32.007	35.172	38.076	41.638	49.728
24	33.196	36.415	39.364	42.980	51.179
25	34.382	37.652	40.646	44.314	52.620
26	35.563	38.885	41.923	45.642	54.052
27	36.741	40.113	43.195	46.963	55.476
28	37.916	41.337	44.461	48.278	56.892
29	39.087	42.557	45.722	49.588	58.301
30	40.256	43.773	46.979	50.892	59.703
31	41.422	44.985	48.232	52.191	61.098
32	42.585	46.194	49.480	53.486	62.487
df = degrees of freedom. Refer to chapter 9.3 to determine its value in a chi-square test					

Appendix F4

Mann–Whitney U table, p = 0.01, two-tailed

Critical value of the Mann-Whitney U statistic. The calculated value of U must be smaller than or equal to the table value for significance. Two-tail; p or α = 0.01

n1→ n2↓	1	2	3	4	5	6	7	8	9	10	11	12	13	14	15	16	17	18	19	20
1																				
2																			0	0
3									0	0	0	1	1	1	2	2	2	2	3	3
4						0	0	1	1	2	2	3	3	4	5	5	6	6	7	8
5					0	1	1	2	3	4	5	6	7	7	8	9	10	11	12	13
6				0	1	2	3	4	5	6	7	9	10	11	12	13	15	16	17	18
7				0	1	3	4	6	7	9	10	12	13	15	16	18	19	21	22	24
8				1	2	4	6	7	9	11	13	15	17	18	20	22	24	26	28	30
9			0	1	3	5	7	9	11	13	16	18	20	22	24	27	29	31	33	36
10			0	2	4	6	9	11	13	16	18	21	24	26	29	31	34	37	39	42
11			0	2	5	7	10	13	16	18	21	24	27	30	33	36	39	42	45	48
12			1	3	6	9	12	15	18	21	24	27	31	34	37	41	44	47	51	54
13			1	3	7	10	13	17	20	24	27	31	34	38	42	45	49	53	56	60
14			1	4	7	11	15	18	22	26	30	34	38	42	46	50	54	58	63	67
15			2	5	8	12	16	20	24	29	33	37	42	46	51	55	60	64	69	73
16			2	5	9	13	18	22	27	31	36	41	45	50	55	60	65	70	74	79
17			2	6	10	15	19	24	29	34	39	44	49	54	60	65	70	75	81	86
18			2	6	11	16	21	26	31	37	42	47	53	58	64	70	75	81	87	92
19		0	3	7	12	17	22	28	33	39	45	51	56	63	69	74	81	87	93	99
20		0	3	8	13	18	24	30	36	42	48	54	60	67	73	79	86	92	99	105

Appendix F5

Mann–Whitney U table, p = 0.05, two-tailed

Critical value of the Mann-Whitney U statistic. The calculated value of U must be smaller than or equal to the table value for significance. Two-tail; p or α = 0.05

n1→ n2↓	1	2	3	4	5	6	7	8	9	10	11	12	13	14	15	16	17	18	19	20
1																				
2								0	0	0	0	1	1	1	1	1	2	2	2	2
3					0	1	1	2	2	3	3	4	4	5	5	6	6	7	7	8
4				0	1	2	3	4	4	5	6	7	8	9	10	11	11	12	13	13
5			0	1	2	3	5	6	7	8	9	11	12	13	14	15	17	18	19	20
6			1	2	3	5	6	8	10	11	13	14	16	17	19	21	22	24	25	27
7			1	3	5	6	8	10	12	14	16	18	20	22	24	26	28	30	32	34
8		0	2	4	6	8	10	13	15	17	19	22	24	26	29	31	34	36	38	41
9		0	2	4	7	10	12	15	17	21	23	26	28	31	34	37	39	42	45	48
10		0	3	5	8	11	14	17	20	23	26	29	33	36	39	42	45	48	52	55
11		0	3	6	9	13	16	19	23	26	30	33	37	40	44	47	51	55	58	62
12		1	4	7	11	14	18	22	26	29	33	37	41	45	49	53	57	61	65	69
13		1	4	8	12	16	20	24	28	33	37	41	45	50	54	59	63	67	72	76
14		1	5	9	13	17	22	26	31	36	40	45	50	55	59	64	67	74	78	83
15		1	5	10	14	19	24	29	34	39	44	49	54	59	64	70	75	80	85	90
16		1	6	11	15	21	26	31	37	42	47	53	59	64	70	75	81	86	92	98
17		2	6	11	17	22	28	34	39	45	51	57	63	67	75	81	87	93	99	105
18		2	7	12	18	24	30	36	42	48	55	61	67	74	80	86	93	99	106	112
19		2	7	13	19	25	32	38	45	52	58	65	72	78	85	92	99	106	113	119
20		2	8	14	20	27	34	41	48	55	62	69	76	83	90	98	105	112	119	127

Appendix F6

Wilcoxon table, p = 0.01, 0.05 and 0.10; one and two-tailed

Critical value of the Wilcoxon *t* statistic. The calculated value of *t* must be smaller than or equal to the table value for significance.			
	p values or level of significance two-tailed test; note one tail read up from below		
n ↓	**0.1**	**0.05**	**0.01**
5	1		
6	2	1	
7	4	2	
8	6	4	0
9	8	6	2
10	11	8	3
11	14	11	5
12	17	14	7
13	21	17	10
14	26	21	13
15	30	25	16
16	36	30	19
17	41	35	23
18	47	40	28
19	54	46	32
20	60	52	37
21	68	59	43
22	75	66	49
23	83	73	55
24	92	81	61
25	101	90	68
26	110	98	76
27	120	107	84
28	130	117	92
29	141	127	100
30	152	137	109
35	214	195	160
40	287	264	221
45	371	343	292
50	466	434	373

Appendix F7

Related t-test table, p = 0.01, 0.02, 0.05 and 0.10; one and two-tailed

Critical value of t; for the related t test. The calculated value of t must be more than or equal to the table value for significance				
	p values or level of significance two-tailed test; note one tail read up from below			
df ↓	**0.1**	**0.05**	**0.02**	**0.01**
1	6.314	12.706	31.821	63.657
2	2.920	4.303	6.965	9.925
3	2.353	3.182	4.541	5.841
4	2.132	2.776	3.747	4.604
5	2.015	2.571	3.365	4.032
6	1.943	2.447	3.143	3.707
7	1.894	2.365	2.998	3.499
8	1.860	2.306	2.896	3.355
9	1.833	2.262	2.821	3.250
10	1.812	2.228	2.764	3.169
11	1.796	2.201	2.718	3.106
12	1.782	2.179	2.681	3.055
13	1.771	2.160	2.650	3.012
14	1.761	2.145	2.624	2.977
15	1.753	2.131	2.602	2.947
16	1.746	2.120	2.583	2.921
17	1.740	2.110	2.567	2.898
18	1.734	2.101	2.552	2.878
19	1.729	2.093	2.539	2.861
20	1.725	2.086	2.528	2.845
21	1.721	2.080	2.518	2.831
22	1.717	2.074	2.508	2.819
23	1.714	2.069	2.500	2.807
24	1.711	2.064	2.492	2.797
25	1.708	2.060	2.485	2.787
26	1.706	2.056	2.479	2.779
27	1.703	2.052	2.473	2.771
28	1.701	2.048	2.467	2.763
29	1.699	2.045	2.462	2.756
30	1.697	2.042	2.457	2.750
40	1.684	2.021	2.423	2.704
50	1.676	2.009	2.403	2.678
100	1.660	1.984	2.364	2.626

Appendix F8

Spearman's rho table, p = 0.01, 0.05 and 0.10; one and two-tailed

Critical value of Spearman's *rho* (ρ). The calculated value of ρ must be more than or equal to the table value for significance			
	p values or level of significance two-tailed test; note one tail read up from below		
n ↓	**0.10**	**0.05**	**0.01**
5	0.900	1.000	
6	0.829	0.886	1.000
7	0.714	0.786	0.929
8	0.643	0.738	0.881
9	0.600	0.683	0.833
10	0.564	0.648	0.794
12	0.506	0.591	0.777
14	0.456	0.544	0.715
16	0.425	0.506	0.665
18	0.399	0.475	0.625
20	0.377	0.450	0.591
22	0.359	0.428	0.562
24	0.343	0.409	0.537
26	0.329	0.392	0.515
28	0.317	0.377	0.496
30	0.306	0.364	0.478

Appendix F9

Pearsons's r table, p = 0.01, 0.05 and 0.10; one and two-tailed

Critical value of the Pearson's r. The calculated value of r must be more than or equal to the table value for significance.			
	p values or level of significance two-tailed test; note one tail read up from below		
n - 2 ↓	0.10	0.05	0.01
2	0.900	0.950	0.990
3	0.805	0.878	0.957
4	0.729	0.811	0.917
5	0.669	0.754	0.875
6	0.621	0.707	0.834
7	0.582	0.666	0.798
8	0.549	0.632	0.765
9	0.521	0.602	0.735
10	0.497	0.576	0.708
11	0.476	0.553	0.684
12	0.457	0.532	0.661
13	0.441	0.514	0.641
14	0.426	0.497	0.623
15	0.412	0.482	0.606
16	0.400	0.468	0.590
17	0.389	0.456	0.575
18	0.378	0.444	0.561
19	0.369	0.433	0.549
20	0.360	0.423	0.537
25	0.323	0.381	0.487
30	0.296	0.349	0.449
35	0.275	0.325	0.418
40	0.257	0.304	0.393
45	0.243	0.288	0.372
50	0.231	0.273	0.354
60	0.211	0.250	0.325
70	0.195	0.232	0.302
80	0.183	0.217	0.283
90	0.173	0.205	0.267
100	0.164	0.195	0.254

Index

abbreviations, list of 5
abstracts 4–5
academic phrasebank 42
accreditation 8
acknowledgements page 4
adverse events during research 21
aim of research 70–71
alternate hypothesis 81–82
alternating poles on statements and
    questions 127–128
analysis 10–14
analysis of data 137
ancillary variable (AncV) 83–88, 186–190
anonymity of participants 20–21
appendices 252, 255–256
appendices, list of 6
audio data 168

bibliography 58–59
    definition 58
building information modelling (BIM) 34

carbon dioxide (CO_2) as a variable 102–103
categorical variables 93–95
    examples 96, 98
chapters of a dissertation 3–4
chart representations of descriptive
    statistics 190–191
chi-square test 205, 210–226

consequence of spread 216–218
formula 214
more complex form 218
one row 218–219
spreadsheet calculation 216
citation styles 54
citing other authors 53, 54–55
    author names 57
    common mistakes 59
    page numbers 57–58
    paraphrasing 56
    plagiarism, avoiding 62–64
    secondary citing 56–57
    verbatim citations 55–56
    web pages 60
    year of publication 57
client satisfaction 13
closed questions 122–123
COIR mnemonic 93
conclusions chapter 252, 253–255
conferences 45
confidence intervals 247–249
confidentiality for participants 20–21
Constructing Excellence 13
content analysis 170
contents page 5
contingency table 212, 213
continuing professional development
    (CPD) 8–9

Writing Built Environment Dissertations and Projects: Practical Guidance and Examples, Second Edition.
Peter Farrell, Fred Sherratt and Alan Richardson.
© 2017 John Wiley & Sons, Ltd. Published 2017 by John Wiley & Sons, Ltd.
Companion Website: www.wiley.com/go/Farrell/Built_Environment_Dissertations_and_Projects

correlation coefficients 243
correlation test 206
correlations 236–237
   scatter diagrams 237–238
   Spearman's rho 239–241
count 182
covering letters 19

data analysis 137, 168–169
   assembling data 170
   coding data 171–172
   content analysis 170
   example 174
   management of data 169–170
   narrative 175
   procedure 172–173
data coding 171–172
data collection 110–112
data management 169–170
data measurement 112
   populations and samples 112–116
   response points 118–120
   single- and multiple-item scales 116–117
data sources 44–45
data types 198
   hard or soft 100–101
   objective or subjective 100–101
   primary or secondary 99–100
datasets 10
deception 20
declaration 4
degree programmes 9
dependent variable (DV) 13, 77–81
   no relationship with IV 88–89
descriptive statistics 177–178
   calculation 182–186
   calculation using spreadsheet 186
   charts 190–191
   examples 178–182
disability support 29
discussion chapter 252–253
dissenting voices from the predominant
   view 42
dissertations 1–2, 31, 146–148
   appendices 255–256
   conclusions chapter 148, 253–255

description of the problem 35–37
discussion chapter 252–253
embedding theory 49–53
extra support 29
house style 22–23
incorporating qualitative data 175
introduction chapter 33–35, 37–38
literature review 149
nomenclature 2–3
objectives 37
proofreading 27–28
structure 3–6
summary of process 258–259
terminology 2–3
viva/viva voce 30–31
writing style 23–26

electronic surveys/questionnaires 123–124
emotional harm 17
English as a second language 29
environmental variables 76
epistemology 10
error bars 148
error in results 147–148
ethical considerations 17–20
   confidentiality and anonymity 20–21
   general points 21–22
ethnography 111
examiner's perspective 256–258

figures 6
   list of 6
findings from research 137–138
finite element method (FEM) 161
flat distribution 198
frequency histograms 190, 191
   mean scores 229, 230

Gantt charts 160
glossary of symbols 5
glossary of terms 5–6
grammar checking 27
grounded theory 111

hard data 9, 100
Harvard citation style 54

health and safety considerations 149
  housekeeping 150
  induction 149
  risk analysis of procedures 152
  risk assessments 150–151
  student questionnaire 150
histograms 190
house style 22–23
housekeeping 150
hypothesis, stating 81
  integrating the null and the tails 82–83
  no relationship between IV and DV 88–89
  null hypothesis or alternate hypothesis 81–82
  one or two tailed hypothesis 82
hypothesis setting 12
hypothesis with one variable 75–77
hypothesis with two variables 77–78
  cause versus relationship 80–81
  identifying the IV and the DV 78–80
  manipulation or observation 80

impact factor of journals 44–45
independent variable (IV) 13, 77–81
  no relationship with DV 88–89
inferential statistics 177, 204–206
  chi-square test 210–219
  difference between means 223–225
  parametric or non-parametric
    datasets 220–223
  probability values 206–207
  summary of tests 248
information sheet 19
International Standard Book Number
    (ISBN) 55
International Standard Serial Number
    (ISSN) 55
interpretivist research 11
interval variables 93, 94
  examples 96, 97–98
intervening variable (IntV) 83, 84
interviews 166–167
  recording data 167–168
introduction chapter of a dissertation 33–35,
    37–38
  description of the problem 35–37
  objectives 37
irregular distribution 199

journals 8
  availability 45
  civil engineering journals 46
  Construction Management journals 47
  impact factor 44–45
judgements versus opinions 43–44

key performance indicators (KPIs) 91
Kolmogorov–Smirnov test 220

laboratory experiments 141–142
  health and safety 149–151
  recording procedures 145–146
  reliability and validity of
    findings 143–145
  sample size 145
  sourcing test materials 143
  supervisor, role of 151–153
  test methodology 142–143
  writing up 146–148
levels of measurement 93–95
Likert item 116
  response points 118–120
Likert type scale 116–117, 118
  examples 120
line diagrams 191
list of abbreviations 5
list of appendices 6
list of figures 6
list of tables 6
literature review 39–41
  civil engineering journals 46
  Construction Management journals 47
  data sources 44–45
  finding literature 48
  funnel analogy 41
  judgements versus opinions 43–44
  style and contents 41–43
loaded questions 128
log books 145–146

Mann–Whitney statistic 204, 205, 224,
    225–230
  consequence of larger sample size 230
  frequency histogram of mean scores 229
  spreadsheet calculation 228–229
maximum 184

mean 179, 194
measurement instruments 89–90
　high or low variable values 90–91
　measurement levels 93–95
　measurement scales 91–93
median 183, 194
methodological triangulation studies 13
methodology chapter 107–110, 138
　analysis, results and findings 137–138
　data collection 110–112
　data measurement 112–120
　other analytical tools 131–132
　questionnaires 120–123
　ranking studies 129–131
　reliability and validity 132–137
minimum 184
missing data 183
mixed method studies 13
mode 184, 194
money as a variable 102–103
multiple-item scales 116–117

narrative data analysis 175
narrative enquiries 111–112
negative results, reporting 89
nominal variables 93
　examples 95, 96
non-parametric datasets 220–223
　how calculations differ 222–223
non-parametric unrelated t-test 224
non-probability sampling 115–116
non-technical research work 9–10
normal distribution 191–194
null hypothesis 81–82

objective data 9, 100–101
objectives, provisional 37
objectives in research 71–72
　literature review 73
　objectives unaddressed by work
　　undertaken 73–74
　wobbling 72–73
objective research 11
offensive behaviour 19–20
one tailed hypothesis 82
ontology 10
open questions 123

ordinal variables 93, 94
　examples 95–97

page numbering 4
paradigms of research 11
parametric datasets 220–222
　how calculations differ 222–223
parametric related t-test 232–236
paraphrasing 56
Pearson's chi-square 204, 205, 210–219
　consequence of spread 216–218
　formula 239–240
　more complex form 218
　one row 218–219
　spreadsheet calculation 219
Pearson's correlation 241–242
Pearson's product 204
percentage 183
performance variables 76
permission-to-do-research form 17–19
phrases/language useful for literature
　　reviews 42
physical harm 17
pie charts 190
piloting questionnaires 125
plagiarism 53
　avoiding 62–64
planning software 17
populations 113
positivist research 11
preliminary pages 4–5
primary data 99
probability (p value) 206–207
　$p \leq 0.05$ 207–208
　significance level 208–210
probability sampling 114–115
problem to be addressed, identifying 35–37
professional bodies 8–9
proofreading 27–28
psychological harm 19–20
purposive sampling 115–116

qualitative analysis 10–14
qualitative data analysis 165–166, 175–176
　analysis steps 168–176
　data collection 166–168
　steps 176

quantitative analysis 10–14, 177–178,
   204–206
   ancillary variables 186–190
   charts 190–192
   chi-square test 210–219
   confidence intervals 247–249
   correlation coefficients 243, 244, 245
   correlations 236–242
   difference between means 223–225
   Mann–Whitney test 225–230
   normal distribution 191–194
   parametric or non-parametric
      datasets 220–223
   parametric related t-test 232–236
   probability values 206–210
   summary of tests 250
   test selection 243–247
   two variable relationships 197–201
   Wilcoxon t-test 230–232
   Z score 194–197
questionnaires 12, 120–123
   administration 123–124
   alternating poles on statements and
      questions 127–128
   dos and don'ts 128–129
   piloting 125
   reliability 133–134
   statements or questions 126–127

random sampling 115
range 184
ranking studies 129–131
ratio variables 93
   examples 97–98
recording laboratory procedures 145–146
refereeing of papers 45
references 58–59
   as evidence of reading 53
   checking 27
   citing other authors 55–58
   common mistakes 59–60
   definition 58
   plagiarism, avoiding 62–64
   software 60–62
referencing styles 54
related t-test 204

reliability 132–133
   examples 134–136
   laboratory experiments 143–145
report numbering format 4
research areas to consider 7–8
research continuum 11
research goals 68–70, 104–105
   aim 70–71
   ancillary variables 83–88
   carbon dioxide (CO_2) as a
      variable 102–103
   Castle Museum in York analogy 69–70
   categorical variables 95
   data types 98–102
   hypothesis with one variable 75–77
   hypothesis with two variables 77–81
   hypothesis writing 81–83
   interval variables 97–98
   measurement instruments 89–93
   measurement levels 93–95
   money as a variable 102–103
   no relationship between IV and DV 88–89
   objectives 71–74
   ordinal variables 96–97
   questions 71
   ratio variables 97–98
   three objectives with IV and
      DV 103–104
   variable selection 83
   variables 74–75
research objectives and findings
      examples 154
   comparative performance of fibre
      reinforced polymer (FRP) and steel
      rebar 154–156
   concrete with crushed, graded and
      washed recycled construction
      demolition waste as coarse aggregate
      replacement 157–159
   equating steel and synthetic fibre concrete
      post crack performance 156–157
   pull-out performance of chemical anchor
      bolts in fibre concrete 161–163
   surface coating of traditional construction
      materials using microbially induced
      calcite precipitation 159–161

research paradigms 11
research proposals 29–30
    examples 153–154
research topics 153
results of research 137–138
review of literature 39–41
    civil engineering journals 46
    Construction Management journals 47
    data sources 44–45
    finding literature 48
    funnel analogy 41
    judgements versus opinions 43–44
    style and contents 41–43
review of theory 39–41, 49–53
risk assessment grid and template 151
risk assessments 17–20
    analysis of procedures 152
    laboratory work 150–151

sample size 145
samples 113–114
samples of convenience 115
scales of measurement 91–93
scatter diagrams 201, 137, 245
secondary citing 56–57
secondary data 99–100
semi-structured interviews 167
SI units 75
significance level 208–210
single-item scales 116–117
skewed distribution 199
snipping tool 6
soft data 9, 100–101
sourcing test materials 143
Spearman's rho 204, 239–241
    formula 239–240
spell checking 27
standard deviation (SD) 185
    normal distribution 191–194
statistics, descriptive 177–178
    calculation 182–186
    calculation using spreadsheet 186
    charts 190–191
    examples 178–182
statistics, inferential 204–206
    chi-square test 210–219

difference between means 223–225
    parametric or non-parametric
        datasets 220–223
    probability values 206–210
    summary of tests 250
stepped distribution 199
structure of a dissertation 3–6
student/supervisor relationship 14–17
style guide 22–23
subject variables 83–88
subjective data 9, 100–101
    use when no objective data 101–102
subjective research 11
sum 182
supervisor, role of 151–153
surveys 111
symbols, glossary of 5

tables 6
    list of 6
technical research work 9–10
terminology 2–3
terms, glossary of 5–6
test methodology 142–143
test programme 144
theory, spider's web analogy 49–50
theory review 39–41, 49–53
time management 18
transcribing interviews 168
treated data 182
t-test 223
    unrelated or related data 224–225
two tailed hypothesis 82

unstructured interviews 167
U-shaped distribution 198

validity 133–134
    examples 134–136
    laboratory experiments 143–145
Vancouver citation style 54
variables 74–75, 105
    ancillary or subject variables 83–88
    carbon dioxide 102–103
    environmental variables 76
    hypothesis with one variable 75–77

variables (*cont'd*)
  hypothesis with two variables 77–81
  melting into one another 79
  money 102–103
  performance variables 76
  selection 83
  time frame 79
verbal examination 30–31
verbatim citations 55–56
viva/viva voce 30–31

web pages, citing 60
weight of evidence 44

weighting 118–120
Wilcoxon's statistic 204, 205, 230–232
  frequency histogram of mean
    scores 232
  spreadsheet calculation 232
workplace-based projects 7–8, 35
  data collection 110
  sampling 115
writing style 23–26

Yates' correction 214

Z score 194–197